Science, Learning, Identity

NEW DIRECTIONS IN MATHEMATICS AND SCIENCE EDUCATION
Volume 7

Series Editors
Wolff-Michael Roth
University of Victoria, Canada
Lieven Verschaffel
University of Leuven, Belgium

Editorial Board
Angie Calabrese-Barton, *Teachers College, New York, USA*
Pauline Chinn, *University of Hawaii, USA*
Brian Greer, *Portland State University, USA*
Lyn English, *Queensland University of Technology*
Terezinha Nunes, *University of Oxford, UK*
Peter Taylor, *Curtin University, Perth, Australia*
Dina Tirosh, *Tel Aviv University, Israel*
Manuela Welzel, *University of Education, Heidelberg, Germany*

Scope

Mathematics and science education are in a state of change. Received models of teaching, curriculum, and researching in the two fields are adopting and developing new ways of thinking about how people of all ages know, learn, and develop. The recent literature in both fields includes contributions focusing on issues and using theoretical frames that were unthinkable a decade ago. For example, we see an increase in the use of conceptual and methodological tools from anthropology and semiotics to understand how different forms of knowledge are interconnected, how students learn, how textbooks are written, etcetera. Science and mathematics educators also have turned to issues such as identity and emotion as salient to the way in which people of all ages display and develop knowledge and skills. And they use dialectical or phenomenological approaches to answer ever arising questions about learning and development in science and mathematics.

The purpose of this series is to encourage the publication of books that are close to the cutting edge of both fields. The series aims at becoming a leader in providing refreshing and bold new work—rather than out-of-date reproductions of past states of the art—shaping both fields more than reproducing them, thereby closing the traditional gap that exists between journal articles and books in terms of their salience about what is new. The series is intended not only to foster books concerned with knowing, learning, and teaching in school but also with doing and learning mathematics and science across the whole lifespan (e.g., science in kindergarten; mathematics at work); and it is to be a vehicle for publishing books that fall between the two domains—such as when scientists learn about graphs and graphing as part of their work.

Science, Learning, Identity
Sociocultural and Cultural-Historical Perspectives

Edited by

Wolff-Michel Roth **Kenneth Tobin**
University of Victoria, Canada *City University of New York, USA*

SENSE PUBLISHERS
ROTTERDAM / TAIPEI

A C.I.P. record for this book is available from the Library of Congress.

Paperback ISBN: 978-90-8790-080-9
Hardback ISBN: 978-90-8790-090-8

Published by: Sense Publishers,
P.O. Box 21858, 3001 AW
Rotterdam, The Netherlands

Printed on acid-free paper

All Rights Reserved © 2007 Sense Publishers

No part of this work may be reproduced, stored in a retrieval system, or transmitted in any form or by any means, electronic, mechanical, photocopying, microfilming, recording or otherwise, without written permission from the Publisher, with the exception of any material supplied specifically for the purpose of being entered and executed on a computer system, for exclusive use by the purchaser of the work.

CONTENTS

Preface   vii

Aporias of Identity in Science: An Introduction
Wolff-Michael Roth, Kenneth Tobin   1

A. IDENTITY IN URBAN SCIENCE   11

   Introduction   13

1. Structuring New Identities for Urban Youth
   Kenneth Tobin   15

2. Science Learning, Status, and Identity Formation in an Urban Middle School
   Stacy Olitsky   41

3. Learning and Becoming across Time and Space: A Look at Learning Trajectories within and across Two Inner-city Youth Community Science Programs
   Jrène Rahm   63

4. Urban Science Education
   Kenneth Tobin, Jrène Rahm, Stacy Olitsky, Wolff-Michael Roth   81

B. GENDERED IDENTITIES   97

   Introduction   99

5. Learning to be Engineers: Welding together Expertise, Gender, and Power
   Karen Tonso   103

6. Outsiders Within: Urban African American Girls' Identity & Science
   Katherine Scantlebury   121

7. Gendered Identities
   Karen Tonso, Katherine Scantlebury, Wolff-Michael Roth, Kenneth Tobin   135

C. IDENTITY AS DIALECTIC   147

   Introduction   149

CONTENTS

8 Identity in Scientific Literacy: Emotional-Volitional and Ethico-Moral Dimensions
   Wolff-Michael Roth — 153

9 Dis/Continuity of Identity: "Hot Cognition" in Crossing Boundaries
   SungWon Hwang, Wolff-Michael Roth — 185

10 Identity in Activities: Young Children and Science
   Maria Varelas, Christine C. Pappas, Eli Tucker-Raymond, Amy Arsenault, Tamara Ciesla, Justine Kane, Sofia Kokkino, Jo E. Siuda — 203

11 Activity, Agency, Passivity
   Wolff-Michael Roth, Maria Varelas, SungWon Hwang, Kenneth Tobin — 243

D. DISCURSIVE CONSTRUCTIONS OF IDENTITY — 257

   Introduction — 259

12 A Beautiful Life: An Identity in Science
   Yew-Jin Lee — 261

13 When Clarity and Style Meet Substance: Language, Identity, and the Appropriation of Science Discourse
   Bryan Brown, Gregory Kelly — 283

14 Identity Performances in a Science Book Club for Young Children
   Nancy Brickhouse, Pamela Lottero-Perdue — 301

15 Discursive Constructions of Identity
   Yew-Jin Lee, Bryan Brown, Nancy Brickhouse, Pamela Lottero-Perdue, Wolff-Michael Roth, Kenneth Tobin — 325

Identity in Science: What-for? Where-to? How?
   Kenneth Tobin, Wolff-Michael Roth — 339

Index — 347

PREFACE

This book about science, learning, and identity, as any book, is the result of a cultural-historical process that we, as any individuals, produce and are subjected to. Much in the same way that history is not made by individuals who act independently but by individuals who concretize cultural possibility, this book is not merely the outcome of two editors getting together and deciding to do it. Rather, there is a point in the cultural history of a field where realizing such a book that a particular concept becomes a general possibility, which is then realized in concrete form by particular scholars. Over the past seven or eight years, it has become increasingly apparent that the study of identity, which has had a decades-old history in other disciplines, also comes to be an important issue in science education. With this book, we introduce major ways of theorizing and studying identity and attendant issues that currently exist.

The book has three major objectives: (a) introduce science educators to the various dimensions of identity in science; (b) develop a new form of scholarship that is based on the dialogic nature of science as process and product; and (c) achieve the two previous objectives in a readable but scholarly way.

We have planned this book as both very readable and very articulate about all matters of identity concerning science, science education, science learning. We also designed this book as going beyond a simple collection of chapters that look more like journal articles with little connection between them. All through the production process, our concept has been to create a forum in which leading scholars present and interact over and about issues arising from the identity concept. To achieve this goal, we have brought together eleven chapters by leading scholars in the field, who combine an interest in both identity and sociocultural or cultural-historical perspectives. These scholars not only contribute a chapter but also engage in one or more interactive co-authored pieces in which the salient issues of the chapters are discussed. Grounded in different types of empirical situations, the contributors to this volume articulate aspects of identity and how these pertain to learning in science. To contravene a reductionist approach, which places questions such as those at the core of this book into the heads of individuals, the contributors frame the issue of identity in terms of sociocultural and cultural-historical theories.

Victoria, Canada
New York, USA
March 2007

WOLFF-MICHAEL ROTH, KENNETH TOBIN

APORIAS OF IDENTITY IN SCIENCE

An Introduction

The literature over the past decade has shown that identity is increasingly becoming one of the core issues in the study of knowing and learning generally and knowing and learning in science specifically. Although it may appear that the question of "who" someone is can be answered easily, the notions of "self" and "identity" continues to be full of riddles (Mikhailov, 1980). The problematic nature of identity arises from the fact that there are at least two aspects to identity. On the one hand, a person appears to have a core identity, which undergoes developments that are articulated in autobiographical narratives of self. A thirty-year-old person can point to a photograph and say, "This is me at the age of five," and we recognize a resemblance; more so, anything the person says to have done provides us with resources to know who this person is, her identity. In this perspective, events in our lives may provide us with resources to understand ourselves differently, leading to changes in our biographies. This aspect has been articulated in terms of the narrative construction and reconstruction of Self, which is a function of the particular collective with which we identify. Second, in contrast to the contention of identity as a (relatively) stable phenomenon that is constructed in biographical narratives, the experience of the different ways in which we relate to others in the varying contexts of everyday life has led postmodern scholars to conceive of self in society as something frail, brittle, fractured, and fragmented (Giddens, 1991). In some situations, we feel powerless: observers and we might say that we are less powerful or attractive than others; in other situations, we are the focus of attention and wield a certain amount of power. Thus, from one setting to the next, our identities, as revealed by our transactions with others, change. We have to ask, "How can our identities simultaneously be continuous and discontinuous, context-independent and situated, stable and frail, or adaptive and brittle?" and "Why are there differences between the self in narratives and in ongoing, concrete daily life?"

The contributions to *Auto/biography and Auto/ethnography* (Roth, 2005) provide us with a first answer to this question, as they suggest a dialectical relationship between individual and collective. Thus, individual lives are concrete realizations of possible lives, where possibilities always exist at a collective level. More so, biographies and autobiographies never are singularities but both in content and form produce and reproduce culturally available contents and forms. If the content and form of a narrative truly were singular, they would be written in a private language, which constitutes an irresolvable contradiction—a completely personal language would only be understood by the person speaking it and therefore would not constitute a language at all.

In the following, we articulate a general framework for approaching identity and human experience. This will allow us to better understand the relationship between identity, activity, and auto/biography, on the one hand, and provides a context for the different studies in this book, on the other hand. We begin with a phenomenological framing of the different problematic issues, aporias, that those face who attempt to come to grips with the phenomenon and concept of identity across a variety of human experiences. This framework contains several dialectical relations that—in turn—can be grounded in dialectical relations flesh, body, same, and other. This framework requires us to extend agential approaches to identity (and activity) to include passivity as an essential component at the very heart of agency. Similarly, at the very heart of identity is continuity over time, which is produced and maintained through memory and narrative, both of which have, in the same way as the general framework, the human communal experience (i.e., the *with*) as their fundamental condition of being.

CONDITIONS OF/FOR IDENTITY AND THE APORIAS OF BEING

Etymologically, the term *identity* derives from the Latin term *idem*, the same. *Identity*, therefore, means identical with itself, across time and space. But anyone looking back saying "this is me at the age of five" will recognize that she is different today, as a thirty-year-old, than she was twenty-five years earlier. The fragility of identity precisely is its difficult relation to time and the question, what is it that is the same? To get out of this aporia, or rather, to reframe it, the term *ipse* identity has been introduced (Ricœur, 2004), which draws on the semantic field of the same as *ipse*, *Self*. Whereas idem and idem identity refer to permanence in time, *ipse, Self*, does not imply such an unchanging core of a person (Ricœur, 1992). The two terms, idem and ipse are dialectically related, as at any one point in time, a person is identical with itself in terms of idem but is also a Self (ipse), with very different temporal properties. This temporal Self obtains its temporal cohesion, as shown below, in the form of auto/biographical narratives in which the uniqueness (identity) of a person is captured in a unique auto/biographical trajectory. This trajectory, as it will turn out, is not so unique, because narratives make use of language, plots, and characters that are cultural possibilities, and therefore also expresses a Self generally possible and available. The uniqueness is in part achieved in the dialectic of the Self and the material body, which is a source of passivity and being-affected.

A second moment of fragility of *identity* derives from its confrontation with others, or rather, with *the other* generally. The *same* thereby comes to be confronted with the *other than the same*, and, in fact, stands in a dialectical relation with it. The complex play of the same|other and oneself|another dialectics stands out quite clearly, for example, in the child who has lost a limb and comes to school with prostheses. Materially, these additions clearly are other than the body parts that they come to replace; clearly the body of the child has changed dramatically,

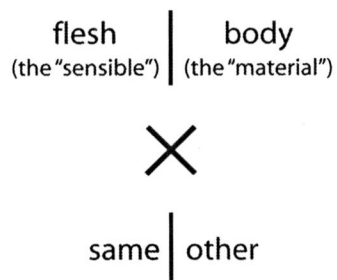

Figure 1. Identity involves the relation of two dialectical relations, flesh\body and same\other, leading to additional dialectical relations when one dialectical relation is conjugated to terms in another dialectic, and associated aporia.

which is perceptible especially when the artificial limbs have been taken off. But at the same time, the child can look back, remember the games played with the original leg, and, perhaps, the moment of the accident that damaged it. He can talk about *himself*, who he is, and, perhaps, how he has changed as a consequence of the trauma, all the while assuming that there is a constant element across the sequence of events: the *Self*. There therefore exists yet another dialectic, the one involving the person with senses, memory, and the material body that constitutes the substrate for the former (Franck, 1981).

All of these dialectics can be visually expressed in a simple schema, whereby a flesh|body dialectic comes to be conjugated and iterated with a same|other dialectic (Figure 1). The *flesh* is a phenomenological term denoting the body with all its sensual properties. Thus, whereas the *body* refers us to the mere material, the *flesh* refers us to the very possibility of being, agency, and passivity. The *flesh*, seat of agency and senses, is the mediator between Self and world. It is through the flesh that we are open and exposed to the world, the generalized other, subject to being affected and fashioned by the cultural and material life conditions: from the beginning, we are (in flesh) "subjected to a process of socialization of which individuation is itself a product, the singularity of the 'me' being forged in and through social relations" (Bourdieu, 1997, p. 161). The contradictory identity experiences described by Hwang and Roth (chapter 9) derive precisely from this openness to the world of the fleshly nature of the human Being, who can no longer (or not easily) make sense when physically moving from one cultural context into another. The very source of difference in physical experiences lies at the heart of the difficulties African American students experience, when their sense of rhythm, temporality, and proximity are confronted with the different forms of physical-material relations typical for the white middle-class culture that governs U.S. schools. The emotional-volitional and ethico-moral dimensions of identity Roth describes in chapter 8 also derive from the fact that the flesh constitutes a condition for human nature, Self, and identity. Without the experience of the flesh, itself the condition of the possibility to experience, there would not be emotionality or the intentionality it enables.

We may now conjugate each term of one dialectic with the opposing dialectic to yield dialectical relations that are at the heart of the troublesome nature of *identity*. For example, when we conjugate the *flesh* with the same|other dialectic, we obtain the mentioned oneself|another dialectic. This dialectic, which means that oneself and another presuppose each other, embodies some interesting features. Thus, to articulate but one of these, the Self forms in the image of the generalized other, it is a possible self within the context of the culture of the person, who concretely realizes one of these possible selves as her personal Self. Below, we further articulate how a Self is realized in this way in and through conversations about a person's auto/biography; this approach also is at the heart of Yew-Jin Lee's "A Beautiful Life" (chapter 12).

When the *body* is conjugated with the same|other dialectic, we arrive at the opposition between material bodies; it is this opposition that irrevocably separates each person from all other persons. It is the source of the unbridgeable otherness of other persons and the world. But it is also the condition for anything like human Being, which requires signs as a form of communication; and signs—sections of the material continuum [traces, sounds] standing for other sections of the material continuum [objects, events]—there would be nothing like cognition, knowing and learning. Without signs, there would be no memory, no culture, no learning from others, no human forms of society. This is the origin of the link between science, learning, and identity invoked in the title of our book. Phenomenologically speaking, the same–other distinction is made on some surface, for example, the skin, the locus where self and other rest in *proximity*. Not surprisingly, therefore, the notion of proximity has become central to those attempting to construct a *first* philosophy, that is, a philosophy that takes into account anthropogenesis, the coming into being and the emergence of everything that makes human nature possible, including the distinction between self and other, thinking, (cultural) learning, memory, and so forth.

AGENCY AND PASSIVITY AS SOURCES OF AND FOR IDENTITY

Most scholarship not only in science education but in many other disciplines as well focuses on agency at the expense of its correlate, passivity, which can be theorized such that it becomes the privileged attestation of otherness (Ricœur, 1992). This comes with a theoretical advantage in the sense that it leads to the agency|passivity dialectic, which decenters the self to the extent that it no longer serves as the exclusive anchor and foundation of identity. No longer, therefore, can the theoretical concept of identity be derived from intentionality, so that our identities no longer are at our will. We do not simply construct identities, but our identities emerge from the agency|passivity dialectic that grounds human nature as such.

The agential production of identity underlies much of the work conducted in the social sciences, including science education. Thus, for example, Nancy Brickhouse and Pamela Lottero (chapter 14) show how, through their forms of discourse, boys and girls *position themselves* in and during discussions of books in their book-club meetings. Maria Varelas and her colleagues, too, feature reading sessions, this time in class, where students produce and reproduce particular identities. In both instances, the authors draw on the theoretical notion of *positioning* to theorize the ways in which individuals contribute to producing their identity. Similarly, Bryan Brown and Greg Kelly (chapter 13) show how minority students produce differentiated identities in and through their talk about the physics of baseball. Discourse also plays a role in Karen Tonso's account (chapter 5), in which she articulates the production of identity through the discursive assignment of individuals to particular categories (e.g., nerds, curve-breakers, brownnosers). The attribution may derive both from self-attribution and other-attribution. Although identity production involves discourse in all three cases, the third account differs from the two preceding ones. Whereas identity is produced and reproduced in and through talking about science in children's books and the physics of baseball in the former two instances—even without talking about it—it is the topic of talk in the Tonso study. In the former two instances, identity is a by-product of talk, whereas in the latter instance it is its main object.

An over reliance or exclusive use of agency as the source of identity—which is an approach that constructivism leads us to—leads us to an aporia: how does a constructing agent construct its own beginning? How does the conscious self, which is said to construct its identity, construct its own beginning? *Passivity* and *participation in social relations that precede consciousness* are the answer—both from the perspective of anthropogenesis (becoming human) and ontogenesis (becoming a person). Preceding anthropogenesis and the first instance of consciousness, pre-humans lived together, hunted together, used and learned to use tools together, related to each other, expressed affection to individuals and collective— the *with* is the condition for anything like human consciousness to emerge and exist (Nancy, 2000). Similarly, from the very moment they are born, babies participate in social relations even though they do not experience themselves as separate Beings and prior to any form of consciousness. In fact, parents change their behaviors (practices) as and because they interact with their babies, who therefore contribute to transforming cultural practices of child rearing prior to being conscious of themselves as separate Selfs.

Ultimately, then the world comes to be comprehensible, is immediately endowed with sense, because the incarnate person, with its senses and mind, has the capacity to be present in the world outside of itself. A simple experiment with the sense of touch provides us with evidence of this form of experience: sliding our fingers along some surface provides us with a sense of its characteristics, its roughness or smoothness, ripples and cracks. But these characteristics are not felt to be inside ourselves, and even less within our minds—we truly sense these characteristics to lie just outside of the skin of our fingers that slide along the surface. Precisely because the flesh is outside of itself in this way it is open to be impressed and lastingly modified by the (material, social) world (Bourdieu, 1997). At the very moment that we touch something, this something touches us in return, which we, shifting our attention to the sense of touch, experience on the inside of our skin.

MEMORY, NARRATIVE, IDENTITY

Bodies are singular; identities are not. Experiential trajectories of bodies and flesh are singular; accounts of trajectories are not. Narrative forms, language, and grammar are structures people draw on when providing accounts of who they are; and because these resources are general, existing at the collective level—from the other, for the other—they inherently embody and encode patterned ways in which auto/biographies and identities can be recounted. Each auto/biography therefore simultaneously is particular and general, singular and plural. This capacity to recount a life and therefore individual and collective memory, essentially derive from the relation of Self and Other (Franck, 1981); the social is the condition for temporality and the memory that bridges the distance between then and now.

Narrative forms (genres), too, are resources that can be transformed into new forms at the very moment that they reproduce an aspect of culture, producing and communicating narratives. When we talk to someone else about who some third person is, her identity, we presuppose that our interlocutor already understands the particular *type* of identity that our descriptions is to evoke. That is, biographical and identity narratives presuppose that the specific individual whose (auto-) biography is being articulated in the interview is a particular *type* of person. Biographical accounts and identity accounts are concrete realizations of presupposed, generally intelligible *plots* and *characters*. It is precisely this relationship between auto/biography and narrative in terms of plots and characters that undergirds the conversion experience from being an alcoholic to being a reformed alcoholic during membership in alcoholics anonymous groups. One is reformed and an accepted, core member of AA at the moment that one can tell one's life story in the form of the typical AA narrative form, its typical plot and characters (Lave & Wenger, 1991). Plots and characters, therefore, are cultural ways of organizing the memory necessary to maintain an auto/biography, individually and collectively. To know who someone is, we do not have to remember isolated bits and pieces of facts, but we remember plots and characters, which are filled with more specific details that make up the particulars of specific lives and identities.

Let us take a closer look at a concrete example. The simultaneous production and reproduction of culture, narrative genre, and language is exemplified in the following episode and analysis, drawing on a conversation between Michael (Roth) and a high school student, who later in the project became co-researcher and co-author (Roth & Alexander, 1997).

01 Michael: How do your parents combine that? Do they believe? Are they regular . . . ?
02 Todd: My mother is very similar to myself, she doesn't think about it as much as I do, because for her it is a difficult question and she was brought up not to think about it. Because she said, she was told that she was just a girl and girls don't answer such questions. And she is a very pacifist person, and she doesn't like that, the conflict. Neither do I, she doesn't like conflict and that's why she doesn't fight all that much. And therefore she– I mean she thinks about it now, and she combines the two. She also brought me up going to church. But very much also because she is the head of nursing, she's got a master's degree in nursing and studied a lot=

03 Michael: =at McMaster?
04 Todd: No she got her master's at U of T ((University of Toronto))=
05 Michael: =and she works at?
06 Todd: She teaches nursing at Sheridan College, and she's been head nurse in the hospital, very bright, very able she coulda been a doctor because she is a scientifically minded woman, who has very very strong beliefs very similar to mine, I mean, we disagree more on the social truths.

As part of a study concerning scientific and religious discourses, Todd and Michael have talked a lot about science and religion. Michael's opening question extends the conversation concerning Todd's acknowledged Christian faith to those he understood his parents to have. Their conversation follows a familiar pattern—see, for example, Lee's chapter 12—whereby the identities and biographies of individuals, such as scientists, are told in terms of their relations to, and influences from, parents and siblings. This type of plot, which we may gloss as "being influenced by parents and siblings and therefore becoming like them," constitutes a resource that has evolved as part of Western cultural history and is available to anybody a little familiar with biographical accounts in Western cultures.

In turn 02, Todd begins his response stating that his mother is very similar to him, but also different in that she does not think as much about the relationship between science and religion as he does. He defines her in terms of who he is, where he presupposes that his interlocutor knows who he is—a reasonable assumption given that at the two had known each other for several years in don|resident and teacher|student relationships. That is, Todd shares with his mother certain aspects, and thereby aspects of their identities are common. Yet there were differences as well. Thus, as Todd continues, he describes her biographical influences in terms of a familiar pattern: "Girls don't answer such questions" as the ones about the relationship between science and religion, especially if there are differences in the discourses of the two with respect to some subject. Here the relationship between the specific person, Todd's mother, and her taking the role of a character in a familiar *type* of plot is not just implicit but made the topic of talk. Todd continues with the discursive fitting of his mother's identity into familiar if not stereotypical character traits for a woman. She *is* a very pacifist person, a claim substantiated by the statement that she does not like conflict. Saying that his mother does not like conflict is already a generalization about the ways in which she goes about her everyday life. Saying that she is a pacifist generalizes one step further by actually using a cultural repertoire for characterizing and classifying individuals and, thereby, attributing particular forms of identity.

Todd does more in and with his response. Using language, he makes available his own subjectivity. Thus, he says that he is very similar to his mother. He experiences himself in the way he perceives his mother to be. That is, although Todd and Michael know that he is a unique individual, having his unique experiences, Todd articulates his subjectivity in terms of someone else—identity here is a character. At the very moment that he expresses his mother's or his own uniqueness, he transgresses the isolation of his personal experience, attributing both lives to a form (type) of life and identity. Language is the means to do it. This language is not their own; it is already presupposed to be those of others. The language both interlocutors use more-or-less existed in the same form and was used when each was born. That is, at the very moment each interlocutor articulates his most private thoughts, questions, and experiences they have to make use of means that are not their own (Derrida, 1998). Our own subjectivities are intimately tied to intersubjectivity; we know ourselves only through our relationship with others, our Selves in fact or the Selves of others, a relationship Paul Ricœur captured in the title of a book, *Oneself as* Another.

The use and presupposition of general identities to develop who a person is continues in the next couple of turns. At the time of the interview, we are in Ontario. When I query Todd whether his mother went to McMaster University, it is not just a question about any university: McMaster University is well known beyond Canada for its medical and nursing faculties and facilities; its problem-based learning approach is a reference point both in the theoretical and practical literature on teaching and learning in the medical profession. Michael's query therefore can be understood as seeking to find out whether Todd's mother is an alumni of a particular university known for the outstanding quality of its program. In turn 06, Todd states that his mother has been a head nurse, which is yet another statement about her identity that draws on a particular *type*. He says and presupposes that his interlocutor understands that she *is* a head nurse rather than stating what she does on any concrete day.

The conversation as captured in the transcript is interesting because it also constructs the identity of a person, Todd's mother, by stating what she is not but what she could have been. At one point, Todd says that his mother is "very bright, very able. She coulda been a doctor because she is a scientifically minded woman." Here an attribute of her identity glossed as scientifically minded is made to work together with the potentiality of being a doctor, supported by the additional character attributions that she is "very bright, very able." The statement that his mother is scientifically minded is used as a resource to support the claim that she could have been someone else, a doctor. As long as the supporting statement itself is not questioned, this potentiality itself becomes an aspect of his mother's identity for the purposes of the ongoing activity.

Choosing a narrative framework, where (auto-) biographical materials are not just about specific persons but also about *types* of persons, we do not have to ponder questions of the truth between what people say and what really happened to them, about their real beliefs and what they say they believe. The discourse analytic framework drawn upon by the authors to Part D of this book allows us to see interview transcripts as establishing versions of the world, versions that have relevance in, and pertain to, the current situation. Different versions, different identities may be evident not only between different situations but also within a single situation, such as the same interview. Thus, all interviews with Todd and his classmates have to be seen in this perspective—participants were oriented to the production of an intelligible text about the nature of science, epistemology, learning, and religion. The interviewer and his interviewees were inherently responsible to one another for producing each meeting and intelligible conversation, and in doing so, drew on culturally and historically available resources. As a result, both interview situation and interview text are concrete realizations of general possibilities—they are dialectical, constituting both particular instances and general cultural-historical possibilities.

There are several different levels of events that occur in this episode, all of which can be traced back to the dialectical nature of culture. First, Todd and Michael produce an interview; and they do so in a way that allows readers to recognize the event *as* an interview. The two participants know this as well as the readers although this particular event, recorded on videotape, is highly singular and occurred only once (in this form). Second, the two participants understand what the respective other is saying, even though they may never have heard a particular question or statement before. Thus, Todd has talked about his mother as having had the potential to become a doctor, the career that he envisioned for himself and eventually realized. In turn 01, Michael asks about how Todd's parents dealt with the issue currently the topic of talk. Todd has an immediate response, which is concerned with the similarities and differences between his mothers and own identity. Third, both Michael and Todd draw on a particular aspect of telling a biography—influence of parents: the latter volunteers information about the influence his mother has had on his going to church. But even at the very moment that one of the two interlocutors begins to draw on the family repertoire in auto/biographical accounts, he presupposes the possibility and intelligibility that family members may play a significant role in autobiographical accounts. The biographical nature of character and plot that are developed in such interviews allow the articulation of learning, development, and change narratives. Rather than being narratives about particular learning, development, and change, these narratives are concrete realizations of possible narratives that exist for and can be drawn on by all members of a culture speaking the same language.

At all three levels, the participants realize cultural possibilities for doing interviews and for constructing auto/biographical accounts of their careers. That is, despite the very singularity of *this* interview and *this* student's identity, we recognize in the event and the narrative produced culturally possible forms of doing interviews and telling identities. Now the singular nature of the event and identity also means that they have not existed before, which means that they are not reproduced but newly produced forms of interview and identity. Yet the very fact that they recognizably *do* an interview and *construct* an identity tells us that they *reproduce* a cultural form.

REFERENCES

Bourdieu, P. (1997). *Méditations pascaliennes.* Paris: Seuil.
Derrida, J. (1998). *Monolingualism of the Other; or, The prosthesis of origin.* Stanford, CA: Stanford University Press.
Franck, D. (1981). *Chair et corps: Sur la phénoménologie de Husserl.* Paris: Les Éditions de Minuit.
Giddens, A. (1991). *Modernity and self-identity: Self and society in the late modern age.* Stanford, CA: Stanford University Press.

Lave, J., & Wenger, E. (1991). *Situated learning: Legitimate peripheral participation.* Cambridge, England: Cambridge University Press.
Nancy, J.-L. (2000). *Being singular plural.* Stanford, CA: Stanford University Press.
Ricœur, P. (1992). *Oneself as another.* Chicago: University of Chicago Press.
Ricœur, P. (2004). *Memory, history, forgetting.* Chicago: University of Chicago Press.
Roth, W.-M. (Ed.). (2005). *Auto/biography and auto/ethnography: Praxis of research method.* Rotterdam: SensePublishers.
Roth, W.-M., & Alexander, T. (1997). The interaction of students' scientific and religious discourses: Two case studies. *International Journal of Science Education, 19,* 125–146.

Part A

IDENTITY IN URBAN SCIENCE

INTRODUCTION

Urban classrooms pose particular challenges to systems of schooling, as the contradictions arising from the reproduction of an inequitable society are most salient here. These contradictions are especially salient to the research at the University of Pennsylvania, and perhaps to other U.S. universities as well, where the Faculty of Business, with annual tuition fees exceeding US $60,000, lies within minutes of walk from the poorest neighborhoods and neighborhood schools, where students do not come to school because they only have one set of clothing or do not have the 25 cents for the bus ride. Within minutes from the university, it is dangerous to walk through the streets at night, as there is a high possibility of getting robbed, beat up, and shot. (Few people nowadays remember that MOVE, a radical African American back-to-nature and anti-technology movement, had its headquarters in the same neighborhood. In 1985, the mayor had explosives dropped on a house and more than 10,000 rounds of ammunition were shot, killing six adults and five children.) Within minutes from what many regard as the world's leading business school, young women may be raped if they walk through the streets at an inopportune moment. To the students of the two comprehensive high schools in this part of the city, this is their 'hood, where they live their "normal" social lives; many of them witnessing shootings and experiencing extreme forms of violence from the tenderest of age.

In such situations, it is not just that the identities of science teachers and students are at stake—identities always are at stake. But making it safely through the day takes a particular fluency in survival techniques, that is, a fluency in cobbling together the resources at hand for making and making it through the various settings in which urban youths might find themselves. Urban schools are places that are only marginally safer, as there is always a chance of "getting rolled" or otherwise assaulted. Many U.S. high schools have weapon detectors and it is impossible to enter the school without passing through the detectors and getting checked in some other way by security personnel. Schools as much as any other setting through which urban youths pass during their day are constitutive and become an integral part of their identities. More specifically, students' existing identities continuously are transformed; their new identities continuously emerge from participating in an activity system focused on teaching and learning. From the virtual identities that the students concretely realize is drawn a potential that the identities appropriate, always in a bricolage fashion, transgressing boundaries, never pure, but always characterized by hybridity, heterogeneity.

Nevertheless, scholars and even the individuals themselves think identity in terms of a constant core Self that remains unchanged through time. The constancy of identity is called into question especially, however, in moments of crisis or when

PART A: IDENTITY IN URBAN SCIENCE

people change from one activity system to another, thereby continuously threatening the sense of a constant self that is maintained over time. In each situation of their daily praxis, students (and teachers) are involved in the struggle of making and remaking who they are, how they understand themselves, and how they are understood by others. Identity definitely is not a stable given that individuals take in and out of situations; rather, identity can be regarded as one of the outcomes of a person's participation in ongoing activity.

In this first part of the book, both Kenneth Tobin and Stacy Olitsky write about the development of identity from the same urban setting, two schools in Philadelphia where there has been a seven-year research program on learning science in urban settings. The two chapters, therefore, are complementary and should give readers at least a kaleidoscopic perspective of the issues arising for development in the kinds of settings that the authors have worked in. Tobin's longitudinal study focuses on Shakeem, a student who, despite all the odds stacked against him, ultimately makes it into college, where he, contradictorily, supported by a drug dealer, eventually makes the dean's list (of outstanding students). Tobin's account is rich, noting all the contradictions one might find in heterogeneous and continuously changing identities cobbled together from a multitude of resources in a constant bricolage that has as its major goal to *make do*. It is perhaps above all his participation as a youth researcher that mediates Shakeem's learning, his appropriation of sociological discourse on Tobin's research squad, and his becoming aware of his own societal position that ultimately allows him to be sufficiently successful to get into college and be successful even there.

Stacy Olitsky provides a fascinating account of identity development in another urban science classroom in the same city and, in the course, develops a theoretical framework grounded in sociology of emotion. She shows how during successful interaction rituals, emotional energies are both reproduced and augmented, as other participants who co-constitute the collective entrain them into solidarity.

Jrène Rahm conducts her studies in the larger Montreal area where she works with poor urban youth, many of whom are immigrants, and especially with girls. Although Rahm does not do so, we can envision the usefulness of the concept of identity as the result of a diasporic praxis that leads to hybridity, difference, and heterogeneity (Roth, 2006). The children Rahm works with come from different countries or are born to parents who recently immigrated to Canada, and who cobble together momentary and continually transitional identities from the various cultural resources—parent culture, French, English, other cultures present in Montreal.

REFERENCES

Roth, W.-M. (2007). Identity as dialectic: Agency and participation as resources and fields in the emergence of hybridized identities that acknowledge science and home culture. In A. Rodriguez (Ed.), *Multiple faces of agency: Innovative strategies for effecting change in urban school contexts* (pp. •••–•••). Rotterdam: SensePublishers.

KENNETH TOBIN

1. TELL ME WHAT YOUR LIFE LIKE ...

Your Life is Dis—Your Life is Dat—Mine's Real

> *Tell me what your life like*
> *Shit, mines is real*
> *Everything signed is sealed*
> *Tell me what your life like*
> *Nigga, mines real*
> – Beanie Sigel, February 29, 2000

Beanie Sigel's rap about the life of African American males in inner city Philadelphia is a shocking rendition of a world I have never experienced—a set of truths involving death, violence, incarceration, drugs, guns, sex, poverty, relationships, rage, loyalty and a dialectical relationship between hope and despair. "Tell me what your life's like. Shit. Mine's real." As I listen to Sigel's "truth" my eyes well with tears. How can I understand the problems of urban education without first experiencing such an ontology—one that is vastly different from my first-person experience? In this chapter I explore the ways in which my experiences of urban life, urban education, and urban youth have changed my identity. I address issues of identity through case studies, the first an autobiography and the second of an African American youth who I refer to by the pseudonym of Shakeem.

EXPERIENCING URBAN LIFE

When I started my research in urban schools I was very much an outsider. Unfamiliar with the youth culture of urban African Americans, my first encounters with them, en masse, was on a train from suburban to inner city Philadelphia. I was residing in temporary accommodation, making the four-mile journey to my inner city office via public transport—a new experience for me. I did not know how to stop a train, pay the fare, change trains at 69th Street, and anticipate the stop where I would get off. I felt incompetent and hesitant. When I boarded the train most of the fellow passengers were white—sitting silently, reading newspapers, sleeping, and working with a variety of paper and digital media. As we got closer to the downtown area my discomfort increased, largely due to the African American youth who swarmed on to the train at consecutive station stops and, since there were no seats, stood in groups of two to many and interacted loudly, often in animated ways as they greeted one another and caught up on the latest news. I sat stoically, dealing with an odd mix of emotions, including anxiety, fear and excitement. At a conscious and rational level it was students like these who had dislodged my professional equilibrium, bringing me to Philadelphia to address the problems of urban education.

Reasons for my discomfort are hard to articulate. Race was a factor, possibly the main one. The students were black and I was white, yet the source of my discomfort was not skin color—more likely a lack of familiarity with the youth culture that governed their ways of being. I was clearly in a strange place without the stocks of knowledge needed to make sense of the urban youths' interactions. Even though I had no plan to interact verbally with these youth, my proximity to them necessitated that I learn to interact nonverbally and emotionally since I was on the train with them—in close proximity. As they enacted their culture in dyads and larger groups I interacted with them continuously, without a conscious effort to do so. I watched them, experienced a range of negative emotions (e.g., anxiety, discomfort, and fear), and registered numerous disapproving thoughts as I judged their otherness through deficit lenses. Also, as their bodies entered what I regarded as my private space, I moved instinctively to retain a distance from them that was comfortable for me.

Even though I can now experience events like this with comfort and a sense of normalcy, at that time I was acutely aware of my cultural otherness and the associated emotions that produced insecurity—I did not feel safe and at ease. My sense of self was shaken by the presence of others I did not understand well and the novelty of being in an unfamiliar place. My lack of familiarity with the culture of these youth was not something I was conscious of and their practices were interpreted by me through cultural lenses I had developed in Australia and the southeast of the USA. I had a tendency to make sense of my experiences in ways that privileged white middle class culture and viewed differences as deficits. I felt a sense of disapproval and would have been more at ease riding a train with youth who were like those I had taught and researched—more like my own kids. My goals of having a quick, comfortable commute in which I quietly read and produced scholarly pieces were breached by the practices of others. Hence my agency was truncated and I developed a strong sense of not belonging and being incompetent.

My emotional responses to riding and alighting from the train were strongly negative. Mistakenly, I associated the urine-stench, discarded liquor bottles, and trash in the exit stairwell with the above ground neighborhood. As I emerged from the stairwell I was in a part of the city where African Americans lived, shopped, commuted to and from, and generally enacted life—it felt strange for me to walk through this neighborhood to my office. I was acutely aware of my whiteness and middle class lifestyle. Like being on the train, being in the streets produced a high level of discomfort. Almost everything seemed strange and the strangeness played with my emotions. I was cautious, tentative and ill at ease. Also, I was alert; on the lookout for signs of danger—making sense of where I was as a potential victim of social and physical violence. If I were to become a successful urban educator I was conscious that my ways of being would have to change, including my frames for making sense of urban life.

Four blocks from the station stop was City High, a comprehensive high school that mostly catered for students in a neighborhood in which almost all residents were African American, from home circumstances of economic hardship. City High was to be the central site for my research in Philadelphia, placement of new science teachers from my university, and for my coteaching and solo-teaching with Spiegel, who had just commenced a career as a science teacher. When I first set eyes on City High it reminded me of a prison, a place that reflected what I regarded as pervasive squalor. Just as my perceptions of the train, youth, and area surrounding the exit from the train stop were saturated with deficit perspectives, so were the neighborhood and the building in which City High was situated—the site for the education of more than 2,000 urban youth.

Producing Relevant Stocks of Knowledge

Within a few months I purchased a home in the southern part of the city, about four miles from my office, which was in the west. At this time I decided to address my cultural and social otherness by undertaking an ethnography of the neighborhood. There were several parts to this ethnography. I went into the streets most mornings at about 6 A.M., walking west from 2nd Street to 9th Street and north beyond South Street, which is one of the hubs for night life in the city. I looked for people who lived in the streets and observed them as they left the doorways, sidewalks, benches and stairwells that had for the night been their homes. I watched where these people went to clean up, how they scrounged food from trashcans and the gutters, used washrooms in coffee shops, and begged for food and money from anyone who was out and about. I also watched them milk coins from parking meters, public telephones and newspaper dispensers. I began to recognize participants and soon knew who was a stayer and who was passing through. I no longer flinched when street people yelled at me or moved toward me in threatening ways. Also, I began to be less conscious of the stench of South Street; smells that inevitably accompany a day and a night of human activity.

Several days a week, and at various times of the day, I'd sit in the window of a coffee shop on South Street, watching the people pass by. Also, I'd observe as I walked through the streets, learning about the different rituals enacted by an astonishingly diverse array of participants. I began to understand the streets of this neighborhood and how the enacted cultures changed in different places as a function of the hour of the day, day of the week, and season of the year. Gradually, my stereotypes about urban street life began to erode and the valence of my emotions changed from strongly negative to strongly positive. I began to enjoy being in the streets and my experiences with diverse forms of culture associated with the buildings, shop and body displays, dress, hair and shoe fashions, and practices of the ever-changing participants.

The next step in building my urban street identity was to walk to my office, choosing different routes for coming and going and varying the time at which I made the journey. These experiences afforded deeper insights into the neighborhoods and the ways in which cultural fields were defined in the streets according to time and place. As I better understood urban communities my own ways of being in the streets changed to afford my movement through them with confidence, albeit cautiously and with a growing sense of becoming streetwise.

My ethnography of the neighborhood allowed me to use my professional stocks of knowledge to understand the agency|structure relationships among the participants in a street culture. At the same time I sought perspectives from the professional literature and colleagues such as Wolff-Michael Roth, whom I interacted with regularly via email. Within my university there were resources I accessed too—scholars in cultural sociology, the sociology of emotions, and African American psychology. For example, Elijah Anderson, an African American sociologist, had written two books (Anderson, 1992, 1999) that greatly informed my understandings of African American youth and the streets. Also, Pierre Bourdieu's writing and frequent conversations with Wolff-Michael Roth about habitus in relation to my experiences in the streets and as an urban science teacher produced fresh stocks of knowledge that greatly expanded my agency and praxis in a variety of urban fields. My being in the streets allowed me to enact culture as praxis, feel failure and success, and reflect on what seemed to work and what did not. Colleagues such as Anderson, Bourdieu, and Roth were sources for new schema, referents for reviewing what was happening in my ethnography and for creating models for why events were experienced as they were. My agency expanded and my practices in the streets changed as my relevant stocks of knowledge increased and my understandings of what was happening deepened—allowing me to meet my goals, develop emotions with positive valence, and build an urban street identity associated with confidence and competence.

In parallel I was developing an identity as an urban science educator, distributed over three primary fields—science teacher education, research in urban science, and teaching urban science. Questions raised in one of my science education classes, about the extent to which research findings were applicable to the urban schools in West Philadelphia catalyzed my teaching of science at City High (Tobin, 2005). Because my efforts to successfully teach students in the low track of a neighborhood high school were quite unsuccessful, I called into question the efficacy of my knowledge about urban science education and launched a program of research that permeated all three fields, thereby changing my agency within those fields and restructuring the fields themselves.

Critical to my emergence as an urban educator was the establishment of an urban street identity and living in the city and doing ethnography helped in that regard. However, also salient were coteaching assignments at City High with a teacher called Spiegel, especially with small groups of students. Spiegel was young, strong, and popular with the students. He had grown up in a neighborhood like the one in which he was teaching and he could interact successfully with most students. They liked and respected him. As Spiegel and I interacted together to prepare small groups of low-achieving youth to peer teach in a nearby middle school, it is apparent from videotapes taken over a period of a year, that I learned to successfully interact with youth, anticipate what they would do, and enact praxis in ways that were timely and appropriate. No doubt the street ethnography equipped me in some ways to be more successful in these small groups, but it is clear that being with Spiegel was an enormous help. I learned to teach those students successfully by coteaching with Spiegel (Tobin & Roth, 2006). At the same time that I was struggling to teach science at City High as a solo teacher, I was building the capital needed for success in small, cotaught classes with Spiegel.

In the science methods class my students were new teachers who were also teaching in urban classes. The methods classes provided all participants (including me) with opportunities to be reflective on our experiences in relation to others' experiences, and the involvement in the class of urban youth, who advised us on *how to teach kids like me*, provided robust student perspectives that had to be considered in our science teaching and research. The porous boundaries of fields enabled me to take what I learned from my involvement in the methods course and enact it in other urban fields in which I participated, affording the attainment of individual and collective goals in those fields, while changing the structures of those fields, and expanding my agency and identity.

The sources for my production of capital were global in extent, involving people from all over the world through the Internet, writing papers for peer reviewed international journals, and participating in international conferences. I did not have to rely on personal experience and local resources to solve the problems that arose or even to frame the problems theoretically. Scholars from many continents and countries interacted with me regularly about my challenges in re-learning to teach science in urban high schools. Hence, I maintained my identity as a prominent science educator and felt very much like an explorer in a new world. Probably because of my status as a researcher and the foregrounding of research, neither others nor I regarded my initial lack of success as an urban teacher as a sign of weakness. Instead, my teaching was a resource for learning more about the challenges of teaching and learning in urban high schools.

Before moving on to explore identity in the context of an African American male student from City High, I address several issues that are salient to identity. My identity changed from field to field, depending on my agency—that is, the power to act in a given field. Even though I considered some parts of my identity as relatively stable across fields—the core parts of who I am—my move to Philadelphia certainly changed what I did and whom I interacted with. In terms of the individual|collective relationship, which are constituents of identity, there were noticeable changes because the collective was radically different—not just on the trains and in the streets and schools, but also in this region of the country. To be successful in most facets of social life I had to produce new forms of social, cultural and symbolic capital, expanding my agency in ways that would allow me to appropriate structures in all fields of my lifeworld, including those I produced through my own praxis. A personal goal was to become fluent and that involved creating new stocks of knowledge, including those associated with knowing the cultural possibilities of others comprising the collectives in which I was a coparticipant. The core parts of my identity, though resilient to change, were not immune from the turmoil in my social life. I could not check my emotions at the door and being in urban fields was a highly emotional experience that disrupted my sleep patterns and hence my ways of being in the home and opportunities to participate in leisure activities. Accordingly, the culture enacted in the home was mitigated by my emotional states as I entered into the home from other urban fields, and my attention focused on resolving the relentlessly unfolding problems from those urban fields. I was determined to succeed as an urban educator and willing to change and push until I was successful—until the negatively charged emotions were positive and a new identity was forged along with new (and appropriate) stocks of knowledge, allowing me to participate successfully in urban fields to attain individual and collective goals.

YOUTH CREATING AND SUSTAINING IDENTITIES IN URBAN FIELDS

In the second section of the chapter I explore the identity of Shakeem, an African American male who has been a participant in our research squad for more than five years. The data resources that support what I have learned about Shakeem's identity include ethnographic accounts produced by various members of our research squad, two video ethnographies and PowerPoint productions prepared by Shakeem and, over the five year period, transcripts of conversations recorded on audio- and videotape, interviews, and field notes.

Campaigning for Respect

As a middle school student, Shakeem lived in a poor, crime-ridden region of Philadelphia. Given the type of neighborhood and Shakeem's physical prowess it is not surprising that he dealt drugs and owned a gun at a young age. According to Shakeem it was a visit to his middle school by President Clinton that caused him to throw away his gun forever—because of the likelihood that the FBI would discover it during their frisking of the students. Another factor that probably contributed to Shakeem being able to walk away from dealing drugs was that his family moved to a different neighborhood, affording the chance for a fresh start. Shakeem's mother, unable to make ends meet financially, moved Shakeem and his brother to live with her mother in a different part of the city. Although Shakeem's mother and brother soon moved away, Shakeem continued to live with his grandmother for his three years at City High. According to Gale Seiler (2002), most people in Shakeem's neighborhood were African American, retired, on welfare, or unemployed. Unfortunately, selling drugs and other criminal activities were rife in this part of the city. As is the case in many inner city neighborhoods two forms of culture are advantageous—decent culture for the home and fields such as the church and the mosque, and street culture for successfully navigating the streets. Elijah Anderson drew attention to the necessity for decent folk to also enact age and gender appropriate forms of street culture Shakeem used a similar dichotomy to describe residents in this neighborhood—decent and *hos* (whores), *alchies* (alcoholics), and *crack heads* (drug addicts). Seiler noted that many of the youth in the age range from 18 to 20 were drug dealers and carried guns. Shakeem confirmed this tendency and his determination never to return to the drug trade or to own a gun. Furthermore, he only knew of one high school graduate from this neighborhood who had gone on to college.

I employ the dichotomous view of culture as a point of departure, especially as Shakeem made it clear through words and other actions that at any moment he would enact any culture, street, decent, and hybrid forms of each. Furthermore, as I make clear later in this chapter, the resources for producing, reproducing and transforming culture are global in extent and are locally accessed and enacted.

Getting to Know Shakeem I first learned about Shakeem in April 2001 from Ryan, a new teacher who identified poverty as a factor that militated against Shakeem's participation in science. Ryan noted:

> Shakeem, a fourteen-year-old African American, walks into his science class everyday. He is big for his age—about six feet tall and two hundred and fifty pounds. He is visibly poor, usually dressed in sweatpants, worn-out t-shirts, and beat-up sneakers. While most of his fellow classmates also are considered poor, he appears more deprived than the others. Under normal circumstances, he would intimidate his classmates or anyone in the school, but he is not viewed as intimidating in this classroom and this school. Despite Shakeem's size he must constantly fight, physically and emotionally, for respect, arguing with others during class. Although these arguments never start over his poverty status, verbal confrontations usually escalate to personal attacks against his appearance as he defends his reputation. Arguments such as these drastically affect Shakeem's performance in the classroom . . . he frequently acts out in class and, at times, refuses to do work.

Anderson (1999) explained that according to the code of the street, respect is a form of currency that can offset poverty. He showed how, in urban streets, youth who are highly respected have greater access to resources than those who are less respected. Being physically strong and a good fighter are obvious physical attributes that earn respect in the streets. Also, street code attributes respect to being attractive, sexually appealing, and well dressed. Because students have to successfully navigate the streets to make it to school, and many of the youth in the school are from the same or similar neighborhoods, the code of the street also can be highly salient in urban schools. That is, the boundaries between the fields of the street and the school are porous and culture originating in the streets often is enacted in the school, especially in peer–peer interactions. Those who command respect in the streets are likely to be given respect in the school because after school the code of the street likely will be enforced on the way home. Though there are obvious ramifications for disrespecting in school someone who commands respect in the streets, this did not mean that the best street fighters spent their time fighting in school. Having grown up in tough circumstances Shakeem understood how to survive in the streets and he also knew that he could not fully enact street code at school with impunity. He commented that, "when you come around my neighborhood and you talking shit I would knock you the fuck outta there. That's just how it is." In contrast, at City High Shakeem was relatively non-violent and recognized the necessity to earn sufficient respect to create and sustain the social networks he valued so highly.

Poverty A first strike against Shakeem was that he was regarded by peers as po (poor). According to Cedric, Shakeem's closest friend in the ninth grade, "there are different classes of being poor. Some people are less fortunate. I am poor, but some people might be po." Cedric's observation suggests that within a group of almost 2,000 students, most of whom are from homes where the annual household income places them below the poverty line, there are recognizable categories of economic status, po being the lowest. One of numerous status indicators is where others rank an individual's access to cash. Clearly, Shakeem was regarded by most (including himself) as belonging to the lowest economic category. Willingly he accepted "the freebie," breakfast and lunch provided by the school district. Even though all students in the school qualified for the freebie, some refused to accept free meals because it is a sign of poverty. Instead they opted to purchase similar food from lunch trucks parked in the nearby streets. Clothing is another example of an artifact used to categorize students economically—owning recognizable brand name clothes (e.g., Rocawear and Guess) and footwear (e.g., Nike and Timberland) are status-earning symbols used as proxies for relative economic standing among peers. In discussing his experiences in school and the ways he gets "talked about," Shakeem explained the reasons why he gets called dirty: "Cuz I don't got like Polo and Guess like everybody else. I don't got the expensive jeans. I don't got the latest sneakers. Stuff like that."

Taking from Others An important part of street code is that youth might take others' material and symbolic possessions to earn respect and create social networks. Customs for protecting material goods that apply in the streets may also apply in the school, evident in students refusing to remove backpacks from their backs, preferring to lug them around so that others do not steal their material possessions. However, symbolic capital cannot be protected in quite the same way. For Shakeem a key issue was how to overcome the strikes against him. Although he was good looking, strong and athletic, and could win fights against most of his peers, Shakeem's dark skin and overt signs of poverty, especially his often-soiled clothing, reduced his status among peers. Shakeem's solution to the problem of having insufficient status among his peers was to earn respect by publicly disrespecting peers and authority figures. In an effort to earn respect by dissing (disrespecting) others, he publicly used his verbal fluency and quick wit to direct derogatory comments at selected peers and to argue against rebuttals. Students who Shakeem did not respect became targets for verbal and sometimes physical aggression. When he succeeded he gained symbolic capital and if his campaign for respect was unsuccessful his status among peers decreased.

Unfortunately for Shakeem his attempts to interact with some of the more attractive females in the class were rebuffed because they regarded his clothing, unkempt appearance, and lack of money as resources to shower him with disrespect. Why choose this course of action? Because the females were attractive Shakeem could earn respect by having them as friends—that is, he could earn respect by expanding his social network to include individuals with high status, and beauty is a status-earning attribute. However, because of Shakeem's perceived low status, the females risked losing respect by interacting with him—that is, they could not risk losing symbolic capital by including a low status individual in their social network. Furthermore, Shakeem's overt signs of poverty and his dark skin were resources for such females to earn respect—by dissing him in public, especially because of his strength and reputation as a good fighter. Hence, Shakeem's initial overture, an effort to create social capital, was a resource for others to rebut his effort publicly and earn respect from peer witnesses. However, Shakeem would not accept public insults from females and their rebuttal became a resource for the trading of insults and an associated build up of negative emotions.

Cascading Insults In a vignette written by one of our research squad, Shakeem's insults were met with swift, increasingly derogatory rejoinders from two female students and, when one of them told Shakeem to "shut his mouth," Shakeem raised the stakes with even stronger language and emotion. Both females then attacked Shakeem's appearance and clothing, eliciting louder and stronger profanity-laced insults from Shakeem. Raising his voice made the interactions more public and using profanity increased the magnitude of the insult. Shakeem raised the stakes and displayed his anger—in losing his cool, Shakeem effectively ceded victory to the females. His anger soon subsided and he became sullen and withdrawn.

Shakeem described incidents like these in the following way. "They call me dirty and everything, but I wash and I am clean. I am one of the people who stand out. That's why people don't like me." The verbal attacks on Shakeem visibly upset him and affected his subsequent behavior and performance in the classroom. "I got into a lot of trouble that day. If that situation would not have happened, I probably would have dealt with it better. They made me mad and then someone else talked about my clothes . . . I was in one of my 'I don't care modes'." Incidents like this were commonplace and every now and then they cascaded out of control—drawing the attention of teachers and school administrators to Shakeem's uncontrollable tendencies, an identity marker that cost him dearly in terms of earning course credit.

Creating and Sustaining Social Networks

The campaign for respect is part of the capital exchanges that occur within fields of cultural production. To succeed in a field it is essential for an individual like Shakeem to have sufficient social and symbolic capital to afford the production of capital—to kick start an upward capital production cycle. Often lack of respect was a problem for Shakeem reaching his goal of creating and sustaining large social networks, a squad of individuals who would respect Shakeem and support his participation in the different fields of his lifeworld. In the home and streets, just to take two examples featured in Shakeem's videoethnography, his social networks were cross-age and he had the verbal skills to interact successfully with people ranging in age from babies to the elderly. At school a particular challenge was for Shakeem to interact successfully with adults because those who were in authority and wanted to establish control over students usually did not permit his playful oral exchanges. Shakeem's wit and sarcasm challenged established power hierarchies and his efforts to include adults in his social networks, as was the case in the 'hood, were unsuccessful in most cases. He was regarded as being too familiar and teachers felt a need to structure classrooms to prevent Shakeem from interacting in the ways that typified his being in other fields—that is his oral fluency, a trademark for Shakeem, was not a useful resource in most classrooms.

Shakeem's dispositions to interact fluently with all participants were suppressed by structures, often teachers endeavoring to enforce class rules about verbal interaction. These structures became resources for Shakeem's agency and a frequent way for him to earn symbolic capital at school was to violate rules about interacting with adults, thereby disrespecting those in authority. He chose his targets judiciously and did not take on administrators or others who had the clout to suspend or expel him from school. However, relatively inexperienced female teachers, especially new teachers like Jen (who is the teacher in the chapter by Kathryn Scantlebury and Sarah-Kate LaVan), were pushed to the limit by Shakeem who used his oral fluency and quick wit to gain an edge over them—showing his peers that teachers could not control him.

Teachers were not regarded as authority figures in the same way as a principal. For example, female teachers were often treated disrespectfully when they challenged Shakeem for his lack of focus in class or his failure to do homework. A tendency to disrespect female teachers led to Shakeem being thrown out of class frequently and under-performing in key subjects like science, mathematics and English. Hence, teachers and administrators regarded Shakeem as a student likely to fail and drop out of school. This was a contradiction for me as I regarded him as having an awesome intellect that could propel him to graduate school in a leading university.

Through his actions in class and his conversations about teachers Shakeem showed that he regarded teachers as people who had to earn his respect and always show him respect. He did not automatically assign status to teachers and those who made demands of him or endeavored to admonish him for shortcomings were almost certain to be unsuccessful and catalyze retaliatory actions from Shakeem as a demonstration of his agency. For the most part, Shakeem's primary goals in classrooms were associated with maintaining his autonomy to pursue his goals (i.e., to act agentially), earning respect, and not being disrespected in public. These priorities and Shakeem's insistence that all teachers had to earn his respect and their status as teacher led to problems with some teachers, especially Alex.

Alex, whom Shakeem regarded as cool, expected him to be respectful and enact decent culture in the classroom. Alex did not want Shakeem using street culture in the classroom and neither did he want an entirely level playing field. As an adult, parent, and teacher, Alex expected Shakeem to interact with him according to the established adult-child norms, unaware (or perhaps unwilling to accept) that these were not the norms that applied in Shakeem's 'hood. Alex enacted middle class culture in the classroom. He encouraged high levels of interactivity and expected control over his students. This was a major sticking point as Shakeem would not cede the power to Alex to have control over him. Accordingly, the relationship between Alex and Shakeem was like a tug of war, the unfolding environment not conducive to maintaining Shakeem's regular participation, interest and success. Shakeem was up and down in science classes, sometimes very involved, and often subdued. When he was "popping," he listened to the questions, called out answers, and often asked questions. As was his practice in many other fields, Shakeem was often playful, articulate, outward going, and involved. Also, he was curious and liked to connect what was happening in class to his lifeworld, especially his interests. Just as he was when he rapped, Shakeem used metaphor in creative ways to connect to the science he was learning. How can someone as smart as Shakeem struggle to succeed in science? He seemed to have little interest in passing, rarely doing homework or committing to high quality work in class. Accordingly, he failed science on numerous occasions, despite having what I regarded as a high aptitude for success.

My first interactions with Shakeem were playful. He'd tease me as I entered the science classroom in which he was a ninth grader. His taunts took several forms and always included light manhandling—"you gotta get in shape doc," he would often say as he grabbed my love handles or patted my gut. At other times he and Cedric would shadow box with me, making out that I was a punching bag at the gym—throwing flurries of punches but never making contact.

One day Shakeem was involved in an altercation with Jen, a new teacher with whom I was coteaching. He decided to leave the classroom and I intervened, requesting that he speak to me outside. When he declined I gave him the choice of seeing me outside or coming with me to see the principal. My heart was in my mouth as he towered over me, and then after a moment of hesitation (that felt like a lifetime), he agreed to speak with me outside. As we entered the hallway his demeanor switched dramatically to decent and he agreed to return to the classroom and work hard. Later he explained that he realized I must be important, deserving of respect, and therefore he should cooperate and get to know me. Notably, from this moment onward Shakeem stopped teasing and taunting me and in many ways became my teacher and protector within the school. Symmetrically his actions were reciprocated and I frequently intervened on his behalf to assist his successful navigation of the school.

A Gangsta and a Gentleman

Shakeem used the title Gangsta and a Gentleman for a QuickTime movie he made and a PowerPoint presentation. He emphasized the contradictions of being both a Gangsta and a Gentleman, categories that correspond to those used by Anderson—street and decent. However, unlike Anderson who presented these systems of culture in an either or sense, Shakeem clearly embraced a both/and perspective in his uses of gansta and gentleman culture—he was always both and, as necessary, could switch from one to the other. Shakeem's depictions of self are subtle and sophisticated, making the case that he is not one or the other, but always consciously both. In the following subsections I explore how Shakeem represented his identities in the numerous fields he included in the videoethnography and in his interactions with me.

The videoethnography shows Shakeem with others enacting culture in a variety of fields including the home, community garden, streets, basketball court, and staging arenas for youth subculture, including stairwells and sidewalks. He also shows how fields have porous boundaries and culture that does not belong in a specific place can be enacted there—for example, the living room at home when no adults were present. Analogously, decent culture is enacted with a street vendor selling chips, pretzels and water ice—an illustration that it is not just the location that structures a field and elicits one form of cultural enactment rather than another. In the streets Shakeem shows his capability of speaking respectfully and appropriately with a peer of about his own age and an older gentleman who was "Just chillin'. Tryin to make a dollar." The interactions show how Shakeem affords respect to an elder from his neighborhood. Throughout the videoethnography there are numerous examples of Shakeem speaking respectfully and with a sense of playfulness to adults from his neighborhood. Shakeem's oral fluency, quick repartee, humor and good nature are apparent in all fields, though their expression depends a great deal on the community. It is clear from Shakeem's actions in those fields and his oral accounts about what he does that it is always a priority for him to maintain and earn respect while pursuing his goals.

A critical contradiction arises in that Shakeem's transactions with adults in the home and neighborhood are mutually respectful and successful, whereas in school, his transactions with adults are often unsuccessful. In the out-of-school fields Shakeem is respected—others interact with Shakeem respectfully and his interactions with them (including adults) are similarly respectful (measure for measure). In school, teachers act to gain control over Shakeem, endeavor to shut down much of his cultural capital (especially orality and expressive individualism), and thereby inflict social violence on Shakeem, who feels disrespected. Experiencing the teacher's efforts to shut him down, Shakeem's fluency is breached and he has to maneuver, using alternative forms of culture, to maintain his respect among peers and earn back the respect that the teachers appear to be taking from him. Accordingly, the campaign for respect emerges inside the school and the classroom, and is evident in transactions between Shakeem and those teachers who seek to gain control over him.

Decent Culture Largely due to the way his mother raised him in the home and then the influence of his grandmother, Shakeem enacted decent culture throughout his lifeworld—presumably when his interests were served by enacting "decent" stocks of knowledge. He was articulate and modified his voice to retain interest and attention. He also modulated pitch and amplitude to fit the occasion, being very gentle with children and friendly yet jocular with adult family members, such as his female cousin. The videoethnography begins with Shakeem's young niece and his grandmother on the front steps of her home. A garden is shown to the side of the steps where the young girl, colorfully dressed, is playing, doing what Shakeem asks her to do. He speaks softly and modulates his voice to establish a non-threatening and enjoyable set of transactions. The camera then switches to an infant in a carrycot whom Shakeem addresses in a loving voice and then playfully introduces his cousin, the "best cousin he's got and the mother of these two children." Although Shakeem's representations in the video may be colored by a desire to project self-flattering images, everything that Shakeem did and included in the videotape was consistent with our experiences of him in similar fields for more than five years. His representations of self in the home field are highly consistent with the ways in which we have experienced him in across-age environments with friends and family.

Shakeem shows little Steve hard at work weeding the grass adjacent to a flowerbed in a community garden. His interactions with little Steve seem premised on an assumption that playin' would allow the child to experience street code in a safe place. In all of his interactions with little Steve, Shakeem was a role model, a teacher and protector—making sure that his protégé was prepared for the streets. As he commends little Steve for "trying to save the world" he begins to discus rap in a soft, caring voice that is laced with sarcasm; acknowledging a need to ease "his son[1]" into street culture.

Street Code   The camera sweeps down 52nd St, showing neat dwellings with cars parked on a street that contains a maze of power lines. The videoethnography then depicts an African American youth, without his shirt, standing behind a metal fence on an elevated deck, several feet above Shakeem who, with the camera rolling, playfully taunts him. The youth's voice is rhythmical and he uses wide gestures as he and Shakeem banter back and forth, with Shakeem being aggressive in a playful way and the youth answering him without escalating any tension by taunting back. A sample of the conversation follows.

```
Shakeem:  But you pretty nigga?
Youth:    No. Pretty Tony. Pretty Tony.
Shakeem:  Pretty Tony what they call you?
Youth:    No.
Shakeem:  What's up with the with the equipment you got work-
          ing on? What's that?
Youth:    It's a back brace. I was moving stuff around the
          house and I hurt my back.
Shakeem:  I'm sayin that pretty boys don't hurt their back.
          They just stay pretty all the time. Am I right?
Youth:    Ah no.
Shakeem:  About how long you stay in the mirror man?
Youth:    Huh?
Shakeem:  How long you stay in the mirror?
Youth:    I don't stay in the mirror. I ain't look in the mir-
          ror?
Shakeem:  So how you know you're pretty?
```

```
Youth:     Girls say. Girls say.
Shakeem:   Girls say? Then I'd like to find me girls. Are they
           desperate?
Youth:     (inaudible) they're fairly desperate ((gestures to-
           ward Shakeem))
Shakeem:   So, uhm. If some girls don't go see this tape too, I
           want to see how pretty they think you is.
```

The two youths enact a joust routine in which there is asymmetry in that Shakeem is older, much bigger, and verbally aggressive. The youth accepts Shakeem's taunts and the short, sharp exchange has a playful dimension to it. Because no peers were present, the disrespect shown by Shakeem toward the youth likely reaffirmed the existing status differentials within the 'hood. Probably this is another example of Shakeem being a role model for younger *bols* (friends) in his *'hood*, playin' with them, helping them to learn street code. Even so, Pretty Tony probably breathed a deep sigh of relief when Shakeem and his camera moved down the street.

The videoethnography shows a similar set of interactions at a basketball court. Male youth at each end of the court compete in one-on-one competitions. Though he does not compete himself, being fit and strong are important to Shakeem and in his talk he constantly emphasizes that even though he is big, he is not fat and can run fast and jump high. As he videotapes the basketball exchanges he is sarcastic, but receives no rebuttal from his peers who are absorbed in their contests.

A clue into the across-age dissemination of youth culture can be seen in Shakeem's presentation of two youth performing rap. "I do the rap thing. He do the rap thing. So for right now, he gonna spill those sounds for you. You wanna do that?" Shakeem was announcing little Steve as the first performer in a two-rap performance in a stairwell leading down to a basement level apartment. The performances are evidence of the ways in which youth subculture is produced, reproduced and transformed. Little Steve, probably no more than six years of age, is asked if he'd like to perform a rap. Readily he agrees and his performance is identifiably coherent with *gangsta rap*—including genre-appropriate profanity, body movements, prosody, and gestures such as crotch grabbing. The lyrics contain reference to guns, misogyny, and machismo. Moments later, on Shakeem's cue, a fifteen-year-old peer, smoking weed, also performs a rap that depicts an even more authentic reproduction of freestyle gangsta rap. Shakeem requested a specific rap. The performer thought for just a moment and then adapted a freestyle performance of a rap by "hot nigga Cas." The three youth seemed to enjoy the rapping experience and each regarded performing rap as a key identity marker. According to Shakeem he always had a beat in his head and he performed raps by himself and with peers and listened to hip hop on the Internet, radio, television, cell phone, and an electronic CD player. Artists like Cassidy and Beanie Sigel, from Philadelphia, are performance role models for Shakeem and his friends.

The videoethnography also shows Shakeem with a group of his friends in the living room of his grandmother's home. The group is multi-aged and includes one female and four males. Even though these are Shakeem's friends, the videoclip depicts the living room as a staging arena for earning and maintaining respect. The conversation is fluent, interactive and shows evidence of comfort in using and accepting profanity. Humor and the exchange of insults are pervasive. Shakeem's verbal interactions with two male peers show him establishing an edge with his oral fluency, using his wit to put them down. His peers seem reluctant to get involved and the female does not speak at all. Perhaps there is a pecking order based on the degree of orality, expressive individualism, verve, and physical strength possessed by each of the participants. Challenging Shakeem might result in humiliation at best and a beating at worst, whereas silence simply maintains the status quo.

Consistent with all of his interactions with little Steve, Shakeem spares him from his aggressive sarcasm and turns the camera on an empty chair where his grandma usually sits. He states that he is only doing what he is doing because his grandma is not present. If she were present she would "kick ma black ass." As he uses profanity in relation to his grandma little Steve protests as if such language has no place in the home. The porous boundaries of the field have allowed street culture to be enacted in a place where decent culture is the norm, but not without some discomfort from little Steve and Shakeem whose actions suggest profanity does not belong in this place.

Shakeem as a Youth Researcher

From the moment I met him I had a strong interest in working with Shakeem as a student researcher. Accordingly, when I received a four-year grant from the National Science Foundation, Shakeem was one of five student researchers invited to join my research squad. As soon as the research started the student researchers made it clear they did not want the research experience to be too much like schoolwork. In subsequent conversations about why they liked their work with us Shakeem made it clear that he felt valued and enjoyed being treated as an equal. In the research squad Shakeem found the autonomy that was denied him at school where adults' endeavors to exert control over him, eventually pushing him out of school as a virtual failure. As a researcher, Shakeem quickly caught on to the sociocultural framework we employed, using ideas such as capital exchange, agency|structure relationships, fields, and culture to make sense of teaching and learning urban science, schooling, and social life. He had a flair for technology and was adept in using the Internet, preparing PowerPoint presentations, video editing, and editing photographs and images. Just as his oral fluency was a large part of his expressive individualism, Shakeem's impressive technological fluency equipped him well for doing videoethnography in our research on the teaching and learning of science in urban schools. Similarly, his broad interests enabled him to make connections between science and the world outside of the classroom and he gave us advice on adapting the curriculum for use in urban schools. Finally, his insights into teaching and learning were sophisticated and he interacted successfully with new and experienced teachers seeking a master's degree, showing them how to do research in their own classrooms, incorporating ethnography and microanalysis.

Toward the end of our research in Philadelphia, Shakeem's science teacher (Alex) requested that the youth be removed from the research squad until he had a passing grade in science and an exemplary attendance record. Alex regarded Shakeem's participation in his science class as unsatisfactory and his attitude as too egalitarian—Shakeem seemed unwilling to accept the authority structures of the classroom. Although I was opposed to terminating Shakeem's participation as a student researcher, I wanted to support Alex too. When the teachers and administrators at City High cried "Enough!" I reluctantly agreed to a *three strikes and you are out* rule. The writing was on the wall. It was only a matter of a short time before Shakeem earned three strikes, was off the research team, and he was already on a pathway to failing chemistry. From Shakeem's point of view he failed because he would not do the homework and from his science teacher's point of view he failed because he did not learn enough chemistry. The contradiction is that Shakeem knew what he needed to do, had the resources to learn and do well, and opted not to do what it took to pass the course. He would not be threatened with the loss of his job as a researcher and would not submit to Alex's control, even though Alex was a black male for whom Shakeem had affection. It was not that Shakeem wanted to fail science, just that the social costs of passing were too great in terms of him losing respect within his peer group. Also, it seemed as if there was an expectation that Shakeem would learn to interact successfully using middle class culture. At the same time there were no sustained efforts to permit him to use his cross-age ways of interacting with elders producing hybrid forms that would be accepted as decent and appropriate, thereby allowing him to fully deploy his stocks of knowledge—especially those pertaining to orality.

On reading the above paragraph one of our research team commented that: "Shakeem, felt like the behaviors we were suggesting he would do so as to pass Chemistry and remain a student researcher were too difficult and unnatural for him. Specifically, I remember that after you and I met with Shakeem after school in our office space and explained the 'ultimatum' with him. We came up with a list of suggested ways of acting in the classroom and with Alex and wrote these down on paper to make the abstract more concrete—but Shakeem found this list overwhelming. Also, around the time all of this was happening, Shakeem was smoking quite a bit of weed—which often resulted in his cutting class. On more than one occasion, he appeared high during school hours when I saw him." I include this perspective because my changing identity obviously shapes the ways in which I represent Shakeem in the research. As I was becoming and continue to become more understanding of the cultural resources of urban youth there is a tendency for me to move away from white middle class perspectives that may well have been in play at the time in which the three strikes provisions were formulated. It was not just Alex and the teachers at the school that expected Shakeem to accept and align with white, middle class perspectives: we did too.

At the end of the school year Shakeem rejoined our research squad for the intense summer session and then moved to Atlantic City where his mother and brother had established a home in a low-income neighborhood where housing was subsidized as part of a national welfare program. After moving to Atlantic City

Shakeem continued as a youth researcher to collaborate and contribute significantly to our learning through research.

Shakeem's participation in the research squad allowed him to forge new identities while contributing enormously to our intellectual growth. One of our research squad put it succinctly: "I benefited manifold from opportunities to interact with Shakeem and his generosity in teaching me about how to be more like him." Shakeem was allowed to be who he wanted to be—to be himself and succeed within a new collective and new field. As he had done so well in his home and in the neighborhood, Shakeem was successful in adult and cross-age interactions and he reciprocated in giving us the respect we gave him. We valued him highly and provided opportunities for individual expressiveness in the videoethnography and other projects he undertook as a researcher, curriculum developer, and teacher educator. As a researcher he interacted with scholars on a daily basis, traversed the university, taught in undergraduate and graduate classes, and also participated with success in local, regional and international conferences. Ironically, despite his aptitude, he was failing high school—when he left City High after three years he had not accrued enough credits to be a junior in his new school.

Starting Over

Shakeem adopted a fresh approach to high school in Atlantic City. He was determined to get through and get on with his life. In an interview with me he described his school experiences in Atlantic City in the following way:

> I know that I have to graduate high school soon, very soon so I talked to the people I was supposed to talk to and got the grades that I needed to, to where as though, I'll graduate next year, rather than the year after cause we both know that I messed up, at "City" but that's in the past to me. When people ask me what grade I'm in, I don't hesitate to tell them the tenth grade and they say well you big and you got all that facial hair, why are you in the tenth grade, you look all old? I am old and I have no business in the tenth grade but I'm here because I messed up and what you should do is what you have to do, therefore you won't mess up and be in the same situation I am.... I mean I do the average thing that every other kid does in school, go to class, laugh and joke and stuff like that but when I see kids I hang out with and I associate with outside the classroom, they not going to class or they cutting class, I tell 'em like "yo" you gotta go to class man. You in the eleventh grade, you might as well keep going and it's like why stop now, why are you in the hallway cutting around if you don't, what are you here for? If you don't want to do anything, you cutting class all day, you might as well get a job because while you're cutting classes you're gonna eventually either graduate dumb or drop out early and get a crappy job so you might as well get a head start getting that crappy job or you can get your stuff together, you can go to class and you can learn and you can.... Even if you don't learn as much as you think you should or you know you could, you could still get the grade to pass.

> If you go to class and average a "D," you'll still pass and get outta high school and you can get a job because you graduated high school. It's like there's, of course there's a step after high school, that's like the minimum to get a job now you need is like a college degree, if you want a good job but if you don't even have a high school diploma, what are you planning on? . . . I'm going to summer school this year and take chemistry and next year I'll be a senior and I can graduate.

Although life for Shakeem never seems to work out exactly as planned, eventually he graduated from high school and then realized that his high school diploma was virtually worthless as far as gaining entry to a good job or to college. In an email message he wrote to me soon after graduation Shakeem sought assistance.

> i want to get a new job, but not just any job, a good job. i also want to get away from home, but am afraid to go to the service. unfortunitly my grades in school havent been consistant and college is pretty much out of the question. i dont know what to do. can you help?

The email message from Shakeem galvanized action on my part and I pulled strings to have him admitted to my university, but to no avail. At that moment Shakeem was out of touch with the research squad, having reconciled to a life of unemployment and welfare. What a waste! One of the most talented youth with whom I had worked was virtually pushed out of school and, having finally graduated, his credential was for all practical purposes hardly worth the paper on which it was written.

After about a year of flipping burgers and working in a variety of other minimum wage jobs, an adult neighbor of Shakeem's suggested he enroll at a local community college, offering to drive him to the college to make an appointment with someone to get him registered. Shakeem wanted a break and seized the opportunity to begin a certificate course in massage therapy—a qualification that would lead him to employment in which he was interested and to interact with lots of females in his classes—also something he enjoyed. Recently one of the research squad received an email from Shakeem—an email that in many ways said it all.

> Oh yeah before i forget, I MADE THE FUCKIN DEAN'S LIST FOR ACADEMICS, now aint that some shit. I told yall this school shit aint no joke no more, this aint city or atlantic city this is big business for me!

To paraphrase, Shakeem's goals were aligned with the achievement ideology and the socially transformative potential of education. He acknowledged that, unlike his time in two high schools (*City* and Atlantic City), now he was serious in his pursuit of success, as it was defined within mainstream life, gaining recognition through his inclusion on a merit scholar list. In a way his communication wrapped up many issues. His use of the Internet reflects his technological fluency and within the text of the message is strong evidence of hybrid culture, a blend of street discourse dissolved within a matrix of mainstream ideology. The message was not sent directly to me, but to one of the more youthful female researchers. Shakeem's use of profanity in his interactions with the younger female researchers is consistent with his ways of interacting with younger female science and mathematics teachers at City High. His tendency to repeat this practice in different fields is a contradiction to the more common pattern of Shakeem using field appropriate discourse as he enacted social life. This contradiction is one to be raised with Shakeem in our ongoing efforts to make sense of the lives of urban youth.

Communalism

Recently I drove through the backstreets of Philadelphia with Rowhea Elmesky and Shakeem. Now twenty years of age and earning a college certificate in massage therapy, Shakeem was visiting a homie in his old neighborhood before catching a bus back to Atlantic City. During our drive through the inner city Shakeem called *homies*, including two *ol' heads* (i.e., older males whom he regards as mentors), insisting at one stage that we speak to one who was paying for his college tuition. Using speakerphone Shakeem announced proudly that he had told us all about how the *ol' head* was keeping him off the streets by putting him through college. Shakeem was grateful and through public acknowledgement showed his respect for the actions of an older friend he regarded a role model and teacher.

When we met him on this occasion Shakeem's first utterances were about not being in jail, not doing drugs, and not carrying a gun. These affirmations mirrored the first words Shakeem's *ol' head* uttered on the phone. As a drug dealer, the ol' head was emphatic that he would prevent Shakeem from living his life in the streets. He had Shakeem's back and stressed the centrality of education in escaping an underclass—characterized by unemployment, poverty and crime—that entraps many African American males like Shakeem and him. "Thanks my man. Holla at cha boy. See ya n'a coupla hours." Shakeem disconnected from the call and proudly announced: "He's ma man."

In a twenty-minute ride Shakeem demonstrated repeatedly the centrality of social and symbolic capital in creating and sustaining a community that supported his lifeworld. *Ol' heads* and *homies* were essential—never taken for granted, relationships were continually refreshed with highly respectful interactions. As mindful as ever of the salience of social capital, Shakeem initiated all the calls and visits. In his interactions with others Shakeem's praxis exemplified capital exchange theory. He nurtured relationships by showing respect to those he wanted in his social networks and he used the networks to produce respect, loyalty, solidarity, an array of positive emotions, and forms of cultural capital.

The ride through West Philadelphia is memorable for many of the contradictions it raised for me. Shakeem still carries the signs of poverty and yet he has commodities that in some ways point to economic resources, including a cell phone, CDs loaded with rap, a CD player, cigarettes, and weed. As he interacts with Rowhea and me he initiates calls in and out of state and in so doing refurbishes his social network, showing respect to those who will have his back. He is succeeding at last in his academic goals and the credential he earns will likely earn him a job that will propel him toward the middle class. Yet his college studies are supported by an ol' head, a drug dealer with a social consciousness and social transformation as a goal. The *ol' head* will do what it takes to make it possible for Shakeem to have an improved social life—better than it is now and better than his own. The *ol' head* fights against the determinism of social reproduction, even though it involves self-sacrifice. Although he sells drugs to make money, he uses the money to support his family and youth such as Shakeem.

Shakeem's social network is robust; bonds are sustained by showing respect continuously to those who are in it, honoring others by having their backs and intellectually engaging with social life. Multi-tasking and fluency are ever present, occurring without conscious awareness as he switches between decent and street, and creates hybrid forms of culture when it is appropriate to do so. Riding with Shakeem is indeed an opportunity to learn.

One of Shakeem's favorite hip hop artists is Beanie Sigel, a Philadelphian whose raps provide deep insights into the social lives of many of the youth from whom we have endeavored to learn through our research. Sigel performed raps that were salient to Shakeem and his peers, presumably because of the lyrics, which were focused on growing up as a poor African American in Philadelphia. Rap is a central part of Shakeem's being in the world. I have studied Sigel lately, especially the raps on his album *The Truth*. Without claiming the same status as Shakespeare, making sense of what Sigel has written in his raps is about as intellectually demanding for me as making sense of the Bard's famous works. The intellectual engagement of youth with rap extends beyond rote representation and to my way of thinking is analogous to poetry, literature and drama. Until recently I have failed to acknowledge the intellectual engagement with rap. I have tended to dismiss rap as social violence, works of resistance that target modern-day colonizers. Yet I am struck by the themes of love, loyalty, respect, solidarity, compassion, loss, and communalism that coexist with themes of crime, violence, sex, misogyny, and futility. Out of this dialectical mix arise hope and a sense of expanded possibilities. Any of Sigel's raps from *The Truth* would serve as an illustration, but to end this section I choose a verse and a chorus from *Ride 4 my* . . . I choose these excerpts because Shakeem performed *Ride* often and with pride. Just as urban youth are feared and mistrusted, a rap like this may be regarded by many and perhaps most in the middle class as a shocking example of what must change in America—and yet I find the text intellectually engaging, challenging, replete with deeply emotional representations of the contradictions of youths' lives in inner city USA.

> *Aiyyo, from Boyz II Men to the End of the Road*
> *Yo we boys to the end never been in the cold*
> *You got me, I got you till our souls grow old*
> *He shot you, he shot me, how we supposed to roll*
> *I take a shot for my nigga gimme two to the ribs*
> *Run a spot for my nigga, while you doin your bid*
> *Fuck a step pop these c's gonna know who you is*
> *I get it through to your wiz that it's due to the kids*

TELL ME WHAT YOUR LIFE LIKE . . .

> *For my dog, I swear to God I'll sit in a box*
> *Gimme three hots and a cot before I snitch to the cops*
> *Six foot ditch pam box covered with rocks*
> *Tombstone ready die for the love of his block*
> *Know what I want in my life I want for my brother*
> *Know what I want for my wife I want for my mother*
> *It aint a question on what I would do for my squad*
> *Ask yourself if you really true to your squad*
>
> *[Chorus X2]*
> *I'ma ride with my niggas*
> *Die with my niggas*
> *Get high with my niggas*
> *Flip pies with my niggas*
> *Till my body get hard, soul touch the sky*
> *Till my number get called and God shut my eyes*
> – Beanie Sigel, February 29, 2000

The Shifting Sands of Identity

In my analysis of Shakeem's social life I found convenience in Elijah Anderson's dichotomy of decent and street as categories to depict two different forms of culture that were applicable to the youth who attended City High and the neighborhoods they inhabited. Especially when we commenced our urban research, the different forms of culture were strident and it was apparent that more than 90 percent of the students at City High brought street culture directly into the school's cultural fields—for example, the entrance to the school, front office, hallways, lunchroom, assembly hall, and classrooms. Efforts to resist street culture were futile, largely because it was like a tidal wave sweeping through the building. If educators at City High had only started with the question of what is it that these students do in praxis they would have observed two viable systems of culture, street and decent—both markedly different than middle class, white culture and, for that matter, middle class combined with any ethnic culture.

What Julius Wilson (1987) wrote about the emergence of an underclass is applicable to Shakeem, who lived in an inner city neighborhood with African Americans having similar racial and social histories to one another. The minority cultures associated with race and poverty were immersed in a majority mainstream culture that tended to reproduce inequity for African Americans. Grounded in a shared history that predates and includes enslavement of Africans and the Civil Rights movement of the 1960s an underclass can emerge with a shared and characteristic culture. A key idea is that culture is produced in praxis and individuals within a community produce and enact similar forms of culture by being with others and having first-person experience with their praxis. In high density inner city neighborhoods the chances of an underclass culture developing seem likely. In the case of City High, most students lived in close proximity, hung out in the same streets, attended the same church and school, and participated in the same sports and hobbies. Poverty limited mobility and restricted the variety of pastimes that might characterize middle class neighborhoods. Accordingly, participants had opportunities to learn from one another consciously and unconsciously as they interacted with people having similar racial and economic histories.

Elsewhere Wolff-Michael Roth (in press) has written about the creation of a diaspora when displaced people come together in another place to create a sense of home, cobbling together mainstream and minority culture to create hybrid forms that afford their successes in social life. This framework has more appeal than the ideas of underclass and ghetto culture, both of which are saturated with deficit images. As is evident in the analysis of Shakeem's videoethnography, there are numerous resources to support his cultural production. For example, gangsta rap extends well beyond Philadelphia and the northeast of the United States. Accordingly, the beats and lyrics of gangsta rap are resources for reproducing and transforming hip hop culture. Dissemination of popular culture occurs very quickly and globally through cable television, Internet, cell phones and CDs. In this way new raps, associated clothing and body fashions, and body moves can saturate youth culture across time and space. Recall that children like little Steve were performing and talking about raps from an early age and cross-age interactions encourage widespread dissemination and participation. Accordingly, cultural brokers such as Shakeem can disseminate culture produced locally, as a reproduction|transformation of culture from elsewhere in the world.

Stuart Hall (2006) also has highlighted some of the strengths of the diaspora construct in taking account of the ways in which displaced peoples come together to create hybrid forms of subculture within mainstream culture. This idea has salience in the ways in which Shakeem has produced diverse cultural resources to support his social life. Over the five years in which I have known Shakeem, the resources he has accessed to produce culture have expanded significantly and include the Internet, cell phones, and cable television. Movies, books and increasing global commodification of race, ethnicity, sex and hip hop are resources likely to be salient, especially in a context of increasing levels of immigration, in disseminating culture into cities like New York and, to a lesser extent, Philadelphia and Atlantic City. Because of the pervasive and readily accessible communication networks the boundaries for macroculture are global in extent rather than national, state or city. It is not that the culture produced in the 'hood is no longer viable, just that there is an increasing array of resources for cultural production—even in neighborhoods where there are high levels of poverty and welfare recipients. Accordingly, culture flows across national borders and sows the seeds for the creation of hybrid cultures and that are enacted locally. Within large cities like Philadelphia and New York, neighborhoods no longer lock in characteristic forms of culture. Instead popular culture penetrates the fields nested within mainstream culture, providing common elements from which hybrid forms of culture grow—bases for the emergence of shared culture that can be used for successful interactions across previously difficult to penetrate boundaries thought to contain macrostructure.

SUMMING UP

> Cultural identities come from somewhere, have histories. Like everything which is historical, they undergo constant transformation. Far from being eternally fixed in some essentialized past, they are subject to the continuous "play" of history, culture and power. Far from being grounded in a mere "recovery" of the past, which is waiting to be found, and which, when found, would secure our sense of ourselves into eternity, identities are the names we give to the different ways we are positioned and position ourselves within the narratives of the past. (Hall, 2006, p. 19)

I began this chapter with an analysis of a period in my professional life that was more dynamic than most others. However, looking back over my life it is clear that my lifeworld has been constantly changing according to my goals, the associated fields in which I live my life, others with whom I interact in those fields, the accessibility of structures to support my agency, and the stocks of knowledge I have to appropriate structures in pursuit of success. These factors do not constitute a complete set and nor is there a hierarchy among them. In fact there are many dialectical relationships that are salient to my identity, including first-person|third-person perspectives, individual|collective, and agency|structure.

Who am I? Well, it depends on where I am, whom I am with, what I want to accomplish, and what I can do. Who I am is contingent on others, the resources available to support my agency and theirs. But there are emotional parts to who I am as well—I am honorable, I care about the welfare and interests of others, I respect others and want them to respect me, I enjoy success and like to belong to communities. Also, I am aware of my strengths and limitations. My stocks of knowledge equip me to be good at some things and not so good at others. Accordingly, I have interests, often in doing things I am good at and also in rising to the challenge of producing new knowledge to allow me to develop new strengths—that is, to succeed in the pursuit of fresh goals. Hence, I am intellectually engaged in my lifeworlds.

"I" incorporates a first person perspective of self, an unfolding first-person identity associated with a conscious|unconscious dialectic. As culture is enacted fluently and successfully "I" is reinforced and is not an object for reflection and possible changes. "I" changes through interaction and the associated successes and failures. Capital exchange supporting cultural production (reproduction|transformation) occurs continuously and in the process identities are constructed, reproduced, and transformed. "ME" is a manifestation of objectifying "I," a call to reflect on identity and all this may entail. The conscious manifestation of identity can occur because of failure; a rupture in the fluency of cultural enactment, a call for adjustment, perhaps requiring the application of different stocks of knowledge or perhaps the production of new culture. Failure can occur because of the intended and unintended actions of others, but in each case identity is inscribed. A third person perspective of identity, "YOU," comprises the collective and individual constructions of "ME" within a macro|meso|micro dialectic. Accordingly, identity is a tri-part relationship of I|ME|YOU—unfolding continuously as social life is enacted.

Who is Shakeem? Who is HE? Obviously there is not only one ontologically viable way to represent Shakeem. As he navigates life, as a first-person singularity (see Roth, this volume), his identities change as a function of space, time and the ways in which he appropriates the structures of his lifeworld. I/ME/YOU are changing vectors, dynamically interconnected, but in ways that allow for constant change. From a third person perspective the ways in which I represent Shakeem reflect my changes in identities. With these caveats, I conclude that there are many ways to represent Shakeem and other urban youth. In this research I have listened to and endeavored to learn from many voices. In a period of six years I have continuously looked forwards and looked backwards recursively in an effort to make sense of urban youth, as they have lived their lives, and especially as they have participated in science education and the teaching and learning of urban science. With a dialectical perspective on social life I have prepared my portrait, but is it really Shakeem? Is this the truth? Shakeem, what your life like?

My answers to these questions are from my heart. This is how I have found him to be. This is how I want science educators to think of him as they plan science education for urban youth. This is no single authentic portrait for anyone—not for Shakeem's mother, Little Steve and Pretty Tony. This representation is not for the hos and crackheads from 52nd Street. As an African American youth, Shakeem is what he knows, what he can do, what he can learn to do, and who he becomes. Shakeem has a potential for growth that is staggering, but how will the collectives of his social life afford his goals and accelerate their expansion? These are the challenges for today's science educators with their eyes on equity.

History brought Shakeem and me to Philadelphia. In my case I decided to become an immigrant and I left a well-paid position in Australia to assume a similar well-paid position in the United States. Then, as a result of becoming increasingly concerned about equity and the plight of urban education I decided to come to Philadelphia. Shakeem did not choose to come. He was born and raised in the Northeast and his ancestors were brought to the United States as slaves. Raised without a father, Shakeem, like so many other African American children, was raised by a young mother who did not have the fiscal resources to escape from subsidized housing and other forms of welfare. She was constantly faced with many of the issues Sigel includes in his raps about urban life—poverty, crime, violence, drugs, sex, abuse, and struggle. Shakeem grew up in turbulent circumstances and in the process built forms of culture associated with his successes and failures as he enacted life. He moved from this house to that, from this dad to that, and from his mother to his grandmother. He learned to survive and in some senses to thrive in his lifeworlds and as he grew older and advanced from elementary to middle to high school he participated in education—the great hope to escape from poverty and the underclass into which he was born. Yet, from what we experienced of his schooling, there was collective resistance to him using his culture to thrive in classes like science, to build hybrid forms of culture in a process of learning a creolized science. What he did so well in cross age situations in his world out of school he was prevented from doing in high school. Not surprisingly, the violence of being shut down repeatedly and continuously takes a toll and can be experienced as disrespectful—catalyzing a campaign for building respect in the school fields—not through academic achievement and collaboration with teachers, but through resistance to authority, and earning symbolic capital by enacting street code.

Shakeem, tell me what your life like. How different might it have been if someone had taken the time to find out about Shakeem's life in the first few days he entered City High? Shakeem's life, like Sigel's is real. A talented youth with a flare for life, a personable and gentle giant, Shakeem was often portrayed as a monster and, ironically, he was pushed back out onto the streets by a system that was blind to the necessities for urban youth to be what they have got to be and do what they have got to do. Where was the agency for youth like Shakeem? Why the obsession with control over youth? Alas, the collision of macrostructures with schema from a subculture produced many of the contradictions experienced by Shakeem and youth like him. An ideology of control whereby those in authority say what will and will not be done, who will get what and what they will do with it, and when and how accessible resources will be accessed is at the root of an educational system that is structured to reproduce disadvantage, oppression and failure.

Shakeem, tell me what your life like. I have no response from Shakeem, mainly because he has fallen into line and is working his way upwards and to use his own words "this school shit aint no joke no more . . . this is big business for me!" My eyes are welling up with tears again. Are you sure Shakeem? "Everything signed is sealed." I am not sure what Sigel meant when he sang these words, but my great worry is about what is signed. My journey has taken me some distance into finding ways to respect urban youth culture and to appreciate the awesome stocks of knowledge that they can deploy in living their lives. Unshackling the what, where, how, and when, ceding control to the collective rather than to just a powerful few—can this be the new reality for urban schools? Let's hope so. For me the future is not in colonizing the minds of urban youth, beating them down with failure, pushing them out to the gutters until they are ready to acknowledge the worth of white, middle class culture. I do not see the continuance of the status quo or efforts to mend it as fully employing the capital of urban youth—instead it allows them to get through, to earn the certificates to get certain sorts of jobs, perhaps even in massage therapy. For me the hope is in unleashing the potential embodied in the "I" of urban youth. Shakeem could have been a theoretical physicist on Monday, a nuclear engineer on Tuesday, a leading business entrepreneur on Wednesday and a pro-basketballer on Thursday—but he was held back, prevented from creating hybrid forms of culture to pursue his dreams in his ways. "Your life is dis. You life is dat. Mine's real." The third person view prevails and this view of education in urban schools simply must change. At a time when Western society is bedeviled there is a dire need to unleash the potential of a diverse and talented citizenry—listen to the "I"s and they build "ME"s—that's real.

NOTES

The research in this book is supported in part by the National Science Foundation under Grants REC-0107022 and DUE-0427570. Any opinions, findings, and conclusions or recommendations ex-

pressed in the book are those of the authors and do not necessarily reflect the views of the National Science Foundation.
1   Shakeem referred to little Steve as his son—reference to his acceptance of being a role model for him.

REFERENCES

Anderson, E. (1992). *Streetwise: Race, class, and change in an urban community*. Chicago: University of Chicago Press.

Anderson, E. (1999). *Code of the street: Decency, violence and the moral life of the inner city*. New York: Norton.

Collins, R. (2004). *Interaction ritual chains*. Princeton, NJ: Princeton University Press.

Elmesky, R. (2005). Playin on the streets—Solidarity in the classroom: Weak cultural boundaries and the implications for urban science education. In K. Tobin, R. Elmesky, & G. Seiler (Eds.), *Improving urban science education: New roles for teachers, students and researchers* (pp. 89–111). New York: Rowman & Littlefield.

Roth, W.-M. (in press). Bricolage, métissage, hybridity, heterogeneity, diaspora: Concepts for thinking science education in the 21st century. *Cultural Studies of Science Education*.

Seiler, G. (2002). *Understanding social reproduction: The recursive nature of structure and agency within a science class* (Dissertation Abstracts International, AAT 3043953). Philadelphia, PA: University of Pennsylvania.

Tobin, K., Elmesky, R., & Seiler, G. (Eds.). (2005). *Improving urban science education: New roles for teachers, students and researchers*. New York: Rowman & Littlefield.

Turner, J. H. (2002). *Face to face: Toward a sociological theory of interpersonal behavior*. Stanford, CA: Stanford University Press.

Wilson, W. J. (1987). *The truly disadvantaged: The inner city, the underclass and public policy*. Chicago: University of Chicago Press.

STACY OLITSKY

2. IDENTITY, INTERACTION RITUAL, AND STUDENTS' STRATEGIC USE OF SCIENCE LANGUAGE

An important component of learning science is developing the ability to think and reason scientifically. The *National Science Education Standards* stress the need for students to acquire process skills such as interpreting evidence, formulating hypotheses, and drawing conclusions (National Research Council, 1996). In order to develop these skills, students need practice both writing and speaking about science, during which they can accomplish tasks such as making predictions about real-world phenomena, developing explanations, and evaluating others' explanations. Engaging in science-related talk is particularly important, as it allows for "back and forth" interaction, during which students can listen to each others' ideas and respond to them. With enough opportunities to engage in science classroom talk over time, outcomes are likely to include not only improved scientific reasoning skills, but also a greater fluency with the content and terms that circulate in science-centered communities, thereby facilitating future participation in these communities.

Many teachers recognize the importance of giving students the chance to talk about science, and strive to facilitate effective whole class discussions during which students make predictions and develop explanations. Sometimes these discussions focus on laboratory experiences, allowing students to come up with their own explanations about experiments they conduct and events they observe. However, many teachers also use demonstrations for similar purposes when it is not feasible to conduct a lab. These demonstrations may not provide students with the same hands-on experiences that labs do, but they still have the potential to be effective in creating a shared experience among the students of observing an event that can serve as a basis for a discussion about the merit of possible scientific explanations.

Like labs, demonstrations can be helpful in stimulating student engagement and interest in science. In the eighth-grade classroom that is the focus of this paper, many students described how what they liked most about science is exciting demonstrations, or in the words of one student, "things that go bang." However, the aim behind many demonstrations is not only for students to like science, but also for students to develop a greater understanding of the science knowledge behind the experiment. Yet this goal can present some difficulty. Students may not connect the science concepts to the demonstration; in fact, they may not see at all what the teacher wants them to see (Roth, 2006). They may not know which features of the demonstration should receive their attention in order to support the content they have been learning, and which features are not relevant (Roth, McRobbie, Lucas & Boutonné, 1997). They may view the demonstration as a diversion from the content, rather than as supporting it. They may not develop their own explanations for the event they observe. They may talk about the experiment, yet may not use science vocabulary. They may remember the demonstration, yet not the associated science content. From a teacher perspective, students ideally generate a variety of predictions and explanations for a demonstration they observe, and would practice using science discourse and applying science concepts in their predictions and explanations. If a teacher conducts a discussion that does not end up having these features, s/he may assume that students do not have knowledge of such terms and ideas or that the students have not learned to transfer this knowledge to a concrete situation or that his/her teaching of the relevant content had been ineffective. While these ideas may have some degree of explanatory power as to why a post-demonstration discussion fails, they are incomplete, as they privilege cognitive processes and neglect other aspects of the classroom interaction. They assume that students may or may not "have" particular knowledge of language or terms, and they may or may not be able to "use" or "apply" it in a new situation, in the form of talking about the particular demo. An implication of this view is that knowledge is like an object possessed by individuals.

Socially situated views of learning emphasize that knowledge is not just "in the head" and distinct from social practice. Rather, what people claim to know relates to their self-presentations in particular situations. Put another way, what a person knows is part of who s/he is, and if a person does not want to portray him/herself in a certain way, s/he will not express the associated knowledge; expression of knowledge is expression of identity. As an example, people in low-status jobs would often claim not to know the relevant knowledge, because to express such knowledge would be to convey their own low status (Wenger, 2000). As another example, a study of an urban classroom revealed that some students acquired scientific ways of viewing the world yet resisted using science discourse because it involved a conflict between identities associated with science class and their cultural identities (Brown, 2004). Just as people may refrain from using language if it makes an undesirable identity claim, they may also purposefully use particular language to make a claim to identities that they find desirable. For example, if someone wants to be accepted by a group of people who follow football, a person may make a point of learning about football and mentioning the previous day's games.

Whereas conceptualizing students' use of science content in their talk as a function of the desirability of identity-related claims (as opposed to conceptualizing their statements as only evidence of cognitive knowledge) offers an important additional perspective, there are still other issues to consider. In order for students to use science language and content during class discussions about demos, these students would need to believe that the identity claims that they make by doing so are not only desirable, but also are likely to be accepted by their peers and result in solidarity.

Discursive moves that make identity claims can backfire. It is difficult to effectively participate in a conversation on football from just reading the sports section of the newspaper. People may recognize the falseness of a claim, and there may be awkward pauses, shared glances, or other non-entraining social experiences that let that individual know s/he has made a mistake in displaying such knowledge or using a particular language. As an example, a teacher's attempt to use youth slang in the classroom in an effort to appear more approachable may be rejected by the students as "trying too hard" or "using it wrong" and these students may laugh at the teacher, thereby increasing rather than reducing the teacher's exclusion from their social world.

In addition, stereotypes can also interfere with the success of someone's claim to a particular identity. Elsewhere I draw on the concepts of *performance* and *facework* (Goffman, 1959) to describe how interactions in the science class of this study are particularly risky for the African-American students, as their categorical identities do not match with images of "good science students" conveyed through the media and through general stereotypes (Olitsky, 2006). These students experienced negative stereotype threat (Steele, 1997) and a greater risk of having their claims to membership in a science-related community through their use of science language rejected by their peers. I discuss how the consequences of such rejection included a lack of solidarity with peers and lower levels of emotional energy and confidence. I also describe how the greater social risk of participation interfered with some students' abilities to obtain the benefits of participating in classroom discourse.

Clearly there is a variety of factors that can mediate whether an identity claim, such as a claim to being a member of a science community, is accepted. In deciding whether and how to participate in science class, how do students assess what is likely to take place, and make "decisions" about whether or not to use science language or to come up with their own creative explanations relating to science content? Why do similarly positioned students in terms of their categorical identities—class, race, and gender—make different decisions, with some students using science language even if it seemingly conflicts with their cultural identities while others do not? Why are students' self-portrayals in science class, as demonstrated by their use of science language and content, sometimes unstable over time?

Whereas students' utterances may be related to their perceptions of the desirability of a particular identity, such an explanation is not sufficient. Most students generally do not decide once and for all whether they wish to express knowledge related to science and whether or not claiming a science-related identity is a good thing or not. It is therefore beneficial to examine identity in science classrooms not only on the mesolevel, by investigating whether a student desires to identify with science-centered communities, but also on the microlevel, by examining how identities emerge from and can change within specific interactional situations. In so doing I focus in this chapter on how identities both structure the everyday classroom interactions among participants and are the outcomes of these interactions. I draw on a vignette from an eighth-grade class to show that expression of knowledge, and the related identity claims, take place in moment-to-moment encounters that can vary considerably from day to day. I discuss how any student's statement in science class can be considered a strategic move, or a choice, aimed at increasing the likelihood of engaging in successful interaction rituals characterized by entrainment and solidarity.[1]

Drawing on the concept of *interaction ritual* (Collins, 2004), I discuss how students, like any other participants in social settings, are drawn to situations in which they can engage in successful interaction rituals that generate emotional energy. These interaction rituals (IRs) are characterized by mutual focus, physical co-presence, entrainment (or coordination) of body language and noisemaking, boundaries to outsiders, and a common mood. Outcomes of these successful IRs include high levels of *emotional energy* (EE) for the participants, EE invested in the symbols associated with the interaction, and solidarity among the participants. I argue that students will not necessarily speak in class in ways that produce the greatest alignment with their current identities, or with what they had previously viewed as desirable, as there is sometimes a possibility of developing new group affiliations, depending on the situation. Rather, students' decisions in what to say during a discussion can be understood as maximizing the possibilities for successful IRs, and avoiding failed ones. Depending on what transpires, aspects of identity can change. While students come with current group affiliations (such as whether they feel a part of a science-centered community), reputations (such as good student, or popular), categories (such as race, class, or gender), and other aspects of their identities, the success of the IRs that emerge can have identity-related outcomes, such as new group affiliations (new identities), the solidification of some affiliations, or the reduction of others.

Identity can therefore be considered both an input and an output of interaction rituals. It is an input in the sense that it is one factor among various other structural factors that contribute to students' anticipation of success of a particular interaction, thus informing their decision/choice/agency regarding when to speak and what to say. However it is also an outcome in that depending on whether students experience a particular interaction as successful, they may experience solidarity, cementing their identity associated with the group, or a lack of solidarity, decreasing their ties to the group. This perspective on the role of identity in interactions can help explain why students' predictions and explanations for demonstrations (or statements in class more generally) sometimes do not include much reference to science content or language, even if students are "able" to do so in that they know the material. In this chapter I argue that if expressing one's best ideas of science content is thought to be likely to contribute to social solidarity, then students will have a tendency to do so. If not, then they may decide not to speak or to make contributions that do not draw on science language and content.

SETTING

City Magnet began in 1958 as a laboratory and demonstration middle school for grades 5–8, adding the high school in 1976. Both schools are housed in the same building. Students are selected from schools throughout the city to attend the middle school based on their third-grade Stanford Achievement Test scores, which must be in the 88th percentile, and their grades, which must be As and Bs. Many families do not know how to apply to the school or if their child is eligible to apply. As expected for a selective magnet school, City Magnet's standardized test scores differ substantially from those of students throughout the district. While in Philadelphia, 57 percent of eighth-grade students scored below basic in math on the Pennsylvania State Scholastic Aptitude tests, only 1 percent of City Magnet students did. While 42 percent of eighth grade students scored below basic in reading, only 1 percent of City Magnet students did.

Once admitted to City Magnet in fifth grade, a student's position in this school is still not secure. In their eighth grade, City Magnet students, like all students in Philadelphia, submit applications for high schools and are chosen for admission based on their grades in seventh grade, behavior marks, attendance record, and scores on standardized tests. Only about 100 out of the 200 eighth-grade students will be selected to enter the more prestigious high school. The other students either attend other magnet schools, private schools, or neighborhood schools. During the year, the students in the eighth-grade class attended numerous assemblies during which representatives from magnet schools came to recruit City Magnet students. In these and in other assemblies, the principal would praise City Magnet as the "best" school in the city, urging students who "have shown their high potential" to choose to stay. This "choice" process in Philadelphia, in which the school gets to do some of the choosing, was a structure that influenced students' interactions with teachers and peers both in and out of class. The eighth grade students have described the tension they experience in trying to prove to their teachers and to other students that they really belong at City Magnet.

City Magnet's enrollment by race is 52 percent white, 34 percent African American, 9 percent Asian, 4 percent Latino, and 1 percent other. These figures are for the high school and middle school combined, and therefore do not reflect that the middle school has a much larger population of African American children than the high school. Whereas in the school district overall, 80 percent of students are from low-income families, only 39 percent of students from City Magnet are.

The eighth-grade classroom that is the focus of this study had 33 students. Of these, approximately 40 percent were white, 34 percent were black, 10 percent were Asian American, 10 percent were Latino, and 6 percent were multiracial. Some of these students came to City Magnet from private schools, some from elementary schools in middle-class neighborhoods, and some from elementary schools in poor, predominantly African American neighborhoods. The teacher, Linda, is white and came to the city from New Mexico.

While all of the students have been identified as talented academically, there were large variations in academic performance while at City Magnet, particularly in science and math, tending to correspond with whether or not they attended elementary school in a high-poverty area of the city. The observations of student performance in this classroom are in accordance with findings that students from lower socioeconomic backgrounds who attend schools with adequate resources still perform poorly in comparison to students from high socioeconomic backgrounds (Oakes, 1990). Given the residential segregation in Philadelphia, the African American students were much more likely to have attended schools in high-poverty areas with fewer resources, and some of them had to struggle considerably in order to succeed at City Magnet. Some students described how they quickly obtained a reputation in the fifth grade for being "smart" or "not smart" that tended to stick with them and affected their agency in accessing resources for learning.

Not only was there variation in performance on tests and quizzes among the students, but there was also considerable variation in students' class participation, such as how frequently students volunteered answers, asked the teacher and each other questions, and engaged in discussions using science discourse. However, rather than certain students consistently participating more than others, participation varied depending on the activity, the topic being discussed, whether Linda was in or out of field, and other conditions. Given the importance of engaging with science discourse for acquiring school science identities, investigating the classroom conditions that influenced this variation was an important part of this study.

As a relatively new teacher, Linda was a "floater," in that she did not always teach in the same room. However, she always had the eighth-grade class that is the focus of the study in the same room, and therefore was able to decorate the walls with posters and student work. The classroom had no sinks or lab materials, and Linda needed to bring in materials herself when she wanted the students to do a lab or hands-on activity at their desks. Conducting labs required a considerable amount of time spent on set-up and cleanup. Given that the class periods were only forty minutes long, the preparation took a substantial amount of instructional time. In addition, the desks were slanted and therefore were difficult to use for experiments. Given the difficulty of conducting labs, on many days students sat in six long rows facing the front of the room. The class activities were divided between teacher-led discussion of concepts and problem solving techniques, demonstrations, hands-on activities, and group projects.

The study of City Magnet is part of a larger ethnographic study in five participating schools in Philadelphia on science education in urban settings. In the interest of conducting ethnographic research that would avoid exploitative relationships between the researcher and the researched, I drew on Guba and Lincoln's (1989) criteria for authenticity, which include an emphasis on working with participants toward positive change in local settings and increasing participants' understandings of each others' perspectives. Toward this end, I worked in collaboration with a teacher-researcher and the four student researchers. I was influenced by other projects that have involved students as researchers, such as one study in particular, where students made significant contributions by providing insider perspectives, conducting interviews of peers, and serving as teacher educators (Elmesky & Tobin, 2005). In selecting the student researchers, the teacher and I asked students who were different from each other in terms of their academic achievement and their expressed interest in science. Of the student researchers, Ashley, Aileen, and Monique are African-American, and Lisa's father is white and her mother is African-American. All four came to City Magnet from neighborhood elementary schools.

FROM INTERACTION RITUALS TO IDENTITY

Demonstrations and Building Solidarity

Demonstrations can be an effective method for teachers to initiate successful interaction rituals in their classrooms. As students anticipate the outcome of the demonstration, they may become entrained in each other's movements and a shared mood may develop. At the peak of the demo, during the noise, small explosion, or color change, students often react with "ooohs" and "aaahs." This shared noisemaking furthers the experience of intersubjectivity, resulting in solidarity among participants.

While the possibility of a successful IR is a valuable feature of demonstrations and may help lead to a temporary solidarity among the students, in order to accomplish the longer-term goal of students developing a sense of group membership associated with science, it would also be desirable for the corresponding discussion to be a successful IR as well. If such a discussion involved students' use of science language, argumentation, and problem solving, the "emotional charging" might be able to extend to these symbols rather than just to the demo itself. Possible positive outcomes include students remembering the content more than they would have without the demo, and developing a sense of membership associated with science discussions. Whereas this would be ideal for many teachers, in practice the successful IR of reacting to a demonstration may not always be matched by an equally engaging discussion in which students make predictions using the content and vocabulary that they had recently learned. In this physics classroom, some students avoided using science vocabulary in generating predictions or explanations. A few of the students described their fear of how others would react if they used the language incorrectly, which sometimes prevented them from participating in this manner.

In this section, I analyze a transcript of a discussion associated with a demonstration and explore why students' contributions were initially limited in their variability and their use of science content and language, yet increased in their use of such content and language later in the class period. In doing so, I explore some of the situational variability and structural aspects that may make it more or less likely for an exchange of ideas using science concepts to take place.

Can Crush Demo

During one day in October, Linda was teaching the students about pressure. As part of her lesson, she conducted a demonstration during which she heated a soda can filled with water, then, using tongs, had a student turn it upside-down in a pan of cold water. If done properly, the can is crushed because of the resulting pressure differences between the inside and outside of the can. In preparation for this demonstration, Linda had students write down their hypotheses. She said:

> Now, listen up, this is where you're going to be a scientist. I'm going to tell you what's going to happen and you guys are going to make a hypothesis

about what's going to happen. Ready Saul? I'm going to use the tongs and if you've seen this before, please don't ruin the surprise for everyone else. . . . Once we have steam coming out of the top of the can, we're going to pick up the can with the tongs. Okay? . . . You're going to pick up the can off the burner once there's vapor coming out of it. . . . Listen one more time. We have steam coming out of the top. Very quickly we're going to move it over here. We're going to flip it upside down, and put it totally into the water at one time. Now, I want you guys to make a hypothesis about what we're going to see. What do we expect to happen when we flip the can upside down in the water? Think about what will happen.

As students began thinking about their predictions, Linda called on Ashley to come to the front of the room. Her role was to turn the can over after the can was sufficiently heated. As the can was heating up, Linda discussed the experiment with the students, asking students to volunteer their predictions of what would take place once the heated can was turned upside-down in the pan of cold water.

Linda:	Okay, while we're waiting, let's get some of your hypotheses. Ummmm. If you have seen this before, I don't want you to answer because I don't want you to – to ruin the surprise. Okay you stand up here and you tell me when it starts to steam. Okay, listen up! James has a hypothesis. Go ahead.
James:	There will be lots of steam coming from the heat.
Linda:	There'll be lots of steam coming from the heat. Okay, that's one hypothesis. Who else has one? Okay, go ahead, read it.
Jorji:	Okay. (inaud) going to condense and turn into a liquid and while it turns into liquid, it's going to lose energy and the energy is going to become transformed into (one word).
Linda:	Okay, there's an idea. (Laughter and mumbles from the crowd.) Okay, Dora also has a hypothesis. (.) Go ahead.
Dora:	When you put the . . . the can into the water, steam is going to come up.
Linda:	Okay, she also thinks there might be some steam coming up. Okay, go ahead.
Shelly:	I think wh– when you flip the can, it'll make a noise.
Linda:	And why will it make a noise?
Ted:	Tssss.
Shelly:	Yeah that. (Shelly and Ted raise and shake their arms)
Linda:	Oh, that sizzle noise. Okay, good. (Mumbles from the class.) All right, Dan has another one.
Dan:	When the hot can flips upside down into the water, the steam, the steam will probably rise.
Linda:	All right. So lots of people are thinking that there's going to be steam rising. Now, let's– let's make a statement about this. Okay, Carl are you ready?
Carl:	Mmm-hmm.
Linda:	Compare the water that's in the can to the water that's in this tray. Make some sort of statement.
Carl:	The water in the can is hot. The water in the tray is cold. (Giggles from the class)
Linda:	Very good. All right. Lisa, you want to make another comparison?

Lisa: The molecules are faster in the heated up can than they are in the pan.
Linda: Oh, so these water molecules are just hanging out, relaxed. And these ones are starting to get really excited. Not only because they're a part of a science experiment, which is of course very exciting, right? But because they're getting hot; they're starting to speed up. Aileen.
Aileen: Can I make a hypothesis?
Linda: Go ahead.
Aileen: Well, since Lisa said that the can water molecules are. . . you know. And the water in the pan isn't. . .you know, then when you turn the can over into the water in the pan, then the water in the pan will move like the water in the can.
Linda: Okay.
(Laughter, ahhs, and oh yeahs from the class.)
Linda: So everybody so far, shh. Well, not everybody, but quite a few people (inaud). A lot of people are saying that we're having some sort of transfer from the can into the pan.

The teacher's goals for this activity included getting students to think more deeply about the science behind the experiment by applying what they had learned about pressure to making predictions about what will happen when the can is turned upside-down. However, these particular comments about the steam and the noise do not seem to be the direction where she wants to take the discussion. At the end of this segment she provides a more structured question, asking a student to compare the water in the can to the water in the tray. In doing so, she elicits ideas about temperature differences, which will eventually lead to her explanation for the crushed can later during the class period:

> This can was full of water vapor. When I flipped it over, the water vapor that was inside this can, suddenly turned back into liquid. Okay? Turned back into liquid. Now the liquid does not take up as much space as the gas. Right? So suddenly, there's what we call a vacuum inside of our can. Like being in outer space, you know where there's nothing? There was a lot of nothing inside of this can. So, if there's nothing pushing out on the inside, look at what the air around us did to the can. It crushed it. (Gasps from the class.) Okay? That is due to air pressure.

Ideally students would be able to practice using science vocabulary words and ideas in making their predictions, such as by referring to pressure, water vapor, vacuum, temperature, or even liquid; only one of the students did so in this situation. One possible explanation as to why the students who volunteered (other than Jorji) did not use science language is that they do not know the words and concepts, or are not able to apply them. However, such a view makes the assumption that students' statements reflect their level of content knowledge, which may not necessarily be the case. Some students with whom I worked on this project would occasionally say, "I don't know" in class, yet competently explain the same concept to a peer when working in a small group. In addition, later in this same class session, many of the students did use science words in providing explanations for why the can became filled with water after being turned upside down, showing that they did "know" the vocabulary and concepts.

Why would some of the same students use the language and concepts later in the class, yet not when initially giving predictions? The data from this classroom support the idea that a student's contribution in a science classroom can be considered a strategic move aimed at maximizing the possibilities for engaging in successful IRs and for increasing entrainment, solidarity, and levels of emotional energy. In this approach, rather than viewing individuals as the unit of analysis and considering their stores of knowledge, as if they exist somewhere in the head, I view the interactional situation as the unit of analysis. This may seem like a subtle change, but it is has implications for formulating questions related to why students may not refer to science terms and ideas in their statements. Instead of asking, "why do the students not know the material?" additional questions of "why do they not always access their knowledge?" or "which situations are more conducive to their using this knowledge?" become important. Rather than conceptualizing the vocabulary and science concepts only as "in the head," they can be considered cultural resources that students may or may not access and appropriate, depending on the situation. In addressing this question, I consider not only what a student appears to "know" in his/her speech, but also other relevant issues such as how aspects of the student's identity (e.g., group affiliations, reputation, categories) may shape participation, how students have responded to each other's science-related statements over a variety of timescales (in previous years, that year, yesterday, and earlier in the class period), and whether a student would anticipate that making a particular statement would result in an IR characterized by solidarity and high levels of emotional energy, or a lack of solidarity and a loss of emotional energy.

Interaction Ritual, Agency, and Identity

There is a "market" for interaction rituals in that people will seek IRs that offer the greatest possibilities for entrainment and the accompanying high levels of emotional energy (Collins, 2004). For example, in choosing which party to attend, people will consider the physical setting, who else will be there, the activity that will take place, and other issues that impact the likelihood of participating in interactions that they will experience as successful. Whereas people may describe "interesting" or "fun" people as a reason to attend a party, this is desirable insofar as the interactions themselves would lead to the high levels of EE that come from participating in conversations characterized by rhythm, entrainment, and smoothness of turn taking. A successful dinner party with these types of conversations results in participants who are energized and feel a sense of solidarity with the other guests. The symbols that circulate during the successful conversations, be they sports teams, current events, public figures, or whatever else people tend to talk about, also become invested with emotional energy. Shared laughter, which entails mutual focus and joint noisemaking, is a particularly entraining experience and a desired feature of good conversations. Other potentially solidarity-producing experiences (depending on whether a person has had prior successful IRs that involve these activities) include dancing or gathering to watch a game and cheer for a team.

Of course, not all IRs are successful. Some conversations are awkward or contain frequent interruptions. This could occur for a variety of reasons, such as lack of match-up of common symbols, or aspects of the physical setting that make it harder for people to become entrained, such as seats that do not face each other. Some games are not interesting to watch and do not provide an effective mutual focus for the participants. Such experiences are unlikely to result in entrainment, solidarity, investment of relevant symbols with EE, and group membership.

Even a successful IR can result in the loss of EE for someone who is present yet not sufficiently engaged. An important feature of successful IRs is that they denote clear boundaries, in terms of who is participating in the interaction and who is outside. Those that share a mutual focus and are caught up in shared laughter or rhythmic conversation recognize that they are inside a group, and those who do not share the same focus and collective emotions are not. If one individual is present but is not caught up in the flow of the conversation, and does not participate actively, this person may lose EE and not develop a sense of identity associated with the group, even if the interaction is perceived as successful for everyone else. Whether one is caught up in the interaction may relate to issues such as familiarity with the circulating symbols, or having had prior unsuccessful IRs with that group or those symbols.

Further, participants' levels of EE will vary depending on how the person is positioned. Some IRs are "order-giving" rituals, leading to a gain in EE for the order giver and a loss for the order taker, without actually increasing feelings of group membership (Collins, 2004). In addition, even within a dinner party conversation that all would consider solidarity-building, different individuals may come away with different stores of EE. For example, a successful, rhythmic conversation where one individual is clearly the leader may generate EE for all those present, but his/her levels would be higher.

While structural factors (e.g., symbols, physical setting) can impact the success of IRs, there also exists some agency, in the sense that people have some degree of choices of which events to attend, with whom to speak when they get there, and what to say (what symbols to invoke). When making these choices, people will be drawn to those individuals, groups and symbols that have been associated with successful IRs in the past, and will avoid those that have not been (Collins, 2004). In order to make these choices people need to consider issues such as the physical setting, norms for that setting or community, their own stores of symbols, their current levels of emotional energy, and their identities. One's identity, which includes self-perceptions, group affiliations, and others' perceptions of self, can be considered one of the many structures that both enables and constrains these choices. For example, a student's identity as a "quiet student," acquired in previous classes, may constrain them to not speak in class even if s/he has a contribution to make, because of the pressure to match others' expectations.

However, one can also exercise considerable agency in remaking one's own identity by choosing among the various courses of actions available given the structural constraints. Hence, there is a structuring aspect to one's identity, which I refer to as "prior identity," yet within these constraints, people have agency in choosing actions that can change both how they view themselves and how others view them, leading to an "emergent identity." This emergent identity later becomes part of the structure—once again a "prior identity" influencing new series of interactions. As an example, a student who was known as not being good at science, and not an active participant, may over time begin to speak more in class, and if the teacher recognizes his/her contributions as valuable, s/he may begin to view his/herself differently, and perhaps other students will begin to view that student differently as well. Of course there are limits to people's agency in changing how others view them, particularly when categorical identities such as race and gender are salient. For example, Ashley has described how talking in classes in which there are only a few black people can be intimidating for her because people make assumptions that black people are not good at science.

In any interaction, a speaker does not have control over the outcome, even with his/her knowledge of the situation. There is always an element of risk, of one's comments falling flat and breaking others' entrainment, of using the symbols inappropriately, or of experiences that produce shame, such as when one's membership claims are not accepted. Aspects of the situation, which can loosely be thought of as structure, influence these outcomes. In this chapter, I draw ideas on structure in the form of schemas and (both human and material) resources (Sewell, 1992). Schemas refer to norms, ideas, principles of action, and habits. Stereotypes can fall under this category as well. Material resources include the physical setting, the nature of the relevant task, and the resources for learning available, while human resources include students' stores of cultural, social and symbolic capital (e.g., Bourdieu, 1986), which are exchangeable and have a powerful influence on people's social positions and on their ability to attain their goals. Structures both constrain and enable action, but such structures can be changed as a result of human agency.

In a structure|agency dialectic, there needs to be a mediating factor through which structure works to influence agency and vice versa (Archer, 2004). She describes this factor as the "internal conversation," which can take place as an agent reflects on structures as s/he perceives them, examines past experiences with these structures, and decides among the various courses of action that are enabled in order to maximize the possibility of one's goals being met. This idea has some similarities to Collins' view that people are drawn to situations that call forth previous successful IRs, specifying a process by which people are "drawn" to various situations. Through engaging in "internal conversations," people can act based on their assessment of these factors. For Ashley, her categorical identity as African-American might make it less likely that she would ask questions in class; however, it is not deterministic, as other variables such as teacher's responses to students, students' responses to each other, the progression of a particular classroom the discussion and whether she thinks a particular question would place her as an insider, or would break the flow and position her outside.

Identity can be seen as an interactional accomplishment emerging from a dialectical relationship between structures, such as norms, physical setting, and schema, and the agency of the participants (Roth et al., 2004). Classroom research that views identities as emerging from a structure|agency dialectic and continually formed within particular classroom contexts can not only emphasize students' agency in self-definition but also allow greater room for discussion of teacher agency in changing classroom environments to increase the likelihood of interactions that can foster group membership and identity surrounding science. This view of identity suggests that teachers can have a considerable impact on students' identities and sense of group membership insofar as they influence the material structures and schema that circulate in the classroom.

A class discussion about a science demonstration in some ways is like a conversation, though of course under a different circumstance than a dinner party. The participants do not choose to be there because they anticipate successful IRs, but instead are compelled to attend because of their roles as teachers and students. Also, the teacher may have more of an interest than the students in encouraging the rhythm of whole-class discussion, as awkward silences or interruptions may induce students to create their own interaction rituals with their peers, such as by talking to the student sitting next to him/her, thus potentially undermining instructional aims.

Science teaching has been criticized for relying too heavily on dialogue with teacher evaluation, which keeps control in the hands of the teacher (Lemke, 1990). However, it is also possible that teachers are drawn to this type of dialogue not only because they have control of the instruction, but also because rhythm and "flow" is easy to achieve when teachers evaluate student responses, with its call and response structure. Of course it can be argued that this dialogue may produce EE for the teacher but not for the students, because they are constantly deferring to his/her expertise, and it is like an "order-giving" ritual.

For the students and teacher in the science class, they did not wake up that morning and decide to attend based on successful experiences the night before. However, student agency extends beyond choosing to be present in the classroom to choices within the classroom, such as with whom to talk, whether they raise their hand, and what they say if called on. Though aspects of these choices may be the result of habitual action or a repertoire of strategies, in any moment there is still a variety of possibilities for action, and the student ends up saying and doing one thing rather than another. To consider their statements as the best examples of their content knowledge would be to suggest that students always would anticipate that displaying their content knowledge would lead to successful IRs. Yet as teachers and students know from experience, this is frequently not the case. For example, sometimes students can gain more EE from off-task comments. Students' choices regarding speaking can be thought of as emerging from their "internal conversations," during which they can take into account a variety of factors.

While the likelihood of a successful IR can in some ways be predictable, such as when a good party host introduces two people who s/he knows have stocks of common symbols, there is always an element of emergence. People may choose actions that offer a good chance of successful IRs and maximizing EE, but it may or may not work out in their favor, and they may be constrained in how these actions are received. Also, as the conversation progresses people are making split-second decisions about whether or not to speak and what to say, based not only on factors such as their prior identities but also on what transpired in the previous moment. It is important to consider a variety of timescales in thinking about student identities, as students not only come to class with particular identities, but also these identities change over time as interactions progress over the course of a year (Wortham, 2003), a week, or even a class session. The discussion surrounding the can crush indicates the importance of looking at the immediate timescale, as students' use of science discourse changed over the course of the class period.

Discussion about the Demonstration

During the first part of the discussion about the can, excerpted in the vignette above, the students were not particularly entrained in each other's movements and voices, and there was a lot of uncoordinated side-talk. While sometimes the students mumbled and laughed at each other's answers, this took place between small pockets of students, rather than in a rhythmic fashion that absorbed most of the class. The lack of entrainment in discussion may have been related to the experience of a failed IR, as about a minute earlier, the first attempt at the demo did not work, and therefore had not served as an effective mutual focus.

The first student who responded to Linda's request for hypotheses is James, an African-American student who had high levels of social capital and was an average science student. His statement, "There will be lots of steam coming from the heat" uses common rather than science-specific language, and does not reference any of the concepts that they recently learned. The teacher accepts his answer, and students are quiet and do not mumble or laugh in response. His answer is repeated by the teacher, which may not be as strong of a show of support as when there is uptake, in that a teacher incorporates a student's wording into her discussion; but it does at least indicate acceptance beyond just an "okay." His response did not lead to any awkward pauses or uncoordinated side talk among peers, indicating that they accepted the answer as well. His claims to membership in a science-centered community were not rejected, and he does not end up being the focus of an IR that excludes him. While James did not contribute to a successful IR, he did not have a negative experience that would be likely to result in a loss of EE.

The next student, Jorji—a white immigrant to the US whose parents work in the science field—provided an explanation that included some science language. It elicited talk and laughter from the other students, some of whom had their own solidarity-building IR at his expense. Whereas Jorji is accepted as a good science student, some students in the class described how they get frustrated with him because he "contradicts the teacher," "thinks he knows more than the teacher," and "is hard to understand." Therefore, it is likely that his statement may have been not accepted by the students because of its use of "too much" science language, in a way that did not make sense to the students and seemed to be "more scientific" than the teacher's talk. It is also possible that students perceived the teacher's response of "there's an idea" as not valuing Jorji's contribution—there was no uptake and she did not appropriate his language in her next statements.

It is therefore understandable that the next students volunteered predictions that were similar to James'. In the interactions that had immediately preceded theirs, use of science language placed the speaker as an outsider, as indicated through it serving as a focus for some small IRs that excluded the speaker (Jorji). In contrast, statements about the sound and the steam did not elicit any verbal response from the students. While the students making "safe" comments that did not use science language did not get the boost in EE that could come from being the center of a solidarity-building IR, neither did they make statements that clearly placed them outside the group. Their discussion of steam did not put their identities as science participants into question, in the sense that it echoed what the teacher had said earlier, "Once we have steam coming out of the top of the can, we're going to pick up the can with the tongs." They did not exceed the wording of the teacher, they did not bring in different science content or language, and they took little interactional risk.

While Jorji received a negative response to his statement, it was not just because he used science language. There were other aspects of the structure that were relevant, such as the teacher's response to his statement "there's an idea" and Jorji's identity within this class, in the sense of being viewed as someone who questions the teacher inappropriately. It is possible that students would have reacted differently to another student. However, Jorji was the first one to use science language, and therefore other students likely considered Jorji's reception in deciding whether to speak and what to say. It is understandable that the other students who volunteered also did not use science language, even if they "knew it." Instead, they followed the example of the one student who did not elicit a negative response.

Another perspective on why students did not use science language, and maybe could not even think of any at the time, is that the interactional situation was focused on supporting the teacher's authority rather than creatively approaching the task. Collins (2004) discusses some of the reasons why people sometimes choose not to speak in a public forum, writing that sometimes in academic lectures, there is a long pause before the audience offers any questions. He writes:

> The subjective experience of members in the audience at that moment is that they can think of nothing to say. Yet if the pause is broken—usually by the highest-status member of the audience asking a question—multiple hands go up. This shows that the audience was not lacking in symbolic capital, in things to talk about, but in emotional energy, the confidence to think and speak about these ideas not that they had nothing to say, but that they could not think of it until the group attention shifted to the audience. (p. 72)

He describes how it is not a negative reflection on the speaker when attendees cannot think of anything to say, as good speakers are even more likely to elicit that silence, because the speaker is too-well respected to be approached. However, if a high status person does ask a question, the focus of the interaction shifts to the audience, and others therefore can become participants.

Although he is writing about asking questions of an esteemed speaker in public forum, a similar analysis can be applied to thinking about why students may hold back from science language in discussions. The science community, given the high status accorded to science knowledge, may be positioned as distant and somewhat like a sacred object. Like the prestigious speaker in Collins' example, a teacher is also distant from the students. She holds the authority on science and is the gatekeeper to their entry, both formally in her assignment of grades and informally as she evaluates students' contributions in class. As the students respond, they hear that "steam" was in Linda's own statements, and therefore that may be the "right" answer. The students cannot think of anything else, as the interaction may have remained focused on upholding the teacher's authority (with Jorji having been criticized partly for stepping beyond those bounds) and had not shifted to their generating their own explanations.

After Linda shifts the discussion to temperature, students do not mumble after Lisa's response, which uses science language, "The molecules are faster in the heated up can than they are in the pan." No one gives her a negative reception, possibly because she has earned an identity as a good science student who is liked by teachers and does not ever question the teacher's authority. In addition, the teacher's own response may have conveyed to the students that Lisa's answer was worth accepting, as there was uptake. The teacher repeated the word "molecules" and elaborated on Lisa's statement, saying, "Oh, so these water molecules are just hanging out, relaxed. And these ones are starting to get really excited. Not only because they're a part of a science experiment, which is of course very exciting, right? But because they're getting hot; they're starting to speed up." Unlike Jorji's, Lisa's use of the science language was accepted by the teacher in a more substantial way, as she drew from Lisa's words in making her own statements.

Next, Aileen, who was not known to be a good science student, and in the beginning of the year said that she "hates science" and is "bad at science," says, "Can I make a hypothesis?" Interestingly, she does not just volunteer an idea as the other students do, but asks permission before doing so. This can be interpreted as avoiding claiming a face (Goffman, 1959) that others will not accept, in the sense of being a good enough science student to provide valuable ideas. This move can also be seen as her trying out the science word "hypothesis," testing how other students respond to her, as their responses may enter into her consideration as to whether to continue to do so.

After being given permission to speak, she draws on Lisa's science words in making her statement. "Well, since Lisa said that the can water molecules are . . . you know. And the water in the pan isn't . . . you know." While in some ways she is making a risky claim, using science language and ideas when she is not sure how others will react, her response can be considered as strategically reducing the risk, as she qualifies her claim. She refers to Lisa's discussion of molecules, rather than just using the word without quoting someone else. She also says, "you know" rather than giving the whole explanation. By doing so, not only does she temper her membership claim, but also she suggests that others know, perhaps better than

INTERACTION RITUALS AND SCIENCE LANGUAGE

she does. She speaks as a science participant, yet defers to both Lisa and the teacher, and does not claim a face that others are unlikely to support.

She then smiles and says in a clear voice and with rhythm and emphasis in her speaking, therefore appearing confident, "when you turn the *can* over into the water in the *pan*, then the water in the *pan* will move like the water in the *can*." People laugh, but unlike in response to Jorji, where the laughter was uncoordinated, sporadic, and punctuated with talk, in response to Aileen, the laughter is sudden, as if she just delivered the punch line of a joke. She looks around smiling, suggesting that the interaction is taken as not at her expense, but as if she intended to bring on the laughter. In this, Aileen is at the center of a successful IR. Her statement can be thought of as very strategic, as she fulfils the teacher's expectations, is funny, initiates a successful IR with her at the center, and makes some type of a membership claim to science-centered communities (though tempered with her "you knows") through her use of science language.

Rather than change the subject as she did after the steam comments, Linda takes up Aileen's response, and through IRE (i.e., initiate-respond-evaluate) elicits from the students that actually it is more likely that the pan water will affect the can water because there is more water in the pan, rather than the other way around. While this uptake indicates that Aileen's response was not correct, it is a substantial use of her response, incorporating it into the discussion, rather than just an acknowledgement such as "there's an idea."

Soon after, the demonstration works somewhat, and students are able to hear a noise, though still not see the dramatic crush. From this point on, there is less uncoordinated noise, the students are more focused on the front of the room, and there are more attempts at using science language and content.

Perhaps significant is that the science language is not only drawn from what they learned in recent class sessions, but also from previous years. For example, Caitlin mentions weather, although they had not even learned about weather this year, and Aileen uses the word "react" even though they had not learned any chemistry this year. The greater variety of responses, the number of volunteers, and the use of science language and content knowledge suggest that the focus of the interaction has shifted to the students. Whereas before they may not have been able to think of these words or this content, now they have the emotional energy to access this resource. Here is an excerpt.

Linda: You saw the steam? You saw (inaud) coming from the top? Okay, now wait. Guys, I did want to ask you about this. I had about this much water in the bottom of the can. Now, look at how much water is in it now. What is causing all the water to shoot up inside the can? Aileen.
Aileen: The water ... is going up inside the can 'cause it's rea—the cold water is reacting to the warm water?
Linda: (.) Okay. . . . Jerry.
Jerry: Okay, like since, like, um, water vapor evaporates and like goes up, it rises.
Linda: Yeah?
Jerry: The, um, the hot water vapor is looking for a way to get out, so when it goes out, it pushes some water in. That way so the steam can come out.

Erin: Okay. In weather, when you have this area of high (.), well, hot air.
Linda: Okay, hot air.
Erin: And this area of cold air.
Linda: (/) Mmm-hmm. (/)
Erin: Doesn't the hot air push up the cold air? (.) (/) Or is it the other way around? (/)
Linda: Okay. (.) This is – this is related to something in weather. What happens then the sun beats down on the ocean? What happens to the water in the ocean? Maria?
Maria: (/) It evaporates. (/)
Linda: It evaporates. And in that, evaporating water goes up into the atmosphere where it's cooler. And what does it do to the atmosphere?
James: It cools down.
Linda: It condenses. It cools down and condenses. Now, (2) these cans, right now, are mostly full of steam. Because, you admit that when the water turns into a gas, it starts to get excited, right? And it's filling up the can. Now, when we flip it over and we put it into the cold pan (2), what's happening to those little pieces (.) not pieces, but what's happening to those gas (.) molecules that are inside the can? What happens to them? Derrick?
Derrick: (/) They condense? (/)
Linda: They condense. Okay? Those, those little particles are gonna condense. Now here's my question for you. When they condense (.) and the top of this has water, (2) what is – what is the can – what does nature want to do with that empty space? Dave?
Dave: Well, it pulls it up. The pressure pulls up the water.
Linda: Okay. Because it's trying to fill it with something.
Dave: Yeah, just like when you have a – a cup, and you put stuff on it and–
Linda: A straw. Right? You have to be able to use pressure to be able to drink out of a straw. That's right.
Dave: No, no, no, no. Like, like, uh, (inaud).
Linda: All right, so here's my question. What is forcing the water to go up inside the can? I mean there's not a magic button inside the can sucking. It's not—there's no one in there sucking the water up. What is causing the water to suddenly enter the can?
Student: Empty space.
Linda: Well, there's empty space, but what's making something go into the empty space?
Student: Pressure.

This discussion has a lot of uptake, with Linda using the students' ideas, such as Caitlin's mention of weather, and repeating their words, such as when she repeats James' "cools down." Also, the students are repeating Linda's words, such as "condenses." Unlike the earlier conversation, the students seem somewhat entrained. They are not engaging in uncoordinated side talk, and they are attentive to the front of the board.

In addition, the students' use of science language was for the most part accepted by their peers during this part of the discussion, which was not always the case in this classroom. One exception was when a student, Samir, makes a comment "the cold water diffused into the can." The teacher repeats Samir's comment, "the cold water diffused into the can?" Samir then receives a negative response of laughter and some insulting comments from his peers. Based on both classroom observations and on discussions with the student researchers, it seems that Samir's claims are rejected in this instance partly because of the teacher's questioning response, and partly because of aspects of his identity, as he is African-American and has a history of low achievement in science (see Olitsky, 2006). Samir's negative reception in spite of the relative success of this particular discussion in eliciting students' use of science language suggests that categorical identities such as race can have a substantial impact on students' science classroom experiences.

Aileen also encounters some of the same challenges that Samir does, as she is African-American and is not known as a strong science student. Why is Aileen's comment accepted, whereas Samir's is not? One factor is that the teacher responded with uptake to Aileen's contribution, but seemed to question Samir's. Another issue is that while Samir makes his claim to membership in the science community through a direct explanation, Aileen tempers her claim by her use of humor and her qualifying statements such as "Lisa said" and "you know." It is possible that Aileen adjusts her statements strategically based on expectations that her peers may not accept a direct claim from her.

The greater use of science vocabulary and concepts in this part of the discussion came right after the success of the demonstration in producing a noise and filling the can with water. This coincidence of events suggests that as students became entrained in the demonstration, their levels of emotional energy increased, thereby providing them with the confidence to speak and the ability to access the resources of science language and content. Another possible explanation for the increased reference to science is that the uptake of Lisa and Aileen's statements by Linda helped shift students' focus of attention away from the activity of supporting the teacher's authority and toward the activity of generating explanations that would become a part of the discussion. Therefore, the students were better able to think of the science responses. Another possibility is that the teacher's uptake of Aileen's response and the students' acceptance, with their coordinated laughter and shared glances with her (indicating they were laughing with her rather than at her), made the prospect of using science language and referring to content seem less risky for the other students. After Lisa and Aileen's statements, as students engaged in internal conversations regarding whether to speak and what to say, the option of drawing on science content did not seem as likely to result in a lack of acceptance, critical comments from other students, and small-scale IRs from which they would be excluded.

While Samir's claim had been rejected, his statement had occurred after the students had already become entrained in the demonstration, and perhaps was not a strong enough experience to have a silencing effect on the rest of the class. In addition, the teacher did not take up Samir's response, but she did take up the other students' responses. The events in this classroom suggest that teachers can have a strong role in facilitating students' use of science language through incorporating student contributions into their talk.

The progression of this discussion also supports the idea that aspects of identity, in the sense of claims to group membership, can change somewhat even during a class period. Aileen, who has shown a lack of identification with science-related communities, and who did very poorly in her science class the previous year, expresses less hesitation using science words as the class progresses. While her comment, "The water . . . is going up inside the can 'cause it's rea—the cold water is reacting to the warm water?" may show some hesitation, as she stumbles over the word "react," she is acting as participant and does not ask permission from the teacher to give her ideas.

Categorical identities (e.g., African American, female) and other types of prior identities (e.g., reputation as a not so good science student) may be influential in shaping Aileen's participation as she weighs the possibilities for entrainment, solidarity with peers, and increased levels of EE. For example, she has described how she often felt discouraged from talking science in the large group at City Magnet because it included white students who would sometimes make negative comments about others' wrong answers and cared only about their grades. However, Aileen's participation in this particular session suggests there is a level of emergence and unpredictability, and that there are local factors that can support science participation. In this situation, the uptake by the teacher, the acceptance of Lisa's comment, and her opportunity to initiate a successful IR though the use of humor in generating an explanation supported her access of science language and her consequent claims of membership. As she engages in internal conversations regarding when to speak and what to say, it is likely that the successful IRs involving science language and her hypothesis contributed to her being drawn to volunteer and use science language later in the class as well.

INTERACTION RITUALS AND EMOTIONAL ENERGY AS RESOURCES FOR IDENTITY BUILDING IN SCIENCE CLASSROOMS

An important part of learning science is formulating ideas, debating explanations, and talking about science with others. In studying science learning, it is valuable to understand the conditions that might be able to increase students' use of science language and concepts as they speak in class. While changes often focus on making sure students know the content, this approach alone may not increase the quality of discussion, as classroom talk is not just an exchange of content by disembodied minds. The approach that I take to interpreting students' statements in class discussions is not only as evidence of their knowledge at a point in time, but also as a decision about how to act.[2] I view classrooms as particular types of social contexts with particular structural features, in which teachers and students make presentations of their selves to others, accept or reject others' claims, and make choices of when to speak and what to say in the interest of maximizing the potential for participating in successful IRs and increased levels of emotional energy, and avoiding the loss of EE.

Desirable interactions for an individual include participating in a successful IR during which participants share a mutual focus (such as a can crush demo) and emerge with feelings of solidarity and high levels of EE. Even more desirable may be initiating a successful IR, with the speaker positioned inside the boundaries (as Aileen did when making her hypotheses). In both of these situations, a sense of group membership and identity was solidified, and the membership claims of participants were supported. Undesirable interactions include serving as the focus for an IR in which the speaker is positioned as outside, such as when membership claims are rejected (e.g., Jorji uses vocabulary more "scientific" than the teacher's).

The results of this study suggest that a student's knowledge of the subject matter is only one factor among many others that influence his/her anticipation of which types of talk will contribute interactional success. Other relevant considerations include prior identities (how students view themselves and how others view them), which shape whether students can anticipate that their membership claims may be rejected. However, while prior identities are influential, they are not deterministic. On a day to day basis, the success of the various IRs that take place in a particular class session may shape levels of EE for particular students, and therefore the confidence that facilitate their thinking of the science words. Each turn in the conversation, with participants' reactions to the previous utterances can shape the next speaker's decision. In this study, teacher uptake of student responses was particularly important in demonstrating to other students both the legitimacy of that student's membership claim and the type of contribution that was made. In this demo, the uptake seemed to shift the focus of attention away from the teacher to the student's explanations, suggesting that the focus of IRs may also shape whether students have the levels of EE and the consequent confidence to use science language and refer to science content.

An understanding of the process by which students act as agents as they make predictions and explanations in classroom talk could help teachers extend their interpretations of seemingly failed discussions about demos beyond issues of cognitive content and into the social realm. Considering the factors that may contribute to students anticipating successful IRs could help in planning to structure classrooms that more effectively elicit students' actual content knowledge. This process could also help in reducing teachers underestimating students' capabilities with regards to understanding and applying the material. It could be that teachers may not always need to re-teach the content but instead may be able to change other aspects of the classroom conditions to make them more conducive to stimulating discussion.

In the science class, the success of interaction rituals surrounding demos, or any other event, has implications for students' identities as science learners. In some ways, the goal of a discussion should not only be to give students practice applying science content to discussing an observed phenomenon, but to give them practice doing so within classroom environments that are structured in such a way that students are likely to experience solidarity and longer-lasting EE associated with science. Discussions in which the students can draw on science content and experience successful IRs at the same time may increase the likelihood that over time students will identity themselves as members of a community centered on science.

NOTES

1 Whereas I use the term "strategic move" and "choice" with regards to students' utterances in classroom discussion, I do not wish to imply that students necessarily weigh particular options in a ra-

tional way and then act on the chosen option. In moment-to-moment interaction, there may be no time for that process of going through all of the possibilities. Yet clearly in those moments of interaction, people do end up acting in one way rather than another, whether it is a result of "rational" decision-making in the moment, dispositions, habitus, or some other process that can convincingly account for human agency and the indeterminacy of outcomes. My use of terms such as choice is intended to reflect that there is some agency and some variability.

2   By "decision" I do not mean that people always are consciously deliberating, but that they do act strategically. They may not always be successful in their goals, there is still some degree of agency, as in any moment a person can speak or not, say one thing or another, try to use the science vocabulary or avoid such vocabulary.

REFERENCES

Archer, M. (2003). *Structure, agency and the internal conversation.* New York: Cambridge University Press.
Bourdieu, P. (1986). The forms of capital. In J. G. Richardson (Ed.), *Handbook of theory and research for the sociology of education* (pp. 241–258). New York: Greenwood.
Collins, R. (2004). *Interaction ritual chains.* Princeton, NJ: Princeton University Press.
Goffman, E. (1959). *The presentation of self in everyday life.* New York: Anchor Books.
Lemke, J. L. (1990). *Talking science: Language, learning and values.* Norwood, NJ: Ablex.
National Research Council. (1996). *National science education standards.* Washington, DC: National Academy Press.
Olitsky, S. (2006). Out of field teaching and the discursive practice of science. *Submitted.*
Roth, W.-M. (2006). *Learning science: A singular plural perspective.* Rotterdam: SensePublishers.
Roth, W.-M., McRobbie, C. J., Lucas, K. B., & Boutonné, S. (1997). Why may students fail to learn from demonstrations? A social practice perspective on learning in physics. *Journal of Research in Science Teaching, 34,* 509–533.
Roth, W.-M., Tobin, K., Elmesky, R., Carambo, C., McKnight, Y., & Beers, J. (2004). Re/making identities in the praxis of urban schooling: A cultural-historical perspective. *Mind, Culture, & Activity, 11,* 48–69.
Sewell, W. H. (1992). A theory of structure: Duality, agency and transformation. *American Journal of Sociology, 98,* 1–29.
Steele, C. M. (1997). A threat in the air: How stereotypes shape intellectual identity and performance. *American Psychologist, 52,* 613–629.
Wenger, E. (2000). Communities of practice and social learning systems. *Organization, 7,* 225–246.
Wortham, S. (2004). From good student to outcast: The emergence of a classroom identity. *Ethos, 32,* 164–187.

JRÈNE RAHM

3. LEARNING AND BECOMING ACROSS TIME AND SPACE

A Look at Learning Trajectories within and across Two Inner-City Youth Community Science Programs

Researcher: Is there something that is special to you about the program that you cannot find elsewhere?
Irina: Yes, all my friends are here, and I have never been in a place before where we have the kind of science that we got and do here.
Mazela: Everything!
Kamila: Yes, there are only girls here and we learn lots of things!

These three girls, about whom we learn more in this chapter, position themselves as insiders and full members of the afterschool program and as girls who do and like the kind of science they get to engage in, a science they cannot get elsewhere. Such positioning work in relation to science goes past publicly circulating models of identity that would position these black or poor girls as not being interested in science and not learning (Brickhouse, 2000). In fact, such a stereotype is often invoked, yet in reality, it most likely does not fit anybody since it is too homogeneous and does neither account for individual diversity nor for agency. Such a stereotype does not account for local identifications of possible selves in science that the statements underline. In this chapter, I focus on learning trajectories and identity work among these three girls as they are engaged in and talk about science within two different science programs in the nonschool hours. I explore their positioning work in relation to science within each program as well as across time within and across programs. In doing so, the girls' forms of participation are taken as "performance(s)" of who they think they are and hope to become (Goffman, 1958). It is assumed that each opportunity to interact and engage in a social system changes who we are and become. It underlines also that identity development happens continuously and that who we really are and are becoming has to be understood in terms of much more expansive timescales (Lemke, 2000), an issue I come back to in the conclusion.

I examine engagement in science in an afterschool and a community science program, settings that have been recognized by educators, policy makers and the public as important for girls' science literacy development. They can be effective in improving girls' self-confidence and interest in science, while also reducing their often sexist attitudes about science through authentic science activities (Ferreira, 2002). They also help girls explore science in to them meaningful and valuable ways while recognizing their own daily experiences and ways of knowing as resources in such a process, as the quotes above also underline. In fact, many afterschool and community science programs emerged in reaction to lack of, or poor science education marked by cookbook science activities with no space for ownership and deeper exploration (Crane, 1994). Science programs that are valued by youth are inclusive, open and respect youth for who they are. They also offer unique opportunities for science literacy development as well as youth development (Delgado, 2002). In short, quality afterschool and community science programs can serve as "door openers" to the world of science and the world at large. How such happens is explored here through a focus on three girls' identity projects.

Two assumptions guide my analysis: (a) that we become different kinds of people as we learn, and (b) that we learn new things all the time. These assumptions justify my study of learning and becoming as a unified process irrespective of context. They also help us move beyond dualistic notions still permeating much research in education such as the separation between "knowing" and "being" as well as between "formal" and "informal" learning. The heterogeneity of social practices and their inherent identity work and learning trajectories are taken as a given. It is assumed that "all activity takes place at the intersection of different communities, each with their own practices, norms, and values" (O'Connor, 2003, p. 70). Individuals themselves also "bring with them a history of participation in different contexts." Accordingly, the girls' learning trajectories and identity projects at the moment will have to be understood in terms of their histories as well as perceived possible future selves. They have to be understood in terms of their positioning work in science in other practices such as school or home, as well as across time or experienced selves in the past, the present and the future. In fact, an analysis of these interrelated dimensions will make possible the study of the diversity in positioning work and learning and make evident the dynamics of contextualization and identification.

My analysis also allows for a discussion of the vast resources that the girls draw upon to position themselves within and across the practices examined here. For instance, one may distinguish sociohistorical models of identity that are invoked to mark selves in learning environments such as "girls being uninterested in science" from local categories of identity, located in and supported by afterschool and community programs, that may contrast with and contradict historically defined categories, making new future selves a possibility. In essence, "models of identity change as they move across time and space, and they are applied in contingent, sometimes unpredictable ways in actual events of identification" (Wortham, 2006, p. 8)—a sort of interplay I am interested in studying here. In so doing I clearly adhere to a notion of identity as dynamic, assuming that "human beings created, construct, work on, and enact their identities, sometimes creatively challenging the limits of the cultural constraints which constitute both what we call selves and the ways those selves can be crafted" (Kondo, 1990, p. 11). Through the stories of three girls' forms of participation within and across two science programs during nonschool hours, I aim to make explicit some of the complexity yet also the dynamics of identity work over time and space.

ETHNOGRAPHIC CONTEXT

Description of Contexts of Identity Work

I explore identity work in two programs that are part of a multi-sited ethnography with a focus on learning opportunities in science in the nonschool hours for poor youth. The first program studied, *Scientifines*, is the only one I could identify within the greater Montreal area (Canada) as explicitly committed to science and poor youth. Next, I looked for programs that are culturally and psychologically accessible while also entailing low enrollment fees. One of the programs identified was *Jardins-jeunes*, a youth centered gardening program that serves approximately 76 percent poor youth (poverty level based on parental occupations and poverty index of school attended at time of study). A second objective of the larger study was to promote partnerships among programs for poor youth. To initiate this process, I teamed up with four girls from *Scientifines* and made their participation in *Jardins-jeunes* possible for the year of the study by covering their tuition fees while also accompanying them to the garden throughout the summer. Such now allows for the study of their identity projects across two different programs. Hence, I focus on three of the four girls' forms of participation within and across these two programs.

Scientifines is an afterschool science program in existence since 1987, serving girls from two neighborhood schools within a community that is poor, ethnically diverse, and the home of many first generation immigrants. While approximately eighty girls are registered in the afterschool activities, about forty girls show up each day. Typically, they receive a snack, which is followed by a homework period, and in turn, they devote about an hour to an activity in science, ending the afternoon with free play which includes working on computers, playing games, or taking care of some of the pets. The last three years, a group of girls also worked on science fair projects one afternoon a week for the whole school year, projects they then presented to the community and a selection of them also went to local science fair competitions—the focus of the study here. The goal of the program is to get girls interested in science but to also offer them some tools that may help them be successful in the future. Given the high pregnancy rate of girls in that neighborhood which then typically results in them dropping out of high school, the program aims to offer these girls an opportunity to develop the resilience needed to complete high school and pursue future careers that may also help them break out of the vicious cycle of poverty.

Jardins-jeunes is a summer gardening program of the Botanical Garden in Montreal. The program exists since 1938 and offers children and youth the opportunity to grow vegetables in a plot they take responsibility for while also exposing them to relevant concepts of the natural sciences, horticulture and agriculture. The program was modeled after the first gardening program in northern America at the time—the Brooklyn Botanical Garden—and was initiated by Father Marie-Victorin, a well-known Canadian botanist as well as founder of the Botanical Garden of Montreal. The gardening program offers a morning section for elementary school children and an afternoon section for high school students—the focus of the present study. The gardening program is eight weeks long, entails work two afternoons a week along with some seeding activities in the spring and some harvesting activities in the fall. While 60 youth were registered for the afternoon program, I followed a group of 46 youth (31 females and 15 males), ranging in age from 12 to 16 years.

A typical afternoon begins with a period of instruction during which time youth are exposed to a theme relevant to gardening. For instance, they learn about different kinds of pesticides, how to make compost, the difference between natural and synthetic fibers, to name but a few. After the whole group activity, youth then tend to their garden. Each day, they have specific tasks listed on a board, such as the weeding and thinning out of the growing plants so they would bear some vegetables, or the control of insects and pests that were taking over certain plants, etc. Only towards the end of the afternoon are the youth allowed to harvest and then water their garden. At times, they also work in community plots, which at the end of the summer yield considerable crop that is then divided among all youth (such as water melons, squash, peppers and lettuce). Every day during the summer, the youth left with some harvest from their own plot.

Data Collection and Analysis

First, I conducted a one-year video ethnographic case study of *Scientifines* and followed the nineteen girls who worked on science fair projects one afternoon a week from September 2003 to the spring of 2004, completing their projects with an open house science fair (initially, 30 girls signed up, but only 19 completed it, resulting in eight projects overall). The 19 girls that participated ranged in age from nine to eleven years. Questionnaires at the beginning and end of the program year helped establish the girls' general science profile and attitudes towards school and science. Semi-structured interviews that year (spring 2004) and the following year (spring 2005), made possible an examination of their ways of talking about science and identity projects over time. Next, a videoethnography was conducted of the program *Jardins-jeunes* following the 46 youth that consented to participate as they engaged in gardening work. Such data were also supplemented by questionnaires and semi-structured interviews conducted at the end of the summer (August 2005).

For this chapter, I re-read the three interviews, the questionnaire data as well as examined videos, fieldnotes and journal notes pertaining to each case, which then led to the development of three stories summarizing these girls' experiences within and across the two programs. Particular attention was paid to youths' positioning work within each program and in relation to science, as well as across the two programs. I was interested in the manner cultural models of identity as well as local or situated identities were enacted, appropriated and challenged. How youth talked about other contexts such as school and home, as well as their future was also examined and taken into account in attempts to understand their learning and becoming across time and space.

LEARNING AND BECOMING

First, I present each case and discuss forms of participation and identity work within *Scientifines* and in turn *Jardins-jeunes*. Second, I look at commonalities and differences across the cases in terms of their identity work as well as contexts that came to constitute who they were and were becoming. In conclusion, I discuss what the study of identity and learning offers to an understanding of science literacy development while I also allude to some of its challenges.

Instrumental Value of Program for Future Possible Self as an Educated Person

Irina is from Russia and came to Canada with her parents three years ago. As we began the study in 2004, Irina was ten years old and in grade five. She participated in *Scientifines* for two consecutive years. After participation in *Jardins-jeunes*, she began her first year of high school in a private college. Her parents were very eager for her to do well in school and to be successful later on. They had moved to Canada in high hopes for a quality education for their daughter. They highly valued her participation in *Scientifines* and also *Jardins-jeunes*. Her mother also volunteered time at *Scientifines*.

Irina liked the general relaxed climate of *Scientifines* as well as having the opportunity to be with her friends and doing science. Yet, like her parents, she primarily saw the program as instrumental for her future success. This came through even in her summary of science when asked to describe it:

> I would say it is something interesting where you never know what will happen. But when you put forth a hypothesis and it is a good one, you are surprised and it is interesting because you discover what it is in reality and the truth. It is also fascinating because one never knows what will happen. And then, science is like . . . one girl that said, "a little step for a girl, a big step for a woman" . . . because when one goes to *Scientifines* [and then] when you are big, one could even invent things. And so it's important.

Irina saw science as powerful but also mysterious. She refers to a quote by another girl that summarizes her experience in *Scientifines*—she learns about little things of science (a little step for a girl), which in turn may result in "a big step for a woman" (in terms of her achievement and position in the world as an adult). She saw the value of learning about science for the future, not necessarily in terms of a specific career but in terms of making "discoveries" and being successful. When asked specifically about whether she would want to pursue science as a career one day, she mentioned "maybe doctor," but then she also valued becoming "a teacher" and added, "in elementary school, that would be easier" and referred to becoming a doctor as entailing "too much science." It shows that she enjoyed the program and the kind of access it offered to science, but science remained mysterious and somewhat distant to her. In fact, a stereotypical image of science as being about the "truth," as being "difficult" and as a world of its own not necessarily accessible to her as a girl clearly is hinted at here. Hence, while the program gave her access to science, it did not make a repositioning in relation to science possible. In essence, Irina orients to *Scientifines* from the perspective of other contexts that have made her come to see science as something beyond her reach as a career, yet as something of instrumental value on her trajectory towards becoming an educated person (O'Connor, 2003).

Irina greatly enjoyed the science fair projects. She was on a team that initially wanted to understand why chocolate makes one excited. To answer the question, they focused on the ingredients of chocolate and, in so doing, found out that white chocolate contains much more sugar than dark chocolate. In fact, according to Irina, white chocolate "has no chocolate in it, it contains just cocoa butter, milk, and sugar" whereas "real" chocolate consists of "cocoa beans [powder] and milk," information they shared with the public in their presentation as follows:

Excerpt 1

1	Yura:	Today we will talk to you about chocolate.
2	Irina:	How much sugar is there in chocolate? Do you know?
3		White chocolate contains forty percent sugar,
4	Yura:	milk chocolate contains thirty percent sugar
5	Hanna:	and black chocolate contains twenty percent sugar
6	Irina:	chocolate is made up of a couple of ingredients but the four primary in-
7		gredients are milk, sugar, cocoa and butter.
8	Yura:	and in dark chocolate there is milk, sugar, cocoa and butter but less sugar
9		than
10	Irina:	and more cacao
11	Hanna:	wait, there is less sugar than in white chocolate . . .
12	Irina:	and milk chocolate consists of milk, sugar and cocoa and butter. That one
13		is between the other two.

Irina was very knowledgeable and proud of their presentation. Yet, as shown above, she constantly struggled to gain and keep the floor during the presentation, making visible some of the challenges she experienced in that team. Irina liked finding out what "things she eats consist of." For her, science was also something that applied to real life and helped her make sense of many everyday things such as understanding what chocolate consists of. In fact, it was this kind of science that she could relate to and identify with. The connection they valued as a team between their project and the real world was very well apparent in the kinds of resources they sought out for their presentation. As one instructor summarized, they "went to a chocolate factory and asked for all kind of equipment . . . and then they wanted cocoa beans and baking moulds, so they called a chocolate factory and also the botanical garden to get cocoa beans." While that team does look successful at the end, their trajectory was not always that smooth. Initially, they wanted to fabricate candy but since they never found the kind of information they were looking for, they eventually switched to chocolate.

The following year, Irina tried to understand how a car works. Again, it was not a topic she initiated but instead the chosen theme of a group that needed another participant. Initially, they hypothesized that a car needs an engine to work. Through the project they learned that "a car also needs oil and other things." Yet, Irina assured a link with the real world again by seeking out her father who then helped them build a model of a car engine given his expertise in car repair work. Despite such connections with the real world, the project was seen as challenging by Irina, underlining again her image of science as something difficult:

> I learned that you have to put a lot of effort into a science fair project if you want it to be successful. At first, I thought it is just about doing some research on paper, just read. I also learned a lot of things about cars this year like for instance V-6 and what this means, that it has six cylinders and they are placed in the shape of a V. Last year, I learned that white chocolate is not really chocolate [since it has no cocoa powder in it].

Nevertheless, Irina was able to summarize well what she learned. Irina also changed her cultural model of scientists somewhat:

> I think the difference is, now I understand that it's not just somebody with glasses and a white lab coat . . . these are people that do useful things, so for instance, if I would not know that water with gas in it and sugar makes foam, for instance, without wanting to do it, I could happen to put sugar into fuzzy water and it could spill over and make a mess. But now since I learn better about science, I can be more careful."

Scientists were described now as doing "useful things" and knowing science for everyday life was valued. Irina had an explanation about how knowing science made her appear educated. It makes her "be more careful" in life. Participation in *Scientifines* contributed to her perception of self as somebody educated that can make informed choices as a citizen given a certain level of science literacy. In essence, knowing science was of instrumental value towards becoming educated. Such a position also came to mark participation in the gardening program. She talked at great length about a link between gardening and ecology in high school and underlined again the instrumental value of gardening "knowledge" in school science, making her look as "ahead of others." Irina also highly valued the opportunity to grow her own vegetables:

> [W]e know these are non-treated vegetables that are biological, one is happy to eat them since one knows that they will not be bad for our health, and I think it's also great to have our own vegetables. . . . I think one is proud when one thinks about all the things we did to make them grow. We put lots of effort into it and so we are very proud and happy to eat something we grew and harvested ourselves and that is biological too.

To grow her own organic vegetables that are not "bad for our health" was important to her. It underlines how science can be put to use towards a positive end. Gardening also helped enlarge her notion of science. While *Scientifines* helped her understand science as the "proofing of things," as entailing much chemistry and human biology, gardening helped her see science in broader terms even though it was more about "doing." Both programs got her excited about science. "One learns things and one can still learn more and from other people, and sometimes I shared things with my mother and sometimes I asked questions about things I did not know, so all helped me learn more about more things." Both programs were seen as contexts in which one can learn a lot for life in an amusing manner. Yet, learning was always valued in terms of becoming an educated person rather than a full member of these communities of practice. Science per se remained somewhat obscure and beyond her reach, a cultural model she had absorbed and that remained unchallenged over time.

Programs as Treasured Spaces to Experience a Successful Learner Identity

Mazela was born in Haiti but moved to Canada with her family not long after. She was eleven years old when I started the study and in fourth grade at the time, participating for her second consecutive year. Overall, she participated in *Scientifines* for three consecutive years, initially together with one of her sisters. Mazela struggled with the many rules in school and highly valued the more relaxed atmosphere of the two programs. While participating, she also did some babysitting in her community. After the gardening program, Mazela had to repeat sixth grade in the elementary school close to *Scientifines*.

In her first interview, Mazela explained that "everybody always talks about science at home" and that made her come to the program, "I wanted to know what it is, the science, and how it was here [in the program] and I learned." Her sister who participated in the program before her was most likely the one that talked much about science at home. Now that her sister was in high school, Mazela did the reporting, "each time when I get home I tell them what I did and they are very happy." While Mazela positioned herself as knowledgeable of science, she was afraid of science initially. She saw science "as difficult to learn about and to learn quickly." In fact, she claimed that in the past, "I did a lot of science but I learned little, and now, I learn quickly and a lot and I do many different things." She also claims that participating in *Scientifines* helped her in school, "usually, in school, I take my time, but now I don't do that anymore, I carefully examine what I do and how." It suggests that the program may have helped her experience success and in turn, develop much self-confidence. It may have helped reposition herself as a partial insider to science. It appears to also have challenged her image of self as academically weak and incapable of doing well in school.

During the first year of the study, we followed her work in the acid team. She was with two other girls who all shared a strong interest in doing an experiment "with a reaction." Unlike others who researched a topic and then explained different dimensions of it, they wanted to do an experience for their science fair project. Eventually, they decided on a project exploring whether "any kind of acid mixed with a base would result in a reaction." One of the members had previously mixed vinegar with baking soda, which led to a "chemical reaction," something they now wanted to explore further by replacing vinegar with fruits that are acid:

Excerpt 2

1	Adult:	What are you doing?
2	Mazela:	We mixed this, this, this and this
3	Adult:	What is this?
4	All:	Vinegar!
5	Adult:	What does vinegar do, what is it?
6	Nisha:	Vinegar, it is like something that makes something go up
7	Mona:	no, vinegar, it's an acid, a liquid acid, that's all.
8	Adult:	Okay, but why do you add vinegar?
9	Nisha:	because, I think that the vapor will make it go up
10	Mazela:	because it is like the vapor will make it go up a bit.
11	Adult:	What if you do not put it, what if you just put the acidity of a fruit
12	Mona:	because, we will do different acid fruits, so we will put lemon here and then
13		we will observe what happens, if it works we will use less . . .
14	Adult:	Ah, okay.
15	Mona:	It's to show that it works better.

Mazela is actively involved in the mixing and doing of science. Like the others, she cared less about following a scientific protocol and was most interested in "acting like a scientist." They decided to add vinegar to all the fruits to ensure a reaction since it is the "vapor [that] will make it go up a bit" (line 10). Only with adult help did they eventually understand that the goal of their study was to see whether lemon could act as an acid and replace vinegar. As one instructor noted, " this team struggled the most with staying in one place since the three girls are really hyperactive." They also had too many things going on at the same time but were unable to focus on one variable at a time. As shown above, Mazela was like the others, she had some vague idea about what they were doing but was most keen on simply "acting like a scientist" without necessarily becoming it. Her identity as academically weak was supported to some degree by the staff in that they saw that team as "challenging" given their "hyperactivity." Yet, in Mazela's words, the program was a place to do the "kinds of activities I wanted to do." It was also a place where she could "make friends," "learn from them," and that way "know more things." The programs were places where she could experience success, ownership and agency, things not possible for her in school. That is, she always oriented her identity talk towards experienced selves in school, making the programs look like "heavens" for valuable learning and identity work. As Lave and Wenger (1991) discussed, such contextualization of identity work, beyond immediate contexts and practices, is central to an understanding of the processes of cultural production of persons.

The second year, Mazela did a project about the phone trying to figure out how voice gets transmitted. The team initially proposed that it entailed some kind of an "echo." In turn, they learned about the transmission of signals through optic fibers. When asked what she learned from the two projects, she noted, "with the phone, I learned that one can get cancer from it, and with acid fruits I learned that baking soda is a base and not an acid." Being able to explain the science behind her projects certainly made her appear academically strong here.

Mazela's sense of self as a successful learner was also supported in the gardening program. She came to the program as a newcomer who struggled to keep her plot weeded and well taken care of. With help from her peers (and often old-timers), she learned to differentiate weeds from "wanted plants" and to take care of her plot by raking and watering. What she liked most was

> harvesting because I am now more confident in myself because I worked hard and then, it showed me how much I had worked, how difficult it was for me, but then I saw that I had beautiful vegetables.

She discovered that her hard work led to a positive result leading to much self-confidence. But as with the science in *Scientifines*, participation in the program also helped her relate to her family in new ways. Her grandfather had taught her parents how to garden and once in Montreal, they managed a small plot in their community, growing vegetables for themselves. Mazela was now capable to actively participate in that activity along with her parents and "show them," making possible the experience of a self as capable rather than as failing, as in the past. Hence, both programs were sites in which Mazela could experience success and learn, something she claimed was more difficult for her to do in the school context given its structure. "In *Scientifines*, when I was new, I learned how to cooperate and I learned many different things and then, in the *Jardins-jeunes*, I learned to cooperate and respect others, and I loved that, it taught me so many things." As articulated, both programs made the experience of success a possibility, challenging her weak academic identity that the school context seemed to support. Her levels of expertise now made it possible for her to position herself as knowledgeable and capable in front of her peers in the two programs as well as in front of her parents. Despite such experiences, such identity work remained highly situated. Mazela insisted that the structure of schooling made such positive experiences unlikely:

> Because in school, in general, we write, we are seated, in the garden we are doing something, we work, we rest, we can go and drink some water but when in school, one has to ask "can I go to the bathroom," in the garden, if you need to, you just go.

Mazela was the kind of student who needed to be respected, who needed much freedom to learn, and when granted, excelled, as she put it: "at school, I have to wake up and I cannot be late and when I am late, I have a detention and all that, it stresses me, and then, especially the mathematics, I don't like that." Mazela also acknowledged that she does not always "feel like studying" something that made her realize that becoming an engineer or studying veterinary medicine—two possible future selves—may be impossible. Hence, like many youth, Mazela does not dislike learning but struggled with the tight structure of schooling that left her voiceless. In contrast, *Scientifines* offered her opportunities to be successful in something *she* decided on and valued and in a less tightly structured environment. The programs made her experience positive possible selves.

Instrumental Value of Programs for Becoming a Specific Type of Person

> I want to be the most best doctor in Canada, and have many people know who I am. (May 2004)
>
> [I would like to become] a doctor, I would save lives of people; and then [maybe] chemist, I could find new medication, things that would change everything for everybody. (May 2005)

Kamila was in fifth grade when we began the study and participated in the program for the first time together with her twin sisters. She was born in Canada whereas her parents came from Guinea. Her father passed away three years ago. Whereas her mother was a chemist in her home country, she pursued a nursing degree as the girls participated in *Scientifines*. Initially, Kamila was one of the less motivated program participants. Once her twin sisters joined the program, however, she became more quiet and involved as if she wanted to put forth a good example for them. Kamila was very active in her community and often took care of other children, besides her sisters. She also participated in a summer arts program while doing the gardening. Kamila's mother volunteered some time in *Scientifines* and always supported her girls in whatever ways she could to make participation in the programs a success.

Kamila came to *Scientifines* to make her dream of becoming a doctor possible. She valued becoming a doctor, in part because of science, but also because of her high value of caring for others, something she also practiced and experienced as she took care of her father who was ill. Given such a strong sense of future self, it may not be surprising to what extent Kamila also sought out the opportunities in *Scientifines* to learn more about themes pertinent to becoming a doctor. Hence, the programs were of instrumental value and tied to a clear future identity in the human sciences. Despite such a clear sense of future self, Kamila's biggest struggle was not to be "lazy." Sometimes it was hot and she would not feel like gardening, but her mother made it clear to her early on that she had an obligation to go given the unique opportunity that had been offered to her. In contrast, Irina was allowed to stay home at times when it was very hot, and Mazela granted herself some absences as well. Kamila was always present which in the end paid off and made her win a special award for attendance and outstanding performance. Yet, Kamila's struggle between laziness and wanting to do well was always apparent:

> Last year [in *Scientifines*] I played around often and did not take it seriously enough and neglected my work. But I was still the one that knew the most in my group, all we did, it was me who knew most [about it], but even then, I neglected things a bit too, I did not always do [what I was supposed to].

Kamila positions herself here as the leader and most knowledgeable in her group, an identity she clearly lived up to despite the fact that she was not always as concentrated as she should have been. In the following exchange, she attempts to explain how a synapse works based on images they had gathered for their poster board and presentation of their project (Figure 3.1). Again, she does most of the explaining given some guidance by the instructor while Hasina, her teammate, is rather quiet.

Excerpt 3

1	Instructor:	this here, what does this explain? When you will show this, what will it
2		explain? *[points to poster board]*
3	Kamila:	It explains what a neuron is.
4	Instructor:	What's a neuron. And this, what does this explain?
5	Kamila:	The synapse.
6	Instructor:	The synapse. What's a synapse? What's the relationship between a
7		synapse and a neuron?
8	Kamila:	Because here they are all connected so that's. That's the medulla.
9	Hasina:	It's the muscle right?
10	Adult:	Yes.
11	Kamila:	That's the synaptic medulla.
12	Adult:	Yes?
13	Kamila:	So, the synapse is here.
14	Adult:	Yes, yes, the synapse is here. The synapse, what is it? What is that little
15		bundle here for, what comes out of a synapse?
16	Kamila:	Chemical messages
17	Adult:	Chemical messages, and what do these messages here have to say?
18	Kamila:	It says for instance, if your brain here said "move your arm" and the
19		light is here, it will do "move yourself from there." *[mumbles]*
20	Hasina:	*[giggles]*
21	Adult:	Exactly, that's right, that's what it means! So then, we have all these
22		messages that circulate, and where do they go? As we said before, to
23		the muscles. That's where, as you explained before *[points to Kamila]*
24		your arm will move. So these two go together, we put them together *[points to pictures]*
25	Kamila:	Yes.
26	Hasina:	Yes.
27	Adult:	On which poster board? Which color? You choose!

The girls had printed out pictures of a synapse and a muscle and were working on the poster boards for their booths (Figure 3.1). As they were assembling the pictures, one instructor came over to check in on them. She began by asking them about the use of the pictures for explaining (lines 1–2). She then probes further and asks them about the relationship between a neuron and a synapse (lines 6–7). Kamila labels the parts yet it is clear that she also understands how it works despite the fact that her talk is somewhat telegraphic. That becomes apparent once she enacts a chemical message. Yet, the instructor's talk helps to make sense of the embodied nature of the talk of the girl (lines 18–19). As the instructor noted during the interview:

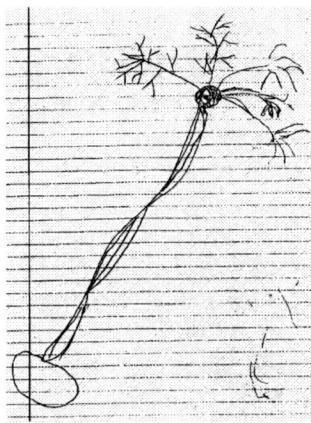

Figure 3.1. Drawing of neuron (left) and synapse (right) that they referred to in talk.

They had a lot of vocabulary, but they had to understand what a neuron is, is it the same as gray matter, white matter, the brain, its parts, etcetera, and at the end they were able, with only a drawing of the brain, to say which part was which and to explain what it does and the synapses, what a synapse does and . . . it's incredible, that's something I learned at the age of eighteen!

That team stood out as having developed a work ethic to put facts together and to understand and appropriate complex scientific terminology. When asked what she liked the most about the program, Kamila sounded somewhat like Irina:

The scientific activities because we do science experiments, sometimes we use our hands or other times we learn a lot and so that can help us in our future, it exposes us to many things and then, one day, we will need it.

Kamila understood the manner in which the program made the world of science accessible to her, a world she also identified with strongly. She also valued the kind of learner identity the program stressed, such as being active while also taking on much responsibility and ownership of a project. She further added in 2005, "I also liked doing work on the computer. I am not sure why, but I like to type, and search for things [on the web]." Through her participation, she learned about facets of a successful learner identity such as perseverance and hard work, as is apparent in her reflection about the science fair project in 2005, "one has to research the topic a lot to fulfill the objective and this year, I even took some work home. I learned a lot of things." Overall, Kamila took much pride in the brain project and was very "happy" with the product of their team "because I put in a lot of effort and ended up finding what I was looking for [and] I like it that we learned a lot of things about what we want." Note her reference to her role in the program. She was considered an active agent of her own learning and given the opportunity to do something that was of value to her. The "curriculum" of the program connected to her everyday activities and possible selves and hence was highly valued and seen as possibly instrumental in her future:

> What is special for me is the fact that we do many activities that are related to the sciences, sometimes we even play games outside but it is always in relation to science, and one learns a lot, and so that can also serve me later in life because I want to pursue a profession in the sciences, I would like to become a doctor or chemist, technician, things like that.

Clearly, there was a strong match here between what Kamila wanted to become and was becoming. Unlike the other girls, Kamila felt at ease to position herself as an insider to science and as somebody seeking out every possible opportunity that would ensure a future self as such. The instructors also talked in positive ways about Kamila and the brain-team's work:

> I thought they were the most scientific ones, as far as science fairs are concerned. I thought they were the ones who had the best vocabulary, who maybe didn't have the greatest structure to explain it and all, but who knew what they were talking about. You could ask Kamila a question, [and she could] tell you [the answer]. What is a mitochondria? She would explain what is a mitochondria, what it's for, they knew what it was for. And details which are really important when you give an explanation, which the other teams did not have. I thought that [for the other teams] the objectives were more superficial, whereas the "Brain" team really got into their subject.

It is as if Kamila along with Hasina managed to develop a work ethic that could then be further refined the following year as Kamila worked on a research project on molds together with her sister, which entailed the following:

> We explain what molds are, what their dangers are, and then also how they reproduce themselves, we explain everything about molds, and then we play a game and also do an experiment to test whether molds develop best in humid contexts, that's kind of it.

When exploring topic choice further, one can also note continuity between Kamila's everyday world, her possible self and the kind of self the program helped her develop. While topic choice the first year had much to do with her possible future self as a doctor, topic choice in the second year was also linked to science things she learned about at home. In fact, the idea to study molds was triggered by a documentary on television about epidemics where they talked about staphylococci—bacteria that typically reside on the skin and entail about 30 species, some of which can result in illnesses in humans through toxin production as well as invasion. For instance, as a toxin it can result in food poisoning. That documentary made Kamila propose the study of molds to her sister. Kamila again identified with a topic in line with her future possible self as a doctor. Even when in the garden program, Kamila made special note of a presentation about medicinal plants, something that may help her "make medication" one day. Note how her sense of self as science literate and as possibly becoming a doctor was reinforced and constituted by positive experiences of self at home and in the programs.

When asked about Kamila's interest in science she noted "I like science since I am six years old and I like it just more with time." While Kamila struggled at times with gardening she learned even more about science there: "I learned a lot of things, really a lot of things, I cannot even list them, I learned about all the parts of a plant, I learned about GM organisms and much more." She also liked to help others in the garden "I like that because I am generous, I like helping others, it's the reason for wanting to become a doctor because I want to help people." Hence, she linked "helping" in gardening with values of her future possible self as a doctor who "helps." Furthermore, she noted that her mother does not have "a lot of money and I bring home vegetables, that helps, she has to buy less, in the store it's expensive, and then also, sometimes there is GM foods and that's bad for you." It shows that in participation in the gardening program, like in *Scientifines*, contributed to her identity development, despite the fact that science was not as transparent in the latter and the timescales for participation varied. The two "fit" with her identity project and nourished it in important ways.

A Look Across the Cases

The identity work of the three girls make visible three different learning trajectories and ways of identifying with the two programs studied. In fact, the analysis makes apparent the fact that by following the three girls for two years, as a researcher I still covered a very small timescale of identity development across few contexts. Yet, some contribution to identity development is clearly taking place. One may argue that the identity projects of each girl made them seek out the learning opportunities offered through the programs in very different ways. For Irina, participation contributed to her development as an educated person whereas for Kamila, it contributed to her sense of self as an insider to science, making her becoming a doctor a possibility in her future. In both cases, the programs were of instrumental value for possibly becoming a certain kind of person. In the case of Mazela, the programs were contexts that challenged her sense of self as academically weak and unsuccessful. Yet, her participation was marked in opposition to other senses of self whereas for the other girls, it contributed to and came to constitute an identity development already in the works due to other past experiences and projected future possibilities. Irina and Kamila's stances made them evaluate and in turn experience the programs in ways that further contributed in positive ways to their own identity projects. In contrast, for Mazela, the experiences in the programs challenged and threatened her previously developed sense of self, yet, whether it was enough to add up to some significant change is difficult to judge if we take seriously the fact that "the formation of identity . . . cannot take place on short timescales." That is, "even if short-term events contribute towards such changes, it is only the fact that they are not soon erased, do not quickly fade—that subsequent events do not reverse the change—that makes it count" (Lemke, 2000, p. 282). We would need to study subsequent events to fully grasp the impact of the two programs discussed here. It alludes to the problematic about the timescale of current studies of learning and identity.

What the analysis of identity projects offers, however, is a means to gain some insights and an appreciation of the diversity of learning outcomes or effects of participation in educational opportunities in the nonschool hours. Each girl approached participation in the programs in different ways, taking away different tools for their future selves. In essence, all three girls were experiencing "a different lesson," which depended upon their "trajectory up to now (and now-in-progress)" (p. 284). Accordingly, when talking about the positive impact of after-school and community science programs, we have to think of such impacts in a variety of ways, and not simply in terms of becoming science literate or an insider to science interested in the pursuit of a science career. In Mazela's case, participation led to the experiences of selves as academically capable and successful, something she had not experienced elsewhere. It may give her the self-confidence needed to pursue an education that will make her successful despite the strong stigma and cultural models of identity in the educational system that tend to categorize immigrants from Haiti as academically weak, failing and a burden for the system (McAndrew, 2001). It may have given her the opportunity to develop a local identity that challenges the persistent negative cultural model she had learned to adhere to, and if supported further in other contexts, may become strong enough to make the experience of success a possibility for her in the future.

CONCLUSION

In this chapter, I illustrate how changes in knowing are always associated with changes in social being, a dialectic that is at the heart of understanding science literacy development and literacy development in general. Through our activities, we become different kinds of people that at the same time make us position ourselves in different ways in relation to the activities we are engaged in. In this chapter, I take a close look at this dynamic among three girls doing science in the nonschool hours. Yet, it is clear that such an analysis is just a beginning. It entails a very short timescale that has to be expanded along with the spatial scale as well (Lemke, 2000). To truly understand how poor urban girls participate in science in the nonschool hours and come to position themselves in relation to science, their forms of participation have to be studied in a variety of activities and contexts over longer timescales. Such could greatly enhance our understanding not simply about their lack of participation in science but most interestingly, in science activities that they value and that are meaningful to them. In so doing we may need to expand our notions of what doing science entails, yet, it would finally move the field beyond the persistent cultural models that define poor girls as failing in science, as not being interested in it, and as not caring. It would move us beyond deficit theories of youth towards theories of possibilities and possibly a respect for a diversity of ways of being and relating to the many sciences that are out there.

The fact that I look at girls only here does not have anything to say about possible gender differences that too often are based on the artificial homogenous treatment of gender as a variable to be manipulated. More important and at the essence of this chapter is an understanding of what kinds of girls they are (Brickhouse, 2000), how they participate in science, position themselves within these programs and "live" science. In so doing I hope to convince the reader that learning and identity have to be studied as a complex system which means the study of its development within "spaces" beyond schools and in terms of broader content areas than solely the academics. As noted at the beginning, we have to move beyond dichotomies, which also suggests that educational contexts can no longer be understood as entailing simply formal and informal educational settings. As I show, their own personal histories and current and future possible selves constitute who Irina, Mazela, and Kamila are and are becoming; a multitude of contexts among which they navigate over time also constitutes them in similar ways.

REFERENCES

Brickhouse, N. W. (2000). What kind of a girl does science? The construction of school science identities. *Journal of Research in Science Teaching, 37,* 441–458.
Crane, V. (1994). An introduction to informal science learning and research. In V. Crane, H. Nicholson, S. Bitgood & M. Chen (Eds.), *Informal science learning* (pp. 1–14). Dedham, MA: Research Communications.
Delgado, M. (2002). *New frontiers for youth development in the twenty-first century.* New York: Columbia University Press.
Ferreira, M. (2002). Ameliorating equity in science, mathematics, and engineering: A case study of an after-school science program. *Equity and Excellence in Education, 35,* 43–49.
Goffman, E. (1958). *Presentation of self in everyday life* (Monograph No. 2). University of Edinburgh, Social Sciences Research Center.
Kondo, D. K. (1990). *Crafting selves.* Chicago: University of Chicago Press.
Lave, J. & Wenger, E. (1991). *Situated learning: Legitimate peripheral participation.* Cambridge, MA: Cambridge University Press.
Lemke, J. L. (2000). Across scales of time: Artifacts, activities, and meanings in ecosocial systems. *Mind, Culture, and Activity: An International Journal, 7*(4), 273–290.
McAndrew, M. (2001). *Immigration et diversité à l'école.* Montréal: Les Presses de l'Université de Montréal.
O'Connor, K. (2003). Communicative practice, cultural production, and situated learning. In S. Wortham & B. Rymes (Eds.), *Linguistic anthropology of education* (pp. 61–91). Westport, CO: Praeger.
Wortham, S. (2006). *Learning identity.* New York: Cambridge University Press.

KENNETH TOBIN, JRÈNE RAHM, STACY OLITSKY,
WOLFF-MICHAEL ROTH

URBAN SCIENCE EDUCATION

Michael: It may be symbolic that two of the three chapters on identity and urban science are based on research in West Philadelphia, near the University of Pennsylvania, in a neighborhood that in 1985 saw the police drop a bomb on a house occupied by the members of MOVE, a radical African American back-to-the-earth movement. As a result of the bombing and the police assault, almost 70 houses burned down and 11 people were killed, including 5 children. What can African American youth expect in terms of their education in a city where the mayor, despite being blamed with "unconscionable behavior" by a special commission, was re-elected for another term? Here, I do not think that the mayor as an individual is to be blamed. Rather, the mayor is but an individual concretely realizing a possibility that exists at the collective level. That is, bombing your own citizens and burning down several dozen homes in your city is an act that a sufficiently large number of citizens consider so normal that it does not make unelectable the politician who gave the orders that led to the incident.

Ken: The story about the mayor and bombing MOVE is complex. A commission appointed by the mayor to investigate the bombing described is as unconscionable. The mayor was the first black mayor in Philadelphia and he grew up in poverty. He shared a great deal with those who elected him and then elected him again. It is interesting that you mention the University of Pennsylvania because in the late 1960s Greene, elected mayor in 1983, earned a master's degree in government administration from there. Throughout his eight years as mayor, Philadelphia continued a decline that led to urban blight in many parts of the large city, especially in the inner city. Rendell, Greene's successor, began a process of urban renewal, however, after eight years of Rendell and now two terms of another black mayor, issues of ghettos, poverty, and a struggling school system are still to be resolved.

Michael: The culture that leads to agglomeration of the poor in slums, ghettos, shantytowns, bidonvilles, banlieus, and cités (e.g., Cité-Soleil, Haiti, where many immigrants to Jrène's Montreal come from) is prevalent around the world. What kinds of identity do children evolve, develop, and construct in situations of continuous precariousness, insecurity, and need? How do the things that they encounter in their everyday lives outside school become resources for acting in school and thereby for expressing identity? Visions of teaching about atoms, molecules, electrons, protons, and neutrons, are they not dreams in the face of the daily struggle for basic necessities?

Jrène: Yes they are dreams, but sometimes it feels good to forget the harsh reality, pretend to be middle class, have access to quality science, play the game, be like them, engage in and author science as the girls could in *Scientifines*. The girls know their position, they know that such is usually reserved for the elite. Yet, it is about trying out an identity in a safe place, even if it is only temporary or can only happen at a place at the margin such as the after-school program or the garden, places that are set apart for them. As Stacy's chapter makes evident, it rarely happens in the classroom—such places are often not safe enough to play with identities. As noted by Stacy, in such spaces, "discursive moves that make identity claims can backfire." Yet, as Ken's own account of his trajectory as an urban science teacher as well as the story of Shakeem makes evident, it is serious time that we begin "to respect urban youth culture and to appreciate the awesome stocks of knowledge that they can deploy in living their lives." It is very much about respect of youth, something the education system ironically seems to struggle with badly.

Ken: Mutual respect is the key to much of what can then follow—as Stacy's study shows. Linda, the teacher in Stacy's study, had earned the respect of her students and she showed them respect too. With mutual respect in place the social networking that can afford learning can then be established.

Stacy: The student researchers described how Linda's treating them with respect on an interpersonal level was one of the most important factors contributing to her effectiveness as a teacher. Quite a few students who described themselves as "not liking science" said that they did like Linda's class. However, Linda is only one teacher working within a larger system in which science is often portrayed as high status and exclusive and in which classroom interactions exacerbate inequalities. Students come to Linda's class with their own previous histories in science that contribute to interpretations of the classroom as a place where boundaries are reinforced and where their emerging uses of science language may not be accepted. While the mutual respect between Linda and the students helped to mitigate some of the students' prior negative experiences in science, and contributed to many classroom interactions that emphasized solidarity and learning rather than division, it was not enough to make the classroom a completely safe space.

Jrène: This all makes me think of another issue. A colleague of mine who teaches high school science in some of the most challenging schools in Montreal wrote a wonderful paper about "le plaisir et les sciences" (pleasure or joy and science; Thésée, 2003) where she explores in what ways the two words, so rarely discussed in science education together, may in fact be related. I would argue that the girls in Scientifines were having fun while doing science; they were given the opportunity to engage in a science that mattered to them, that was in congruence with their lifeworlds to some degree. And it is this kind of science that Shakeem could have enjoyed and would have valued too—a science that emerged from his lifeworlds. As noted by Ken, "what he did so well in cross age situations in his world out of school he was prevented from doing in school." He was not allowed to have fun with science. He was supposed to enjoy a science dished out to him that had nothing to do with who he was and was becoming. Stacy's students had no chance either to own the science they were presented with and not surprisingly, were preoccupied by their strategic use of language rather than enjoyment of doing science.

Stacy: Yes, I agree that enjoyment of science is a crucial goal. I may have a different perspective regarding the path between content relevance, enjoyment of science, and strategic use of science language. I do not think that students in this classroom were hesitant in their use of science discourse because they could not relate to the content and therefore did not enjoy the material. Rather, I think that other structural factors contributed to their perceiving science participation as risky, which then had an impact on their enjoyment of science. From my research in this school, I have come to believe that the competition fostered by City Magnet for scarce spots in the high school, the race and class inequalities in Philadelphia, the high status accorded to science, and certain teacher practices, such as the use of IRE dialogue, contributed to science participation being perceived as "risky," regardless of the content and its connection to their lifeworlds. However, students were not as restrained in their use of science language when the classroom conditions and teacher practices facilitated interaction rituals that promoted solidarity and mutual support, which occurred even during instruction on content that was seemingly irrelevant to their lives (see Olitsky, 2007). Therefore, I think students would enjoy either relevant or traditional science content as long as classrooms fostered solidarity-building interactions that reduce the risk for students to experiment with ideas and language. I suppose my perspective is also informed by the assumption that people are always strategic as they speak or write, as "face-work" (Goffman, 1959) is a component of every interaction. I think that even in situations where people are enjoying themselves, such as in a science activity geared toward their interests, aspects of identities are at stake. Therefore I do not see students "being strategic" as the main problem. The problem is in the details of what particular students have to do to be strategic, such as when African-American students feel they have to be more cautious when using science language than the white students do.

Jrène: That makes me wonder whether some aspects of identity work are always risky, even in a safe environment?

Stacy: Or that there are no safe environments, but only variability in the types and levels of risk within different settings.

Ken: It must be frustrating to be prevented from doing what you are good at—especially if you are trying to learn. For science teachers it is a serious challenge to see the capital in what urban youth do. Similarly, it is difficult for urban youth to get over the feeling of disrespect that comes from being shut down.

Stacy: It seems that there are different "sciences" that we have been discussing.

Michael: For instance, I never had to know the number of protons in oxygen at any point in my life; but other things have been important and they are not taught or available as course options. Thus, being able to repair a lawnmower or being able to do repairs around the house are valuable, especially at times when each visit of a tradesperson will cost you $50 or more.

Stacy: And not surprisingly, being taught about such topics as atoms, molecules, electrons, protons, and neutrons could potentially lead to a perception of science as a collection of randomly chosen facts (how many protons in oxygen?) that may or may not allow youth to be inscribed as good students depending on whether they are able to reproduce such facts or not. But then, there is also the science of collective activities, such as practicing science in a community garden, as in Jrène's chapter, with its own sets of language, skills and knowledge use. In such activities, youth can develop their identities through feeling successful as others recognize them as valuable contributors to the collective activity. Over time, they may experience a sense of solidarity and a sustained interest in science, or as Jrène suggests with regards to one of the students she worked with, a chance for identity transformation that has the potential to contradict previous experiences of failure in academic settings. In which case, learning about science would not be a dream, and in any given moment may not necessarily have middle class connotations or be perceived as irrelevant, but instead would be people enacting positive situational identities in pursuit of common goals.

Michael: Which is why I would like to see schools take an approach where students learn to cope with everyday life situations and learn to expand their own room to maneuver. To me, the research on the negligible correlation between school mathematics and mathematics in the workplace and other everyday situations is a paradigmatic case for arguing against current forms of schooling. I mean, teachers tell us the same: most if not all of what they need to know as teachers, they learned on the job rather than in their formal schooling. Why do we, science educators, not squarely address this issue of the gap between forms of knowing useful to be successful in schools and forms of knowing required to be successful in everyday life outside schools?

Stacy: One danger of course is the creation of two tiers of science knowledge, the practical science that is geared for "everyday life" and the content that is less practical but has been traditionally taught in formal schooling. This more traditional science content is likely to be considered higher status by colleges and employers, giving those who have access to it an advantage. The discussion we are having here ties into a larger ongoing debate among educators regarding whether students should be taught different subject matter depending on their background, interests, and/or aspirations. Either approach seems to reinforce current inequalities, albeit through different pathways.

Jrène: Why two tiers of science? Why not think in terms of multiple sciences that exist out there and that are of value? Important is also the need to think of science as a tool for action rather than an end in itself. If the two ideas could be combined, our students could learn about the many worlds of science and learn to use them for action as informed and critical citizens. Yet, we are still far from such ideals! It would mean that Stacy's students would learn about Shakeem's science as well as garden science, that they would have opportunities to mobilize these ways of knowing as means towards a meaningful goal and action in their community.

*

Ken: In both of your chapters, Jrène and Stacy, you address issues of poverty, gender and the salience of success. Can I ask you each to explicate how each of these categories relates to identity construction in your chapter? Being big, black, poor and male, structured Shakeem's lifeworld, requiring him to enact forms of culture that afforded the formulation and attainment of his goals. I hasten to reiterate these were not the only factors that structured his lifeworld and he produced capital continuously, thereby expanding his agency. Similarly, being in my fifties at the time, white, middle class and male structured my agency in urban fields and the evolution of an urban identity.

Stacy: In my study, I discuss how these categories structure how people respond to each other in social situations, which has implications for their identity work in terms of how people expect others to treat them, how they perceive themselves, and how they act in order to maximize solidarity with others and minimize the risk of shame. For example, in City Magnet, categorical identities such as race had an impact on how accepting students were of each other's attempts at using new science vocabulary (see Olitsky, in press). Categorical identities therefore had implications for whether students were inclined to use science language or claim science knowledge as their own based on their anticipation of the outcome of particular discursive moves. An example from the chapter is when Aileen qualified her use of language by voicing another student, and said "you know," rather than appropriating the language as her own. If students, such as Aileen, do not "talk science" because of the perceived risk, their teachers may develop the misperception that they lack interest or knowledge in the subject. Yet Aileen really does have an interest in the science content, but understandably, does not want to risk having her claims rejected and experience a loss of emotional energy. Over time she may come to see herself as not being an insider to science, which in turn may affect her future participation in science related activities. There is a continual cycling between one's own actions, others' perceptions of those actions, which may be partially based on categories, others' behaviors toward that person, resulting self-perceptions, and an internal dialogue in which the consequences of various future actions are weighed. Hence, the categories laid out by Ken have important implications for identity construction.

Jrène: As is well articulated by Stacy, these categories constitute possible selves of youth. Notions of what one can and ought to be circulate freely within society, and come to position us in certain ways. Yet, youth themselves also reposition themselves through their lived experiences and may challenge ascribed identities such as Kamila, who sought out every opportunity to do science to advance on her trajectory towards becoming a doctor and an insider of science, a trajectory that was not necessarily reserved for her as a poor black girl.

Ken: I am also struck by similar patterns in both of your chapters in regards to the family being salient in affording participation in science and identity work as it pertains to science.

Jrène: I would argue that family is another category that constitutes identity work in important ways here. It makes certain identities possible. Similarly, the classroom described by Stacy made some identity work more salient than others. Family as a category also interacts in important ways with the others mentioned. In Kamila's case, the capital of her family made possible the vision of an identity as an insider to science despite the fact that she was living in poverty, is black and female –categories typical of outsiders to science.

Ken: Salient in all three chapters is the idea that identity structures and is structured by experience. Accordingly, each person experiences life uniquely as he or she participates in social life—consistent with an individual|collective relationship. For me, this raises questions about what individuals regard as truth and real. When we teach science, whether in school or out of it, there may be assumptions about what is true and real that don't hold for the student populations we are dealing with in these chapters.

Jrène: That brings us back to multiple visions of science each constituted by a world that may be distinct from others in terms of its truth and reality. We could have a long philosophical dialogue about truth and what truth in science may imply. Important is the fact that science is historically, culturally and socially constituted, which then also implies a dynamic and situated vision of truth and reality.

Ken: Presenting and saving face while participating in science is also an interesting aspect of identity that arises in Stacy's chapter and is pertinent to both of the other chapters. It is almost as if there is a threshold amount of symbolic capital an individual needs to step forward and "present face." If a person is close to, but below the threshold, he or she might use verbal qualifiers as safety nets—protection against verbal assaults by others that would be regarded socially as disrespectful. In Shakeem's case a similar situation was described in my chapter when the ritual insults from two female protagonists rained down with such consistency that further aggression on Shakeem's part would likely produce less benefit than the crescendo of harms likely as the two females coordinated their verbal assaults, ratcheting up their intensity following every utterance from Shakeem. To minimize the risk of further loss of face and respect, Shakeem opted to withdraw—to cease further participation.

Michael: In such situations, I would like to see some process of raising consciousness. All these students as well as their teachers need to have the resources for recognizing that they are in it together; collectively they have what it takes to modify the situation to the benefit of all. It takes some of the solidarity that already binds us together in society.

Stacy: What would it take for people to perceive that they are "in it together" when the participants in classrooms perceive themselves in competition for scarce resources, and when some stand to gain by excluding others? Students who reject each other's claims to membership may seem to be doing so arbitrarily, yet it can also be seen as being in their interest to do so. When science is seen as an exclusive group, there are those who benefit from associated privileges and others that are set apart.

Jrène: A good question. It gets back to mutual respect, which can only be supported through the creation of "true" communities of learners. In response to Ken's point about saving face, I wonder whether such would be less important in learning communities built on solidarity.

Michael: I would argue that at least theoretically, society is built on solidarity, the idea that by working *together* we enhance the collective power to act, modify our environment, and make it possible to survive even in harsh conditions. It is an inner contradiction, however, that some members live in excessive wealth and many (the middle class) live comfortably, but the remaining, large proportion of society lives in decrepit situations; more so, they have to live in this situation so that those living in wealth or living comfortably can continue to do so. Are we not lying to the children and students when we say that "education" (really, schooling) will give them a better life? This question is especially salient in the light of the fact that (a) resources and wealth are limited and (b) the natural (structural and frictional) unemployment rates in a well-functioning economy lie somewhere between five and ten percent.[1] Are we not culpable of painting images of paradise, giving resources to the dreams of children that they never can realize?

Stacy: You are really getting at a fundamental question regarding the role of education. While some individuals can get a better life from schooling, most people cannot, since as you point out, the realities of the economy do not allow everyone to be wealthy. What then should education, and science education in particular, aim to accomplish?

Ken: Living in squalor and the poverty from which it derives is obviously not what urbanites choose to do. Their lives are tough and yet within those neighborhoods are homes—where people live their lives, enjoy one another's company and together accomplish myriad goals every single day. In so doing they produce, reproduce, and transform culture and the quality of their lives. Being from the middle class it was easy for me to see these neighborhoods totally through deficit lenses and completely fail to see the capital of what they offer their communities. I agree with your point about sharing the wealth and improving social life for those who currently live below the poverty line (especially). My point is, in so doing, we also need to acknowledge what the inner-city poor contribute and are capable of.

Jrène: Ken, you are emphasizing the need to begin with the cultural capital of the inner-city poor and build together, something that was aimed for in part in your classroom and less so in Stacy's but present in the settings I described since the science was somewhat more youth driven and open-ended, as well as embedded in other types of activities driven by broader goals than science per se. If science would more often be co-opted, what would such science practices look like? What possible identities would they constitute? We certainly know little about identity work and learning in such practices.

Michael, you are pointing out the many dilemmas and contradictions of a liboratory pedagogy in science that would give youth such as Shakeem or the ones in Stacy's class and the Scientifines the tools to question their relationship and position within the world of science. Yet, are we, through liberatory pedagogy painting images of paradise? I often ask myself the same question but youth have taught me otherwise. We are giving them access to a world that excludes them. Ken respects Shakeem and tried to use his privileged position as a white male and his cultural capital that makes him successful in that world to make a small space for Shakeem so he too, could gain access, even if it looks like nothing to us, as we examine it from our privileged position. I think another challenge is the fact that even if we learn about the life-worlds of youth like Shakeem, our understanding of its complexity will always be partial, maybe in similar ways as their access to "our" world will always be minimal. Yet, if we are serious about liberatory education in Freire's terms, teachers and students are agents of change and have an important role to play in actively questioning their sense of self and society and in reshaping it together. In other words, if we are serious about an education for all, we have to do it together and not leave anybody out.

Ken: Easy to say and much harder to pull off—as is evident in the *No Child Left Behind* act in the US. Unless teachers and students become cultural brokers it seems unlikely to me that school science can do much more than perpetuate inequities for a high proportion of youth, characterized by discrete forms of culture associated with categories such as gender, race and class.

Michael: But what gets me is the fact that educators' discussions about equity and access and leaving no child behind completely disregard the fact that our economy, to function in the way it does and has to for maintaining current standards of living, requires unemployment rates of seven or more percent. And this does not even take into the account all those who already opted out or opted for (were forced into) welfare. The competition for jobs keeps the labor costs down. From my perspective, then, schools are but a convenient tool to enact the triage necessary to allow some (middle, upper class) to access universities and jobs and to relegate others to the unemployment lines and welfare. These are the economic realities of industrialized societies; and yet, we still tell students that with an education, they can make it out of the morass, out of the neighborhoods, bidonvilles, favelas, and cités.

Jrène: But even if students are continuously told that education is a tool for upward mobility, some may discover that having an education can be worthwhile in its own way and be liberatory, even if it does not shift their position within the system. Many youth I worked with told me how important knowing has become in their lives and how much they value to be doing something in life that keeps them thinking. We also need to remember that a democratic community is always a community in the making. Notions of possibilities, visions of what might and ought to be, keep such a community alive and make us aware of deficiencies and flaws which then lead to action and transformation since "human beings are prone to take action in response to the sense of injustice or to the imagination's capacity to look at things as if they could be otherwise" (Greene, 1995, p. 166). In essence, "action always signifies a new beginning, a new initiative, so that fixed and final frameworks remain inconceivable" (Greene, 1995, p. 197). We also have to move beyond a vision of a democratic community as singular towards democratic pluralism in which multiple diverse perspectives are recognized, valued and explored. If taken seriously, Shakeem, Samir, Jorji, Kamila, Irina and Mazela would have to contribute to such a democracy as much as their teachers, instructors and we researchers.

Michael: But democratic ideals have been sold out to capitalism. There is very little solidarity with the poor. And to keep us all in line, our retirement funds are invested in the economy so that we exploit the very people who we attempt to help out of poverty.

Stacy: In terms of whether education can necessarily give the students a better life, I would agree that communicating such an idea does paint images of paradise, mostly because of its oversimplification of the issue. What type of education? Who is directing the education? Whose idea of a better life? When teachers tell their students, "study hard, get an education, and you can all be lawyers," (as the student researchers have told me that some of their teachers do), such words may be intended to be encouraging but they also promote an ideology that could be harmful to youth if the students, like the Brothers in Willis' study, continually blame themselves for their low socio-economic position rather than question the inequalities in the system. However, education aimed at expanding students' agency, that is collaboratively constructed, and continually interrogated, may have the potential to help students achieve a "better life," by their own definition, by expanding students' cultural tool kits, views of their own possibilities, or knowledge for working for societal change.

Ken: The out-of-school projects seem to connect readily to the students' lifeworlds and shape identity. I want to ask about the take-home message for science educators. Can we adjust school curricula, or is it a case of increasing access to out of school programs—or both?

Michael: I believe that we need to rethink schools in more radical ways, change the sifting functions that they currently have, rethink the relationship between useful forms of knowledge within and outside of schools.

Jrène: Responding to Ken's point, and in addition to the dimensions just outlined by Michael, I would say we need to do all three. The school curricula but also culture of schooling needs to be rethought and in this process, dimensions that make some out-of-school projects successful are worth exploring and may inform change. Yet, out-of-school projects also have an authenticity to them that we should treasure. Most important, science literacy development needs to be seen in broader terms, as not simply reduced to what happens in school, but as also being constituted by these youths' everyday lives. While after-school programs, such as the one I described, can make up for missed opportunities in school, their strength lies in the contribution they make to the development of youth, as youth who have a sense of agency and hope for their future despite their social position in society, who no longer develop a sense of social death but instead, a sense of self as having assets, as being respected by others, and as having a right for an education and a life filled with possibilities and happiness. So yes, we do have to make them more widely accessible since access to quality programs is not equitable at this point. But we as researchers and educators also have to better understand how experiences outside of school, in the community, at work, in the family and in programs like those I described contribute to the making of identity.

Stacy: I think that Jrène has a good point that there is not a specific end point as a goal but a more democratic process of planning. Yet I wonder about the fact that as long as schools serve as sorting mechanisms, they inevitably support the unequal economic structures that Michael discusses. Does it even matter what is taught in the classrooms, when schools are the basis of allocating desired positions in society? On the other hand, is it possible for schools to both serve the sorting function that they do, yet still in ways that contribute to a more just society?

Jrène: Maybe. The crux is that as long as equality is lacking, sorting further perpetuates inequality. Sorting is simply not presupposed on justice as is.

Michael: Recently we have been theorizing identity in terms of (cultural, diasporic) bricolage, where youths pick from their surrounding cultures and cultural expressions (music, fashion, movies) whatever they appreciate as resources in their own ways of being. But does not the success of a few—Michael Jordan, Michael Jackson, Tiger Woods, Ice-T—and the insistence on individual agency ("every one can make it") constitute great delusions that keep the children and students—who use these as resources in identity formation—from engaging in more radical actions to transform society and Self, such as the members of MOVE had done before? Are we, "educators," not culpable of disseminating lies on which the lives and identities of large parts of society are built? Worse, we are hiding from the children the fact that some will have to end up in poverty so that others can live in wealth or comfort and that school is but the first means to put people on the different, divergent and cumulative trajectories to wealth, comfort, or poverty.

Jrène: Again, I am not sure we are hiding it. Youth I have worked with know their position in society. We may pretend to "hide" something from them, but they know better. As Shakeem, they had to develop a tool kit to survive in a very challenging world early on, they have a PhD on the street as a youth worker said. Look just how challenging it is for us, as Ken's story suggests, to participate in that world, even at the margin. One could argue that some youth may be carried away by such illusions, yet such dreams may also allow them to enact fictional identities and help them envision alternative worlds and ways of being in similar ways as the Tij dukha' songs among Nepali women do (Holland et al., 1998). Through such symbolic mediation and play with identities in places that are safe, the students may come to collectively envision a world in which they have a right to do, enjoy and author "science."

Such authored worlds may be dreamed about at moments of distress to lower the pain of living in this unjust world. For instance, Shakeem did experience a world in which he could be an author of his identity and develop an image of science and sense of self that stayed with him during his troubled years and that eventually made him come back and continue on in his struggle towards an education and the becoming of somebody – it gave him a sense of a future with possibilities and kept him going. It is in this way that it may also help the girls in Scientifines keep going as active agents. In contrast, Stacy's paper points out what happens in the absence of such empowerment and shows, in part, identity play of victims of an unjust system.

Stacy: How do we tell the difference between dreams that empower and dreams that serve as illusions and prevent change? Does the difference lie in the dream, the situation, what the person eventually does with it? Can some dreams empower in some ways and disempower in others?

Jrène: For whom is it to tell? Our chapters tell stories of identity work through the lens of the researcher, identity work that would possibly be summarized in different ways by the youth involved if it was up to them to tell the whole story. It underlines another dynamic we have discussed little but that needs to be seriously explored in science education. Who is the creator of knowledge about what works in science education and what does not?

Ken: I agree. It is not just a question of who decides about what is viable, but also where particular knowledge is viable. It is to be hoped that science education is useful in fields beyond the classroom. Is it? The answer to that question is central to the value of science education and cannot be answered meaningfully if one science educators like us seek to do so.

Stacy: Michael's question: "But does not the success of a few—Michael Jordan, Michael Jackson, Tiger Woods, Ice-T—and the insistence on individual agency ("every one can make it") constitute great delusions that keep the children and students—who use these as resources in identity formation—from engaging in more radical actions to transform society and Self, such as the members of MOVE had done before?" is an interesting question—does the success of a few justify that the system "works" and therefore prevents people from seeing the pervasive inequities, or does the success of a few inspire others and serve as a positive force? I suppose the answer is "both," or "it depends"... certainly seeing others succeed, and knowing that it is possible, is important for envisioning one's own future, but conversely, such successes feed into ideologies that leave people blaming themselves for situations that are related to societal inequities. As Michael points out, in capitalist societies not everyone can be wealthy. One issue is whether youth see diversity in success stories. If low-income, African-American youth mainly see people like themselves in certain pathways to success, such as becoming a celebrity because of influences such as media images and stereotypes, whereas middle-class youth see a greater variety of pathways, than the problem is not just that there are misleading success stories, but also that the stories are unequally distributed.

Michael: There are ways to make capitalist societies more equitable, and countries such as Norway show us how it can be done. There, the salary gap between cabbies and doctors is small. PhD students make a salary that is close to that of professors in other European countries. That is, the Norwegian society strives to spread the wealth it produces collectively. It also makes available access to medical and social services for everyone, not like some other countries, including the US.

Stacy: So a simple role that education can play is facilitating students' access to other possibilities, other models for social organization, than those they encounter in everyday life or in the media. Rather than just providing images of individual success, schools can provide images of "societal success," in the sense that there are other ways of organizing societies that could be instructive in planning to reduce some of the problems in ours. At the same time students can explore alternative possibilities for themselves, they can explore alternative possibilities for their society.

Ken: I am thinking about the comment that in a capitalist society not everyone can be wealthy. My impression of Shakeem is that he is highly motivated to obtain money so that he can improve the day-to-day life of those he loves. When she was living his grandmother was someone he looked out for, and even though he never entirely seemed to respect his mother in the same way as his grandmother, he also would share whatever money he got with his family members. He spoke endlessly of middle class symbols, including clothes, car, house and leisure and ample entertainment. In a sense his oral descriptions of artifacts such as these were outcomes he sought from his labors in school. He expected to earn his ticket to the middle class, not to take what he wanted from others through theft or dealing drugs.

Michael: In a way, as a society we have instituted a secularized version of paradise. We provide children and students with images of a better world while hiding from them—and perhaps from ourselves—the fact that there is no paradise and only few will make it out of living hell. Thus, when Ken's Shakeem, Stacy's Samir and Jorji, or Jrène's Kamila, Irina, or Mazela actually achieve to make it, someone else cannot make it. In forgetting economic realities and truths, we create phantasms by making ourselves believe that our work actually makes a difference, when the real difference for those we work with comes at the cost of all those others who do not have similar opportunities. Do we not need to rethink science education in much more radical ways, situating it in a more encompassing effort of raising future generations that are more critical and reform-minded to the extent of changing the capitalist markets and economies? Do we not owe it to ourselves to assist students to build identities that are not built on deceptions and lies?

Jrène: Yes we do, and I think that is our reason for continuing this type of work, for writing a book that uncovers some of the hybridity in identity work and its relevance and importance to science educators. In essence, this section has uncovered the many dilemmas of urban science education that are so often silenced. While it leaves us with more questions than answers, it underlines the urgency for taking these issues seriously. We need to begin building identities of possibilities *with* youth. We are not here to help them, assist them, but to build possible future selves together.

Ken: Though I hasten to add that the identity work is being done as each of us lives our lives. Whether we think of it or not, as we enact culture, identities are being inscribed—in ways that reflect agency|structure and individual|collective —among other relationships, including success and failure.

Michael: Generally, identity work is not work apart from work; identity is co-produced in productive work. This is so because every action is an expression of the person; in acting, subjectivities externalize, objectify, and even estrange themselves. Each action therefore can be used as a resource to construct the identity attributed to the flesh that produced it.

Stacy: While I agree with your earlier point, Michael, that economic realities in capitalist societies prevent everyone from "making it," I am not sure that it necessarily follows that when one person succeeds (in a capitalist-system definition of success), someone else from that same community cannot. While opportunities may be limited, they are not necessarily fixed that rigidly. Also, if Shakeem "makes it," and is able to obtain more resources for himself and his community, such outcomes benefit not only him, but also help to reduce the pervasive inequalities in access to education and wealth based on race in U.S. Society. However, your overall point is well taken. Could science education be aimed at multiple outcomes, a more just society as well as offer better opportunities within current structures for those who have traditionally been disadvantaged?

Jrène: The chapters speak well to the problem of access to science, a problematic not well explored in the literature beyond statistics about the under-representation of poor, linguistically and ethnically diverse youth in the science pipeline. The girls in Scientifines and Shakeem (when with Ken) were given access to science and education, positioning them as insiders and enabling them to co-author their identities. Unfortunately, many poor urban youth learn quickly that science-related identities are not for them. Having worked with poor youth for a while, I have always been struck by "how little young people believe they deserve . . . they somehow believe that they do not deserve a chance to dream" (Nieto, 1994, p. 422). Or as one program director of an Upward Bound Program put it as we talked,

> I hope that through their exposure to our program and being on campus, that they learn they have every right in the world to be here. . . . First generation kids, they're not sure they have the right. They know they are smart enough, but they don't know they have the right to be here, so maybe we can show them.

Ken: Interestingly, it rarely seemed to be so explicit as to raise questions about whether or not students belonged in school science classes. Whether or not they could create emotional bonds that might inscribe science-related identities seemed always to be an issue though. Seldom did I see students engaging with unbridled passion and the love of learning. Instead I saw standard urban schooling being enacted in the worst connotations of what that might entail.

Students intermittently engaged in a curriculum that simply did not connect to their lifeworlds in any meaningful way. The emotional energy related to the learning of science was hardly ever strongly positive, though I believe there would certainly have been positive and strong emotions for Alex, their teacher. The exception arose when Alex broke from the straightjacket of teaching to the state and city standards and enacted a lab oriented curriculum that allowed students to choose what, when and for how long. Elsewhere I have described how intense periods of engagement occurred in a dissection lab that proceeded for more than a week (Tobin, in press).

Stacy: It is important to point out that the students in City Magnet were in a school that was designed to prepare students for college, and allowed teachers little freedom in changing the type of science that was taught. In addition, the students in that class would not necessarily have wanted the content changed to be more relevant to their lives even if that were a possibility. When asked, the student researchers with whom I worked said that they did not think science classes should be changed to be more relevant. While it could be argued that their rejecting the idea of relevant science content indicates their oppression, another perspective is that they knew that they were being taught content that was considered high-status, and they felt they should have access to it in order to leave open the possibility of obtaining the benefits that such knowledge currently brings in our society. Perhaps in this situation what needed to change was not the science itself, but the school structures that contributed to class participation being a risky enterprise.

Jrène: Two of the chapters offer stories of youth who have been given the chance to dream or how one of my youth said in response to his participation in an Upward Bound Program, "I learned about what I can actually do with my life . . ." But then he, like the youth in the chapter, added, "I [still] think it's amazing that people spend that much time on a program that's structured for kids who, you know, don't have any money." It suggests that ironically, we as educators now have to convince youth that social class, race, and gender are no longer roadblocks towards an education and academic achievement.

Ken: Perhaps a large part of this is to ensure that in creating socially relevant science curricula we do not inadvertently project the idea that we are dumbing down the curriculum for these kids because they cannot cut it with the regular curriculum. This is a big risk because of the deficit lenses many policymakers bring to their critiques of what happens in urban science education. I very much embrace creolized sciences formed through processes of hybridization that fully employ what urban youth know, can do, and value. Having said that, I also believe they should have access to the power discourses of social life and one of these can be canonical science—at least being able to pass high stakes tests that test knowledge of canonical science.

Jrène: We also have to convince the system that such discrimination is no longer sustainable. But as the chapters suggest too, we still need to better understand the lifeworlds of these youth. And to do so, new methodological tools are needed, the most promising being youth-led research. In particular, if we take serious that knowledge is socially constructed, "then youth participation becomes essential in the development of a better understanding of youth's perspectives, opinions, needs, and assets" (Delgado, 2006, p. 9), as Ken's chapter underlined well. If we keep objectifying youth and simply legitimize ourselves as knowledge creators, we are not only alienating youth but will also continue to work within an utopian vision of science education that is exclusive, oppressive and has meaning to few.

NOTES

1 An economy actually benefits from unemployment because it keeps inflation from being high. Economists theorize the for a healthy necessary natural unemployment in terms of structural unemployment, which is due to the gap between available work force and required work force, and frictional unemployment, which is the time a person does not work while being between two jobs. The natural unemployment required by an efficient economy may be path-dependent, which means that it increases in the course of economic history as transitory unemployment during a recession may be transformed into structural unemployment.

REFERENCES

Delgado, M. (2006). *Designs and methods for youth-led research*. Thousand Oaks, CA: Sage.
Greene, M. (1995). *Releasing the imagination. Essays on education, the arts and social change*. San Francisco: Jossey Bass.
Holland, D., Lachicotte, W., Skinner, D., & Cain, C. (1998). *Identity and agency in cultural worlds*. Cambridge, MA: Harvard University Press.
Nieto, S. (1994). Lessons from students on creating a chance to dream. *Harvard Educational Review, 64*, 392–426.
Olitsky, S. (2007). Promoting student engagement in science: Interaction rituals and the pursuit of a community of practice. *Journal of Research in Science Teaching, 44*, 33–56.
Olitsky, S. (in press). Facilitating identity formation, group membership, and learning in science classrooms: What can be learned from out of field teaching in an urban school? *Science Education*.
Thésée, G. (2003). «Le plaisir... les sciences». L. Lafortune et C. Solar (Eds.), *Femmes et maths, sciences et technos* (p.114–124). Saint-Foye, QC: Presses de l'Université du Québec.
Tobin, K. (in press). Cultural relevance and alignment in science education. In A. Rodriguez (Ed.), *The multiple faces of agency: Innovative strategies for effecting change in urban school contexts* (pp. •••–•••). Rotterdam: SensePublishers.

Part B

GENDERED IDENTITIES

INTRODUCTION

Despite several decades of science education research and curriculum development (e.g., *Girls into Science and Engineering*; *Women in Science and Engineering*) concerning gender issues, women continue to be under-represented in science, mathematics, engineering, and technology. Some of the reasons researchers in the field use to explain the different retentions of boys and girls in science- and mathematics-related careers include (a) biological differences, (b) differential preparation for science, (c) differential attitudes toward and positive experiences with science, (d) the absence of female role models, (e) irrelevance of science curricula to girls, (f) male-centered pedagogies (e.g., competition over cooperation), (g) "chilly climate" for female students in science classes, (h) cultural pressures to conform to gender stereotypes and roles, and (i) inherently male-centered scientific epistemology (Blickenstaff, 2005). As this review of the literature on gender differences shows, much of the science education research regarding the different numbers of male and female science students and scientists has been conducted from within the very male-centered—or, as Jacques Derrida, combining the terms *phallus* and *logos* (language, logic) evocatively calls *phallogocentric*—scientific epistemology that also figures as one of the mediating elements in the production of difference. That is, there is the possibility that the very differences deplored also are effects produced by the phallogocentric methods that are distinct from women's ways of knowing.

Even more hidden and therefore more difficult to recover is the epistemological ground that presupposes equality (e.g., of gender) and sameness (identity of, for example, A and A in the equation $A = A$) rather than recognizing the inherent plural singularity of each human being. We cannot think of the Self and identity but in ways that are already characterized by otherness: As the French poet Rimbaud said, "Je est un autre [I is another]." He continues, "J'assiste à l'éclosion de ma pensée : je la regarde, je l'écoute [I witness the birth of my thought: I watch it, listen to it]," which means that he is a stranger to himself as he witnesses that which nevertheless characterizes him. This position, which the poet shares with the philosopher Friedrich Nietzsche and with the philosophers of difference during the late 20th century, is radically different from the Cartesian position that begins with the self-identity of the self: "Cogito ergo sum [I think therefore I am]." This position is taken to its limits by the philosophers Immanuel Kant and later Edmund Husserl, who attempted to establish an *egology*, a science of the I, and to show how the individual constructs his or her world. But he failed, having to realize that the other always and already is there: "Je est un autre."

If, on the other hand, we begin with the ontological assumption of difference that exists in and for itself, that is, with the recognition that $A \downarrow A$ (e.g., because

PART B: GENDERED IDENTITIES

different ink drops attached to different paper particles at a different moment in time), then all sameness and identity is the result of work that not only sets two things, concepts, or processes equal but also deletes the inherent and unavoidable differences that do in fact exist. This assumption is an insidious part of the phallogocentric epistemology undergirding science as the method of decomposing unitary systems into sets of variables, which never can be more than external, one-sided expressions of a superordinate unit (Deleuze & Guattari, 1994). That is, the very method of many science educators seeking to understand how gender mediates interest and success in science through and through is gendered, drawing on epistemologies, forms of knowledge, and research methods that inherently are phallogocentric without realizing it as such. Although one can sometimes see authors nod toward feminist scholars such as Evelyn Fox-Keller, Sandra Harding, Donna Haraway, or Helen Longino, all of whom have very critically interrogated the masculine, *phallogocentric* character of science, science education researchers have tended to stay away from similarly critical stances toward science, going as far as leaving out from literature reviews the work these scholars have done (Blickenstaff, 2005). Why might this be?

Feminist critique challenges *science* itself, whereas science educators generally are very close to science and scientists, most frequently concerning themselves with the question about how best to reproduce science as it is by assisting those heretofore excluded to change so that they better fit. That is, rather than changing science, most science educators change students and the conditions of learning but leave science intact (but see Roth and Barton [2004] for an exception). More so, science educators generally are not self-reflexive and self-critical to question their own discourses, which come with concepts that inherently but invisibly are gender-biased so that only a feminist archeology of concepts and knowledge can get at the ways in which such concepts determine discourses, from the outside of the consciousness of their users, so to speak. Otherwise, that is, if scholars do not critically interrogate their own discourses, they tend to reproduce the very conceptual practices of power that both produce and topicalize gender differences (e.g., Smith, 1990).

Differences between approaches are apparent in the two chapters that constitute the brunt of part B of this book. Thus, Karen Tonso shows how the cultures of engineering education, campus, and corporate world produce different levels of access such that some male students with fewer competencies tend to get better, higher-paying jobs than male students with different transactional characteristics or female students. She therefore calls for a cultural change, which, in a dialectical approach, means a change in the possibilities for acting and thereby for producing and receiving an identity. The teacher in Kathryn Scantlebury's chapter, on the other hand, while recognizing the different ways in which female African American youth are positioned in their home and school life nevertheless wants them to develop "science identities," which essentially means—if science remains unchanged—that the flyy girls, othermothers, and their peers have to submit to the science others critique for its phallogocentrism. That is, while these females develop "science identities," which perhaps marginally increase their opportunities to pursue science-related careers, they are in fact subjugated and subjected to symbolic violence from an epistemological perspective.

Both chapters are part of a cultural-historical change in science education to investigate more critically the conditions of science education. To deal with the social injustices that continue to persist, more critical interrogations of *science* and *education* need to be undertaken. For example, why does school have to start at the same moment of the day for all students? Why can students at Charter High School generally and othermothers particularly not start their school day such as to make it possible to attend to other aspects of their lives? (At the Moussac, France, village school, students not only are allowed but also actively encouraged to come to school at their time and leisure.) Why cannot students at City High School—the school where we, the editors (e.g., Tobin, chapter 1), have done a lot of our research—be more implicated in the management of their curricula, programs, and schooling? (In the *Lycées autogerées* [self-managed high schools] of France, [mostly dropout] students and teachers do everything, including cleaning, cooking, managing, and curriculum planning together, leaving it up to the students to decide how to organize curriculum and time.) Why should we not rethink science and scientific literacy to allow alternative epistemologies, such as those that feature more holistic units of analysis rather than the reductionist separation of variables? Why should we not foster communitarianism and collaboration rather than competition as essential aspects of the science process?

To change science education for our students it is insufficient to critique and change *science, education,* and *science education.* For example, few if any science educators challenge schooling, the very purpose of which is to differentiate students along a linear hierarchy—those with high GPA and other symbolic capital and those with less. Those with more symbolic capital tend to occupy higher-paid positions and tend to have greater access to resources than those with less—Karen Tonso provides a compelling example of that. Then, because the present day markets only function at their optimum with structural unemployment around 6 or 7%, it is those with lower rather than those with higher levels of symbolic capital who will have to fill the ranks of the unemployed. Finally, those who drop out or, under the best of circumstances, complete high school but are at the bottom of the scale might never enter the labor market, often female students who end up being pregnant and raising kids, and reproduce the societal strata of those who live on welfare and those who do not even have that.

To deal with these issues generally and with the role of gendered nature of education more specifically, science educators have to become critical of their own endeavors, discourses, and theories, which continue to be phallogocentric right into the genres of science education. For example, one can easily imagine that in an article or chapter that features African American girls' concern for clothing, science educators draw inspiration from écriture feminine that Hélène Cixous has developed throughout her career:

<u>Don</u> du vett
Parfois je me réveille avec 1 désir de vêtement

PART B: GENDERED IDENTITIES

> 1 vêtement très précis. Jouissance
> le pull over, ce pantalon: ce qui nous
> manque à ce moment là –
> C'est quoi? Une force, un mouvement, 1
> couleur, qqe ch. de <u>nécessaire</u> à la vie. J'ai 1
> Instinct animal—pour 1 pull over noir. [1]
>
> – in Sellers, 2004, p. 19

That is, we could imagine Scantlebury's chapter written very differently, for example, in the way bell hooks might write, and with this writing institute a form of critique in addition to articulating critique as the topic. Such alternative ways of "w/ri(gh)ting" classroom research generally and science education research specifically allows us to institute much more substantive ruptures with the current homo-hegemony of the *phal*logocentric genres of science education and better come to grips with the multiplicity and plurality of experiences that exist in science classrooms (Roth & McRobbie, 1999). Such forms of writing transcend the traditional linearity and logic of the ways in which authors traditionally textualize the lives and experiences of their research participants, providing a heterogeneous and kaleidoscopic bricolage that better reflect the epistemology of difference on the basis of which a feminist science education might be constructed.

NOTES

[1] Our translation fills what has been left out in the abbreviated style of writing: <u>Gift</u> of clothing / Sometimes I wake up with a desire for clothing / A specific clothing. Pleasure / the pullover, these pants: what we are missing at the moment—/ It is what? A force, a movement, one / Color, something <u>necessary</u> for life. I have an / Animal instinct—for a black pullover.

REFERENCES

Blickenstaff, J. C. (2005). Women and science careers: Leaky pipeline or gender filter? *Gender and Education, 17*, 369–386.
Deleuze, G., & Guattari, F. (1994). *What is philosophy?* New York: Columbia University Press.
Roth, W.-M., & Barton, A. C. (2004). *Rethinking scientific literacy.* New York: Routledge.
Roth, W.-M., & McRobbie, C. (1999). Lifeworlds and the 'w/ri(gh)ting' of classroom research. *Journal of Curriculum Studies, 31*, 501–522.
Sellers, S. (Ed.). (2004). *hélène cixous: The writing notebooks.* New York: Continuum.
Smith, D. E. (1990). *Conceptual practices of power: A feminist sociology of knowledge.* Toronto: University of Toronto Press.

KAREN L. TONSO

5. LEARNING TO BE ENGINEERS

How Engineer Identity Embodied Expertise, Gender, and Power

As had become common for this senior engineering design team, Pam arrived with work in progress, but no one else had anything to share. Almost immediately, Carson began to grill her about what she had done, calling himself the devil's advocate; after months of this, she became irritated and asked him where his work was. Other team members lowered their heads and studied something—anything—on the table, but soon began to talk about how they spent the weekend, topics they expected on an upcoming test, or other coursework completed instead of work promised to the team. Pam turned to Samuel to go over key chemical engineering concepts and equations that fell outside Carson's mechanical engineering expertise. When their faculty advisor arrived, Pete—who never participated in technical discussion or paid them much attention—took it upon himself to be the team's voice, to present Pam's work as his own, to miscast Samuel as someone not committed to teamwork, and on one occasion to announce that he had "found the problem" that had stymied the team, something he had only moments before been unable to describe to teammates, and in other ways gave the appearance that he was central to the team's everyday engineering work and an accomplished engineer, the opposite impression he created during teamwork. Student engineers "recognized" these students' behaviors—and a wide range of other ways to be engineers—via an array of engineer-identity terms, referring to Pete as a brown-noser or sometimes as anal, to Carson as a hard-core over-achiever, and thinking of Samuel as a nerd. Yet, they had no term for Pam, though someone mentioned she was a "unique individual," clearly marking her as someone not *recognizable* as an engineer on this campus. Evidently, since none of the students on freshman teams used or could explain these terms, but seniors gave long lists of terms and spoke at length about them during interviews, student engineers learned these ways to be an engineer on campus as a routine part of their engineering education. But how did this occur and what did student engineers' sense of the organization of these terms into engineer-identity categories infer about the campus culture's underpinnings? And ultimately, did campus-preferred ways of life become embodied in individual students, and if so how were these embodiments expressed in face-to-face, everyday social interactions on campus; and simultaneously, did the actions and interactions of individuals construct everyday life on campus? Articulating answers to these questions is the central purpose of this chapter.

The approach taken here aligns with that of critical cultural anthropology, which gives pride of place to the knowledge, beliefs, and productions of local communities, while it looks carefully at the ways relations of power play out within communities. That is, critical theorists ask how some forms of practice and their practitioners come to have or be given status over others. And, though such communities have historical persistence and seeming stability, they are by no means static but are instead capable of generating themselves into the future in ways that are sensible to their pasts. However, though local, such communities are not isolable from larger societal forces, but contexts where ideologies of power and prestige seep in and become taken for granted, even as they are taken up in context- or community-dependent ways. Thus, rather than beginning—as many researchers grounded in psychological perspectives seem to do—by asking "What do people make of themselves?," the research reported here asks "What do cultures, communities, and institutions make of people?," and goes on to examine identity productions as a complex process of personas shaped in context, as simultaneously persons shape context. Ultimately, the results here will not suggest absolutes applicable to all sites of engineering education practice, but will instead strive to describe complex, contextual, identity-development processes at play on this campus and illustrate how these are linked not only to campus culture, but also how campus culture took up societal frames of reference (or ideologies) for academic science expertise and gender status. Thus, in time, this chapter will suggest the importance of studying power relations as a way to follow how societal ideologies are taken up in learning cultures, and then how learning cultures reach into face-to-face interactions among individuals, arguing that community-developed identity linked culture to person, that personas embodied relations of power, and persons acting in context through cultural identity produced and reproduced community.

PUBLIC ENGINEERING SCHOOL

Public Engineering School is a well-recognized, stand-alone campus devoted to educating engineers in the U.S. mid-continent. At the time of the study in the early to mid-1990s, women comprised 20–25 percent of 2,300 undergraduates and 11 percent of faculty, representations higher than national levels. Its commitment to reform in engineering education and to women's full inclusion, and its full support of my research project, made it an attractive site

Reform at PES came primarily in one curricular change. In the 1980s, PES began offering explicit instruction in engineering design—a set of courses intended to mimic real-world engineering work. Students took one-semester courses at the first- and second-year levels, and then completed a one- or two-semester capstone design course as seniors. Design courses organized students into teams who completed an everyday project for industry or government clients, and along the way presented (individual and team) oral and written reports for feedback and grades from faculty. At the time of the research in the early to mid-1990s, design teamwork was touted as almost a panacea for changing women's negative experiences in engineering education (e.g., Felder, Felder, Mauney, Hamrin, & Dietz, 1995). Teams seemed well suited as a palliative because of women's perceived organization, writing, and interaction strengths, and because of women's perceived need for more hands-on activities (McIlwee & Robinson, 1992). Earlier research also suggested that peer groups played a key role in women's loss of motivation to study challenging mathematics and science fields, but I was skeptical of an additional finding that classrooms played a less central role (Holland & Eisenhart, 1990). And, much of this research failed to carefully examine men's experiences, which gave the impression that there was a uniformity of experience, and that studying "gender" meant studying women (here Seymour and Hewitt [1997] constitute a notable exception). Thus, student design teamwork at PES provided an entrée to peer group culture that was connected to curricular goals, but was somewhat out of sight of faculty members. Such a study would require carefully organizing the research project to study interactions among student engineers directly, something not possible in earlier research.

The research method aligned with cultural anthropology ethnographic practices: join the community and become a trusted person there; collect field notes grounded in one's participation—paying attention to who is present, what they say, do, and produce; talk to insiders about what is going on and what events mean to them. Fieldwork occurred over the course of four years. For each of the three design-class settings I engaged in a semester of reconnaissance when I collected no data, but became acclimated to each course, then performed active fieldwork for the duration of each class (one semester at first- and second-year levels, two semesters at the senior level)—participating as a design-team colleague on seven student teams (three at the first-, two at the second-, and two at the fourth-year level—a quasi-longitudinal approach), getting to know 33 students well, attending all in- and out-of-class meetings, and formally interviewing students twice and their 11 professors once. Interpretation of events was an ongoing process that began in the field and informed subsequent data collection (for instance, the focus of subsequent observations filled gaps found in analysis of earlier observations and questions asked at interviews depended on analysis of earlier field notes), continued off-site after primary field work was completed, and was bolstered by periods of secondary field work, member checks, and peer review (giving papers at conferences, discussing findings at colloquia) (Lincoln & Guba, 1985; Spradley, 1980). In the main, data analysis tracked patterns of sameness, difference, and affiliation in the data, but the goal was to be able not only to describe what happened, but also explain how things came to be the way they were (a process approach), and why they were this way (in this case an analysis of relations of power—whose sense of the world "stuck"). Reading and re-reading data ruled out competing explanations. Ethnographic strategies were augmented (a) by incorporating items during the first interview to elicit engineer-identity terms and explanations of them, and in the second interview asking each student to sort terms into categories and explain their sorting, (b) by performing an analysis of the curricular structures (how much control the curriculum allowed students to exercise over coursework choices and sequencing), and (c) by surveying students about their sense of the differences between design and conventional engineering courses.

ENGINEER IDENTITY AT PES

Ultimately, in this large-scale cultural study, engineer identity became a central pivot through which societal norms reached into campus culture, and campus culture moved into teamwork via individuals who learned to *embody* engineering practices. To understand engineer identity at PES requires an appreciation of cultural frames of reference for being recognized as an engineer on campus, and an awareness of how student engineers learned and performed engineer selves, as well as, how engineer identities and performances of engineer selves interacted to produce belonging (or not) on campus, which here reinforced campus culture.

Cultural Frames of Reference for Identity

Student engineers used an array of terms to refer to one another as people thought of as engineers on campus. When asked explicitly "what are all the terms you use to refer to one another as engineers here?," seniors usually began with "Well, we're all nerds here." But, they went on to not only list a large set of terms, but also to describe each term in detail, especially answering the question "How would I [the researcher] know a [kind of engineer] if one were on our team?" First-year students, by comparison, gave many fewer terms and only limited explanations of them. At a second interview, as part of the larger ethnographic project, students returned to the issue of engineer identity to sort the 36 most frequently occurring terms into categories and explain their sorting.

Students' understandings of the organization of terms provided a look inside campus culture, the way the world of appropriate engineer-identity was "imagined" or "supposed" to be. The engineer-identity terrain made apparent not only the ways they thought of engineer types, but also which ideologies from the wider society were taken up by campus culture. First, terms coalesced around three central identity categories: Nerds, Greeks, and Academic-Achievers (Figure 5.1). Students thought of Greeks (Social-Achievers) and Academic-Achievers as comprising a super-category, Over-Achievers, signaling the prominence of those who affiliated here, and conversely the lower status of Nerds. Though detailed elsewhere (Tonso, 2006), space limitations in this chapter preclude descriptions of these terms, but suffice it to say that in students' minds these were living, breathing personas that they could identify based on how a person behaved. Each of the categories contained desirable and undesirable terms, or in the words of student engineers "People who are normal and people who go too far." Undesirable identities included those willing to exploit others, cheat, and otherwise cut corners to maintain their standing on campus shown up and to the right among Academic-Achievers: hard-core over-achievers, anal, curve-breaker, and brown-noser. Those who party too much for success in an engineering education exist up and to the left for men among Greeks—BMOC (big man on campus), jocks, frat boy, frat guy, and slacker—and down and to the left for women among Greeks—PES-woman, sorority girl, sorority chick, and betty. Likewise among Nerds, up and to the right are

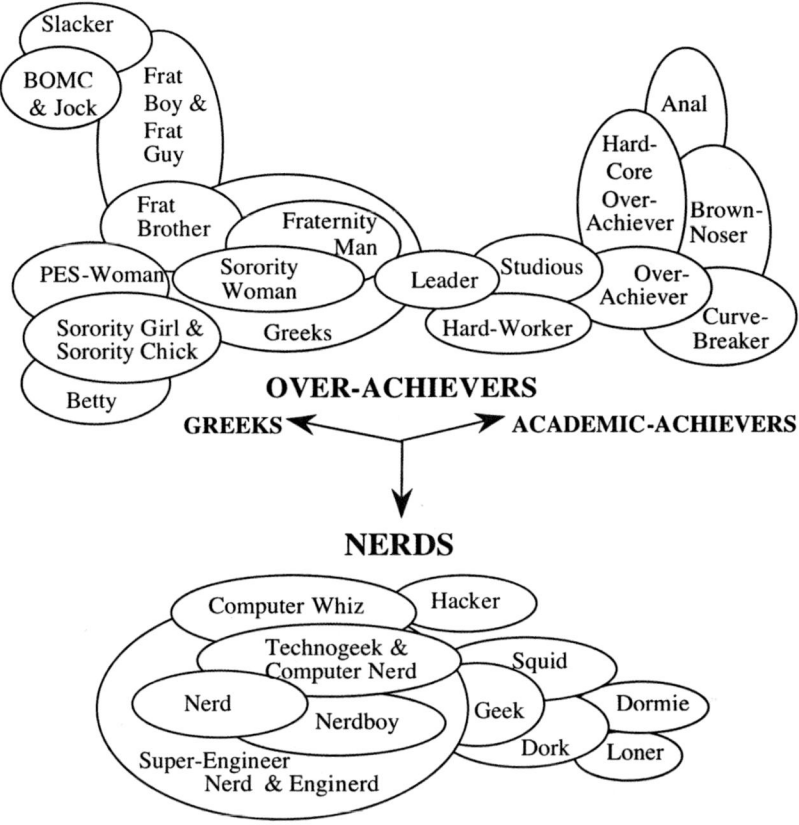

Figure 5.1. Organization of engineer-identity terms

those whose thoroughly asocial behaviors set them apart from other Nerds: hacker, squid, geek, dork, dormie, and loner.

Second, the two strata among engineer identities—Over-Achievers and Nerds—practiced different forms of engineering expertise. Over-Achievers practiced the form of engineering given prominence on campus, answering rapid-fire, one-right-answer, decontextualized, "textbook" problems represented in mathematical equations, using "givens" to seek an unknown quantity. Good grades depended on not only completing long homework assignments comprised almost entirely of these kinds of items, but also taking hour exams throughout the semester and a comprehensive final exam at the end of the term with more of these items. Though it was common to study with other students, some students openly cheated by either sharing homework problems that each had completed, or copying outright from another student without doing any of the items. Students thought this was more common among Greeks than others. And, Greeks had another advantage—the storied "fraternity files." Here, fraternity members accumulated exams from every course, taught by every professor and, to study for an exam, some students only learned the solutions to the narrow set of items used on exams.

Nerds, on the other hand, developed a set of engineering expertise more akin to design engineering. Thus, in one of the more lopsided logics of this campus, Nerd students who understood both the mathematical and scientific principles from their classes *and* could put these ideas to use in specific situations for their clients (the kind of expertise industry urges engineering schools to develop in graduates) did not come to the attention of campus authorities and remained below the prestige "radar" on campus. Thus, in a campus culture that organized evaluation of student work to focus on the easier-to-assess one-right-answer solutions of textbook problems, instead of developing careful ways to assess the more difficult to evaluate—but genuine—engineering products associated with design classes, Over-Achievers gained prestige and status on campus and rose above Nerds. While these two forms of engineering expertise are widely acknowledged (e.g., Bucciarelli, 1994) and the former criticized as inadequate for preparing engineers for industry careers (Dutson, Todd, Magleby, & Sorenson, 1997), rarely is the structure of engineering education directly implicated. But, clearly, campus culture—especially taken for granted sets of ideas about the "right" way to teach engineering, how to assess students' work, and what knowledge counts—should be under scrutiny.

Third, the engineer-identity terrain encoded gender—different forms of masculinity and femininity made salient on campus. Over-Achiever masculinities valorized men who took charge (often described with militaristic code-phrases like "divide and conquer," and "forming, storming, and norming"), took advantage of others, and made a show of themselves through public performances that depended on the work of others and bragging about high grade points and campus standing. Nerds took great care not to reveal any of these markers of over-achievement, treated their colleagues with respect, and practiced egalitarian relations. This "nice-guy" form of masculinity was subordinate to the domineering man of went-too-far Over-Achievers. Some forms of masculinity, however, could not be expressed on campus. For instance, many students told me that the campus was aggressively homophobic, further suggesting that campus culture elevated a prototypic masculinity associated with dominance over others, especially eschewing becoming "feminized." Categories also demonstrated the ascendance of masculine ways of life over feminine in campus culture. Heightened concern about women's intrusion into men's ways of life came from two portions of the engineer-identity terrain. Went-too-far Greeks talked about women in disparaging ways, especially referring to women as sex objects and detailing their physical deficiencies when compared to some men's "ideal" woman. Went-too-far Academic-Achievers described gender relations using expansive stories about reverse discrimination. Respectable Over-Achievers and Nerds did not relate to their women colleagues or talk about them in these ways, but treated them with respect, incorporated women's ideas in the team's work, and gave them credit for their contributions.

But, and this is more telling, terms for women were absent among students known for their engineering work (among Nerds and Academic-Achievers), so woman-terms existed only among students known for their social prowess, only one of which—sorority woman—was a desirable engineer identity (for some, but hardly for all women who study engineering). Sorority woman encoded a femininity associated with social service, being men's normatively heterosexual partners at social events, and being "good" girls. Three other terms—sorority chick, sorority girl, and betty—marked women who only partied on campus and were thought of as promiscuous, an unacceptable sort of femininity, not potential engineers. And the final term—PES-woman (where PES stands in for the campus nickname) encoded the rest of women as "other": "big, easy sluts, who go out too much and flunk out of school, so the women who stay here are ugly." Ultimately, in the student engineer imagination or suppositions about the kinds of people recognizable as engineers, PES-woman constrained the possibility of femininity becoming attached to engineering expertise on campus, and women known for doing engineering that put classroom ideas to work in design projects seemed to fit in only to the extent they eschewed all vestiges of femininity; that is, became indistinguishable from men, or conversely invisible *as* women.

Thus, student engineers used a complicated *cultural* set of terms and orderings to talk about engineer identity and to interpret events and people on campus, that is, to make sense of daily life. Engineer identities were framed by campus cultures having taken up the pre-eminence of academic-science (mathematized abstractions of the real world) over more mundane approaches typical of work in an engineering career and gender status—some forms of men (masculinities) above others, and some forms of women ahead of others (femininities), and within each strata (Over-Achievers or Nerds), men ahead of women. When these ideologies intersected, Over-Achiever and Greek men had more power on campus, sorority women and Nerd men had intermediate amounts, and women not recognizable among, or identifying with, sorority women had the least. These orderings came not from a vacuum but from the way campus practices outside the reach of students framed *becoming an engineer*—especially decisions made about organizing and delivering the curriculum, and about distributing awards and recognition for "excellence." By calling the tune for the piper to which all students danced, campus organizational practices connected a certain kind of domineering masculinity to an academic-science form of engineering expertise and these practitioners ascended.

Student engineers spoke *as if* these terms referred to living, breathing persons, in effect acting as if these crystallized engineer identities, albeit with considerably more variety and complexity than captured in the term "engineer," or expected by some researchers. For instance, in the case of situated learning theorists, a leading identity is conjectured, such as a tailor, and this identity is seen as motivating people to learn the practices of tailoring (Lave & Wenger, 1991). But, identity trajectories are spoken of as if they are somehow linear and lead to the same end result for everyone who can perform the tasks required of tailors. Varieties of tailors, for instance, are not envisioned. Engineer identity variety suggests that science, technology, engineering, and mathematics educational tracks, especially at the post-secondary level, are considerably more complex than a linear trajectory can accommodate. Likewise, research about professional identity in post-secondary education programs (Nespor, 1994) suggests curricular structures—the ordering of coursework, the kinds of choices given students about programs of study, and expectations about appropriate social behaviors to fit into their field—play a key role in producing among graduates a professional identity. But, Nespor, like Lave and Wenger, took for granted that "physicist" or "manager" held the same meaning across graduates, and that being a graduate conferred on all graduates the same kinds of power and standing in their respective fields. Clearly for engineer identity, this was not the case, further suggesting that curricular structures, while they are important to framing the production of engineer identities, may not be sufficient to explain how professional identities operate in day-to-day campus life.

That a set of cultural expectations about acceptable ways to belong as an engineer on campus exists is all good and fine, but several questions about them require answers to understand how they operate. In particular, how did student engineers learn engineer identities, how did they deploy them, and to what extent did the terms capture the performances of engineer selves given by student engineers?

Learning and Performing Engineer Selves

As in other communities where members learn tacit cultural knowledge, student engineers learned engineer identities by participating in a community where others used the terms and implied their taken-for-granted meanings. For instance, one of the students explained that *enginerds* turn ordinary conversation into an engineering analysis, engaging a kind of engineering wordplay. This student told of his hearing the term during a conversation he had with his older brother, a practicing engineer and alumnus of PES, and went on to recall another instance of its use:

> We're talking about gum right, and my sister-in-law said something about [him, that] he chews his gum in about five minutes. And then we started talking about half-life and then we were talking about when half the flavor is gone from the gum, natural log of two over tau, [something like the half-life formula]. And we were watching baseball one day, me and my roommate, and I said that [the pitcher's] ERA is pi over two [about 1.57]. We just do it to be funny. We know we're all nerds. We might as well not worry about it.

Thus, the term "enginerd" named a certain kind of behavior that he and his brother engaged in, and then was used to mark his banter with his roommate. People who exhibited this kind of behavior most of the time came to be thought of as enginerds. But, this form of engineering wordplay also connoted that "we're all nerds"—we belong here and share something that makes us a community. Thus, deploying the terms, and being marked with one of them, meant that a student engineer belonged.

Such terms liberally sprinkled the field notes, and student engineers used them to police the boundaries of acceptable and unacceptable behaviors. For example, a student entered a classroom and a colleague asked, "What did you get on the test?" He replied, "A 93." His teammate grunted, "Curve-breaker." Used pejoratively, curve-breakers did too well on tests, which made others look bad. As I stood in the library waiting for a team to gather, chatting with one of the students, we watched a student from our class, but on another team, roller blade down the asphalt pavement hesitantly, knee pads, elbow pads, wrist braces, and helmet in place. The student commented: "He's a dork, but he provides an endless source of amusement." Being termed a dork inferred social deficiencies, here a lack of athletic skill. On another occasion, a student asked a teammate if he wanted to go for a beer after the team meeting. When the other student declined because he needed to study, the first commented, "Nerdboy," implying that he had just stepped over the line between studying enough and studying too much. I heard students cautioned not to be "such a geek," heard others referred to as slackers, jocks, BMOCs (self-important individuals associated with sports teams but not excelling athletically), and so on. Study too much and risk being characterized as a nerdboy. Do so in a part of the library where anyone looking in could see you and where no talking, eating, or drinking are allowed and become a squid. Toady up to a professor by asking questions after class or walking to the faculty member's office for a conversation and be chastised as a brown-noser. Those whose behaviors settled into these facile grooves came to be thought of as characterizing particular engineer identities. Thus, becoming a certain kind of engineer was a complex interplay of individual action and group recognition of those actions, and it took time to be "recognized" on campus as fitting a certain form.

In addition, student engineers used the terms to characterize other people's fit among the categories, as a way to explain another's actions or predict how they might act in a given situation. For instance, Pam expected her team's climate to be precisely what she experienced (glimpsed in the vignette opening this chapter), because she thought of Pete and Carson in terms of their being went-too-far Academic-Achievers who would either exploit the work of others, in Pete's case, or presume to double-check what others did to protect his standing on campus, in the case of Carson. When a student thought of a teammate as a went-too-far Academic-Achiever, for instance, the student expected that teammate to lobby for a certain interpretation of events and to push for the kind of teamwork that protected their place on campus. Pete's masterful lobbying for meeting as a team as little as possible, for his taking on the least time-intensive tasks, and for his giving public oral presentations (and being seen as a leader by faculty) "since Pam had done so much already," confirmed her interpretation of Pete's engineer identity and influenced how she interacted with him. Likewise, for Samuel on this team, even though his careful explanation of a major dilemma completing their project possessed a wealth of engineering expertise and should have guided the team's changing their approach, Pete's standing on campus gave him the cachet to promote waiting until the last minute before altering their course of action.

Samuel and Pam both told me that it would "not change anything" to confront Pete on this issue, and that Pete would probably retaliate if they did. Samuel had experienced Pete's not-so-subtle controlling behaviors when Pete diminished Samuel's reputation with their faculty advisor. For instance, as Pete, the faculty advisor, and others waited for their team meeting time to arrive, Samuel walked in as the carillon began to chime the hour. Pete commented, for their advisor's benefit, "Finally, here's Samuel, late as usual," miscasting Samuel as a slacker who was not committed to the team's work. These power-deploying events played out on a regular basis—Pete lording it over Pam and Samuel, and Carson grilling Pam about her calculations, but doing none of their own work in their assigned areas. Yet, none of the other team members confronted either Pete or Carson, even though they told me that both sets of behaviors were unacceptable engineering practices, justifying inaction by saying: "It wouldn't change anything." And, if (as I argue) Pete or Carson is the way he is because of campus culture, not because of any innate exploitation proclivity, then student engineers are probably right. Confronting Pete or Carson will not address the campus culture, which grew out of campus practices completely outside the reach of student engineers.

In addition to being used for deciding how to act in this field of power relations, the terms became filters for reducing complex human behaviors to understandable tropes on this campus and these neither provided accurate predictions about unseen behaviors or captured all behaviors. For instance, Pete's being seen as a star in conventional classes, because of his stratospheric grade-point average, brought with it faculty's assumption that he was capable of figuring out the team's dilemma, though he offered no bona fide explanation of it, or any other technical aspects of the project, during their two semesters together. In fact, at the team's final presentation to their client, Pete's description of the condenser he purported to design was all but vacuous, which a company engineer subtly critiqued by asking, "Shell and tube? Cold water running through it?" In effect: "The most basic kind of condenser imaginable took you two semesters to develop?" Because their faculty advisor's expertise lay in electrical engineering, he missed the company engineer's skepticism of Pete's expertise. Pete's conventional-course successes, grounded in academic-science expertise, gave him "star" status, which carried with it the presumption of his being a key player on his team. It seemed that little he did to demonstrate his considerable ignorance during teamwork and oral presentations undercut his image. Ultimately, he had several lucrative job offers, suggesting that on-campus job-search practices further reinforced campus-preferred ways of life and undercut industry desires.

Likewise, aspects of student engineer behaviors that fell outside norms assumed by engineer-identity terms could not be noticed, and might have been penalized had they become known. For instance, though as a computer whiz he was considered the team's expert programmer, he insisted on receiving everyone's feedback on his ideas for software to monitor equipment. Here, Martin practiced a form of leadership that systematically took his teammates' contributions into account, but never took credit for another person's work or coerced them to do work—countering the "normal" kind of leader on campus. Some faculty would have found this unacceptable and berated Martin for needing "hand-holding." Also, Martin facilitated making the team a welcoming place where everyone was included via snack choices (keeping each person's favorite kind of potato chips at his apartment where we met), conversations about rock music, and engineering work. He enacted the role of social hostess, someone who made others comfortable and saw to their needs, a role associated with women in U.S. society. This, too, could not be "seen" by engineer identities, and some faculty and students would find his "acting like a girl" inappropriate. Finally, he actively kept his relatively high grade-point average quiet and gave no impression that he knew more than they did, though his expertise in mathematics and computer science engineering was quite deep. Ultimately, his counter-hegemonic leadership practices, his prototypically feminine social hostess activities, and his deep understanding of scientific and engineering principles learned in conventional classes and his ability to "see" the world in terms of these principles, never came to the attention of faculty and fell below campus prestige radar. Lacking markers of affiliation with campus culture's Over-Achiever ways of life, Martin was assumed to fit among Nerds, which meant he had less access to job-searches than student engineers like Carson and Pete, and fewer, and less attractive, job offers in spite of industry's claim to prefer this kind of graduate.

But, women such as Pam and Marianne—whose skills trumped the engineering expertise demonstrated by Pete and Carson and who were every bit as capable as Nerd colleagues like Martin and Samuel, student engineers able not only to understand, but also to deploy, scientific and engineering principles learned in their conventional classes—had few on-campus interviews and no jobs at the time of graduation. They simply could not be recognized *as* engineers on campus and became invisible there. To be thought of as women, which both very clearly were, meant not being thought of as engineers. Why did their in-team demonstrations of engineering expertise fail to garner recognition as engineers? The best explanation seems to be that the cultural terrain for belonging—engineer-identity terms—filtered them out, much like a polarizing lens filters out glare from the surface of the water. For instance, Martin talked about whether a woman could be considered a Nerd—the sort of engineering self that Marianne performed:

> You don't think of women as that [fitting among Nerds], I guess. 'Cause, at least at this school, I think guys here so much appreciate that a girl chose to come to this campus that you're just like, "Great!" There are so many guys that you can say "this guy's a nerd," the pocket-protector-wearing guy. [But he couldn't say that of a "girl."]

Martin knew Marianne better than almost any other student engineer and was among the most open-minded student engineers when it came to overt thoughts about women's place on campus. Though he recognized, and deeply appreciated, her skill as an engineer, when asked about women being Nerds, he deployed taken-for-granted forms of belonging (cultural knowledge about campus engineer identities) and was unable to place her *as an engineer*: "You don't think of women as [Nerds]." Also, the contradiction between his acceptance of her as an excellent engineer and his inability to see her as fitting among Nerds failed to register; he could not see her as counter-evidence to the way campus culture was *supposed* to work. This is, of course, the dilemma of the engineer identity terms: as cultural knowledge, they are taken for granted, and are rarely open to reflection or conscious thought.

Thus, a campus status hierarchy emerged that was inversely proportional to design engineering expertise; those with the most power demonstrated the least expertise. Women with the best sets of engineering skills—those who could really *do* engineering, who made good use of their opportunities for hands-on experiences, who understood how to recontextualize decontextualixed scientific and engineering principles—were made invisible through campus cultural routines and this was encoded in the identity terrain. Men with extraordinary design engineering expertise, such as Samuel and Martin, never received their due, while watching men like Carson and Pete receive premier job offers and on-campus awards. This thoroughly illogical state of affairs suggests the power of historically salient campus cultural practices to constrain change. And, I argue that the production of a domineering masculinity seemed to be required to achieve at the highest levels on the arcane academic-science tasks that comprised the bulk of the curriculum, and that this concurrently minimized the impact that otherwise progressive men on the inside might have changing such a campus.

IDENTITY DEVELOPMENT

Though there is certainly an engineer stereotype in the US, it fell by the wayside when student engineers thought about being an engineer in PES campus culture. In fact, developing an engineer identity occurred through a complex cultural process for learning to be an engineer. As student engineers went about their everyday lives, they made decisions about how to act relative to affiliating with campus-preferred ways of life. In time, student engineers came to think of engineer types in terms of students being Greeks, Academic-Achievers, or Nerds, and to think of specific people as fitting among these engineer identities. Social control routines used the terms to police behaviors consistent with Over-Achievers versus Nerds, such as a Nerd-affiliate calling a colleague a curve-breaker for his high grade on a test when the average was much lower or a fraternity man cautioning a frat brother not to be a geek. And through these interactions, student engineers actively modified their behaviors and came to be thought of as fitting into niches with which they *identified*.

Each student who ultimately persisted through graduation thought of her/himself as an engineer. Such statements as "we're all nerds" set students apart from the general population and conveyed being a member of the campus community, which student engineers expanded into the larger set of terms for engineer identity. Also, students performed themselves as engineers, whether through persistent shirking of teamwork duties and displaying affiliations with Over-Achiever ways of life, or through demonstrations of design engineering expertise. But, as the cases of the Nerds (Martin and Samuel) and women with design engineering expertise (Pam and Marianne) on the senior teams made clear, being recognized via engineer-identity terms played a critically important role in the extent to which one's development as an engineer counted in this campus culture. For Nerds recognition meant being thought of as second class; for women design engineers, recognition *as an engineer* could not occur sans terms for women among Nerds. To have an engineer identity, it was not enough to *identify* as an engineer, but one must also *be identified* as an engineer. Thus, developing an engineer identity was a dialectic process through which members of the community *did* engineering: demonstrated (a) engineering expertise (of two dominant forms), (b) gender norms, and (c) positions of relative power, as simultaneously they made sense of their student engineer colleagues based on understandings about the kinds of engineers who fit on this campus. A woman might think she was an engineer, might demonstrate extraordinary design engineering expertise, but still not have an engineer identity because she failed to gain recognition—to be *identified*—as an engineer. This flies in the face of wisdom advocating increasing women's numbers in engineering by giving them opportunities to demonstrate their communication and interpersonal skills, and gain hands-on experiences, and deepens the call for cultural change in engineering education (Tonso, 1999).

Engineer identity developed over time through participating in the everyday activities at PES, and encoded ideologies of power aligned with the pre-eminence of academic science and of a domineering masculinity relative to others, as well as of men relative to women. Thus, campus life set the stage for contestations about what it meant to be a "real" engineer. Rather than engineer identity being a prototypic "mature practitioner," as envisioned by Lave and Wenger's situated learning theory, terms students used and their organization of them into three related categories provided ample evidence of contestations about what being a "real" engineer meant on this campus. Rather than students moving along a trajectory toward some shared notion of "engineer," students went about their everyday lives deciding whether to affiliate with Greek life or not, devote themselves completely to their studies or take a more relaxed approach, align with an academic-science form of engineering, or adopt the more expansive engineering form promoted in design classes. In other words, student engineers *differentiated* themselves from one another, but not in open-ended, anything-goes ways, but in culturally salient dimensions. And, for most students, some facets of student behaviors came to be thought of as indicating an engineer type, while some actions and behaviors fell outside institutionalized recognition processes. In time, students assumed these terms (and other things) as cultural fact and deployed them to make sense of how their world was *supposed* or *imagined* to work, even when there was considerable empirical evidence that the world did not in fact work in the way presumed. Thus, at PES identity emerged from, or developed in response to, campus culture, "[t]his collective ability to take imaginary worlds seriously . . . [which] is the magic that anthropologists have tried to capture in the concept of culture" (Holland, Lachicotte, Skinner, & Cain, 1998, p. 280).

At PES in particular, categories of engineer identities served as a refined set of cultural pivots for belonging, and being *recognized* as an engineer seemed to trump either thinking of oneself as an engineer or performing an engineering self. Understanding how academic-science prestige and gender-status ideologies framed belonging proved central to understanding identity productions. Thus, because of the specificity of the knowledge base built into the intended curriculum and of the salience of becoming an engineer to each student engineer's future, one cannot help but wonder about the extent to which the identity-development process that occurred at PES might occur in K–12 settings.

In what we know of high school peer group productions from anthropological studies, a wide range of student identities can exist: Jocks who associated with their school's middle-class-affiliated "corporate" culture and Burnouts who opposed it (Eckert, 1989); jocks, vatos, cheerleaders, band fags, "good" girls, cowboys, and others representing affiliations with ethnic-, gender-, social class-, and school-aligned strata (Foley, 1990); and working-class lads who resisted being made into middle-class ear'oles who accepted school practices (Willis, 1977). Among these are nerds, a term inferring those thought of as out of step with the popular crowd—not as attractive, lacking social skills, and overly interested in mathematics, science, and technology courses, which are notoriously more difficult, time-consuming, and likely to interfere with high grade-point averages and with extracurricular social activities central to being thought of as a popular student. Thus, high schools, unlike PES, produced peer group cultures that encompassed a much wider range of knowledge and of attendant personas, communities not focused explicitly on producing scientist or engineer identities. But, without further study, one cannot rule out a scientist sub-culture being produced among students affiliated with scientific ways of life, and the PES research suggests studying to what extent high school science-learning settings begin the process of students learning academic-science prestige and gender status norms present in U.S. society and given prominence at PES.

When studied in K–12, students' scientist identity appears to emerge relative to a stereotypic "scientist," which the PES research suggests will be inadequate for understanding the next phase of identity development as a scientist. In particular, based on the way that student engineers understood "engineer" in considerably more complexity than an engineer stereotype presages, one wonders if emerging scientists in particular college science majors—student biologists, student chemists, student geologists, student physicists, and so on—eschew becoming a "scientist" of the sort presumed in a scientist stereotype and instead have discipline-specific (cultural) identities produced in similar ways to that at PES. Some evidence exists that this may be the case and that different scientific disciplines take up societal ideologies in different ways. For instance, a pecking order similar to student engineers exists among physicists, which sets particle physicists ahead of others (Traweek, 1988). Their way of life valorizes practices associated with removing oneself as far as possible from the mundane and a masculinity grounded in something akin to a competitive aesthete who aggressively goes after linear accelerator time for conducting tightly controlled experiments harnessing the world of atoms and particles (an academic-science form of practice), and ignores bodily functions like sleeping, eating, and socializing outside one's colleagues (a certain form of masculinity). However, among environmental biologists, a much more collaborative sort of scientist who is connected to the world emerges, even though being a bit too pie-in-the-sky can be pejoratively termed acting like a "flowery bonehead" (Eisenhart, 1996). Here, esteemed practices align with considerable stewardship of, not dominion over, the mundane world and being able to connect with the world, not distance oneself from it, is valued (Eisenhart & Finkel, 1998). These are prototypically feminine ways of life practiced in out-of-the-lab sites, where experimental methods are possible only to a limited extent, practices clearly not aligned with academic science.

In the main, then, identity development at PES aligned most closely with a process approach peculiar to a cultural understanding of identity:

> [Identities] are social forms of organization, [bridging] public and intimate, that mediate this development of human agency ... But the cultural figurings of selves, identities, and figured worlds [cultures] that constitute the horizons of their meaning against which they operate, are collective products ... One can never inhabit a world without at least the figural presence of others, of a social history of person[as]. The space of authoring, of self-fashioning, remains a social and cultural space, no matter how intimately held it may become. And, it remains, more often than not, a contested space, a space of struggle. (Holland et al., 1998, p. 282)
>
> The social and political work of identification, in brief, reveals two closely held interrelated lessons. First, identities and the acts attributed to them are always forming and re-forming in relation to historically specific contexts ... The second lesson is that identities form on intimate and social landscapes through time. The distinction of particular acts as indexical of an identity, and the expansion of identities to include broader or different ranges of acts, depend upon the work of their allies and opponents. . . . [And, this identity production work] cannot simply be dictated by the discourse, the cultural form itself. (pp. 284–285)

Having an engineer identity connoted being recognized through engineer-identity cultural frames of reference produced and re-produced in everyday life. Also, in spite of considerable evidence that the PES campus culture was quite obdurate, it could have been otherwise. Campus could have taken up different societal ideologies and given more credence to design-engineering expertise suited to industry's needs. But, the PES academic community promoted the way of life most familiar to its faculty (who were in an earlier time likely Academic-Achievers going on to graduate studies) and maintained engineering's standing relative to other academic sites of science practice. Taking a critical cultural anthropologic approach to understanding identity made clear that persons could be made to seem more than they are (as when Pete's standing as a star conveyed on him design-engineering expertise he clearly lacked) or less than they are (as exemplified by the lack of recognition given Martin's counter-hegemonic leadership, prototypically feminine social skills, and markers of achievement, or by Marianne and Pam's invisibility as engineers because there was no way to think of them as fitting among Nerds).

Contextualizing individuals not just in the social situation of the moment, but also in the flow of history and imagined reality that was campus culture, illuminated a historically persistent process involving living together as engineers and through this process making the future of engineering. These young adults, actively producing and re-producing science practice through their everyday social interactions, held considerable power because those deemed to belong acted *as engineering practitioners*, and this power made them more than simply students of engineering; it also made them producers of engineering. As such, I am left wondering if it is possible to possess the power that flows with engineer (or scientist) identities (albeit differentially distributed among practitioners) before one achieves an undergraduate degree. Might this power conferred via being recognized—identified—as an engineer be something that sets identity productions of the sort apparent at PES apart from those in K–12 sites of science learning?

REFERENCES

Bucciarelli, L. L. (1994). *Designing engineers.* Cambridge, MA: MIT Press.
Dutson, A. J., Todd, R. H., Magleby, S. P., & Sorenson, C. D. (1997). A review of literature on teaching engineering design through project-oriented capstone courses. *Journal of Engineering Education, 86,* 17–28.
Eckert, P. (1989). *Jocks and burnouts: Social categories and identity in the high school.* New York: Teachers College Press.
Eisenhart, M. A. (1996). The production of biologists at school and work: Making scientists, conservationists, or flowery bone-heads? In B. A. Levinson, D. E. Foley, & D. C. Holland (Eds.), *The cultural production of the educated person: Critical ethnographies of schooling and local practice* (pp. 169–185). Albany: State University of New York Press.
Eisenhart, M. A., & Finkel, E. (1998). *Women's science: Learning and succeeding from the margins.* Chicago: University of Chicago Press.
Felder, R. M., Felder, G. N., Mauney, M., Hamrin, C. E., Jr., & Dietz, E. J. (1995). A longitudinal study of engineering student performance and retention III: Gender differences in student performance and attitudes. *Journal of Engineering Education, 84,* 151–163.
Foley, D. E. (1990). *Learning capitalist culture: Deep in the heart of Tejas.* Philadelphia: University of Pennsylvania Press.
Holland, D. C., & Eisenhart, M. A. (1990). *Educated in romance: Women, achievement, and college culture.* Chicago: University of Chicago Press.
Holland, D. C., Lachicotte, W., Jr., Skinner, D., & Cain, C. (1998). *Identity and agency in cultural worlds.* Cambridge, MA: Harvard University Press.
Lave, J., & Wenger, E. (1991). *Situated learning: Legitimate peripheral participation.* Cambridge: Cambridge University Press.
Lincoln, Y. S., & Guba, E. G. (1985). *Naturalistic inquiry.* Newbury Park, CA: Sage.
McIlwee, J. S., & Robinson, J. G. (1992). *Women in engineering: Gender, power, and workplace culture.* Albany: State University of New York.
Nespor, J. (1994). *Knowledge in motion: Space, time, and curriculum in undergraduate physics and management.* London: Falmer.
Seymour, E., & Hewitt, N. M. (1997). *Talking about leaving: Why undergraduates leave the sciences.* Boulder, CO: Westview.
Spradley, J. P. (1980). *Participant observation.* Orlando, FL: Holt, Rinehart, and Winston.
Tonso, K. L. (1999). Engineering gender—gendering engineering: A cultural model for belonging. *Journal of Women and Minorities in Science and Engineering, 5,* 365–404.

Tonso, K. L. (2006). Student engineers and engineer identity: Campus engineer identities as figured world. *Cultural Studies of Science Education, 1*, 273–307.

Traweek, S. (1988). *Beamtimes and lifetimes: The world of high energy physics.* Cambridge, MA: Harvard University Press.

Willis, P. (1977). *Learning to labor: How working class kids get working class jobs.* New York: Columbia University Press.

KATHRYN SCANTLEBURY

6. OUTSIDERS WITHIN

Urban African American Girls' Identity & Science

Over the past five years, I have had the opportunity to work with female, African American students in urban schools who may not view science as part of their identity. And without a science authority, such as a science teacher, challenging their perspectives, students may remain disengaged from the subject of science, outsiders within the science class. Identity is a combination of one's own construction of self along with how one is constructed by others. Within a given cultural field, identity is simultaneously fixed and changing. Thus, participants' gendered identities are differentiated in different cultural fields (McNay, 2000). And one's gender is a major category that mediates how others construct our identity. Typically, the first question asked after the announcement of a child's birth is "Boy or girl?" In most societies, the answer to that question is the beginning of the unconscious, and at times, conscious, differential treatment of girls and boys by parents, teachers, peers, friends and strangers—beginning with differently colored clothing, the way the baby's room is painted, the form of the clothing provided, and so on.

Gender is a social categorization and as such structures any social interactions, including those that constitute schooling and science education. Jonathan Turner's (2002) theory of emotions explains how gender, as a mesostructure, is inscribed by macrostructures that include mainstream ideologies, some of which are hegemonic. Simultaneously, the mesostructures associated with the category gender are constituted by microstructures that occur within fields as interactions occur, producing success, failure, and associated emotional valences (see, e.g., Tonso, this volume). Formal education occurs in highly socialized settings, and thus, participants have gendered experiences. In this chapter, I focus on how urban, African American female students construct and receive their science identities, and how a teacher used the experience of single-sex cogenerative dialogues—a forum in which participants from another social setting (e.g., the classroom) get together for collective sense making—to promote and encourage her female students' science identities. I use examples from female African American girls attending an urban school to answer the following questions: Why do some girls engage in science class and others do not? Why is learning science part of one girl's identity and not another's? What practices can teachers use to engage African American girls who are the "outsiders within" science?

FLYYGIRLS OR SCIENTISTS?

Jen is an atypical urban science teacher. She is highly qualified with a graduate degree in biology, a master's degree in education from an Ivy League university and, has taught in urban schools for several years. Jen is young, middle class, and white. During her third year of teaching, she perceived that the girls' actions in her class reflected a certain level of disrespect for science. In particular, she interpreted the girls' personal grooming practices as evidence for their disinterest in science. As Jen explained,

> the rigid mold of science that I had in mind and that was preventing them [the girls] from succeeding. Thinking about it now, it was almost a purist idea of science. In my mind, there were appropriate ways of participating in science, and that meant that it looked and sounded a certain way. Ultimately that meant that the girls had to behave and act a certain way in the classroom when doing science. And more often than not, the girls that were in my class did not illustrate *these* practices of science. They were usually in the back of the classroom talking, doing their nails, brushing their hair and doing whatever. And that's when I got upset because I didn't think they were doing science, let alone paying attention and contributing to class. (Jen, Interview 2005)

The girls' practice of unconsciously grooming during science class annoyed their science teacher who perceived this action as an indication that the girls were not serious about learning science or her class. Moreover, Jen's self-report of her frustration with the girls is an example of how an outsider constructs another person, in this case, the teacher constructing the girls as nonscientific because of their non-science behavior. Jen labeled the girls as "nonscientists" because they did not adhere to her "purist idea of science," and did not approach science in an "appropriate" way. For Jen, to be scientific, a person should "act in a certain way" in class. Possibly, the girls did not adopt science practices as part of their identities because they did not relate to the subject matter, but rather placed importance on their personal appearance. That is, the mesostructure of gender mediated the microstructures within the science classroom.

The science classroom is a cultural field where Jen intended for her students to enact a particular form of science. But culture is not contained within fields, and there can be occasions where a different culture is enacted within a scientific cultural field. From Jen's perspective, the girls' grooming practices did not belong in the science classroom but the girls' habitus—i.e., dispositions for acting, perceiving, and feeling in certain ways—meant that those practices were enacted regardless of the field. This example illustrates how cultural fields are structured and have porous boundaries. The girls' unconscious grooming provides insights into high school girls' self-perceptions, that is, foregrounding their femininity and physical attractiveness, rather than their mental acuity and scientific knowledge.

Previous studies of girls in middle and high schools document the importance girls, and others, place on physical beauty and sexual attractiveness. The fictional teenage, female character from Tyree's (1993) novel, Tracy Ellison, provides insight into the important lifeworld issues for African-American girls. As Tracy moves into middle school, her focus is not on academic success, but rather acquiring the clothes and apparel she needs to become a flyy girl[1], that is, a sexually attractive young woman. In Philadelphia, flyy girls dress fashionably, wear large eye-catching earrings and have trend-setting hair coiffures. Elijah Anderson (1999) wrote about the salience of respect to African American youth, including females, in the code of the street. Important ways for female youth to earn respect included their beauty and wearing fashionable clothing. Hence, in urban streets, such as Philadelphia, there is a social force toward females looking their best and wearing the most fashionable clothes they can muster—that is *to look flyy*.

In the course of my research, I have met flyy girls. For example, Rashida took great care with her appearance. She wore different, multi-colored, braided hairpieces. She also insisted on wearing a light pink jacket even though it violated the school policy on uniforms that resulted in her receiving repeated school suspensions. However, African American females often are constructed as matriarchs, "mammies," or sexually active and predatory as either "jezebels" or welfare mothers (Collins, 1991). Although for some girls, being acknowledged as flyy is an important aspect of their identity, it is also possible for others to interpret a girl's efforts to portray herself as sexually desirable as a sign of sexual promiscuity. For teenage girls, being attractive is a significant resource for earning symbolic capital among peers and may be a more important facet of their identity than learning science.

The acquirement of gender identity is an unconscious act governed by the dialectical relationship between individual and collective. Similarly, an individual does not have sole control over enacting a gendered identity since within any field material, symbolic, and social resources structure cultural enactment. As is the case with all praxis, a person's actions are both conscious and unconscious and reflect the structures of the field. That is, persons as much are structured as they are structuring. Hence as girls enact practices in a field in which they are familiar with the other participants and the available resources they do so in ways that are nonreflective, enacting stocks of knowledge as habitus—which is generated over time and reflects gender and other mesolevel categories such as race. At an early age, children learn acceptable, gender-appropriate practices that reflect mainstream culture and the contradictions of being inscribed according to categories such as gender, social class, and ethnicity. In this milieu the experiences girls have lead to the sedimentation of various forms of habitus and other cultural resources that contribute to their identities in the fields of their lifeworld. Anderson (1999) focused on two such identities, which he labeled as *decent* and *street*—highlighting the contradictions youth experience in developing cultural resources for the home and needing completely different cultural resources to successfully navigate the streets. Other fields, such as the *school* and *church* also require distinctive forms of culture, including science culture, which is created in part through formal schooling.

The science classroom is a field where science culture is enacted in age-appropriate forms. As Jen observed, the girls' practices did not reproduce science culture, at least in a form that she recognized as such. From her perspective science remains a masculine endeavor and its structures reinforce gender stereotypes. For example, rationalistic, logical, unemotional, and positivistic stances to objects of activity are masculine attributes associated with science. Conversely, female attributes such as nurturing, emotional, and subjective orientations are not valued within the culture of science. Accordingly, a masculine gender identity is more aligned with science than a feminine identity. For many years, educators have studied the paradox of encouraging girls and women into a culture that is not well suited to a feminine habitus—a field where females may not have a "sense of the game."

When girls come into the classroom, a different field from their school corridors, their neighborhoods, or the math class, their feminine habitus continues to dictate their conscious and unconscious behaviors, rather than "doing science." It is a feminine habitus that the girls are enacting when doing their nails and hair in science class, they have learned and repeated these practices over time with peers like those in the science classroom. The same cannot be said for science practices. Most students do not begin to acquire a science habitus from an early age, as they do with gender. Thus, one can argue that the girls have had few opportunities to develop science dispositions and as such have not developed a "sense of the game" regarding science. Boys are more likely than girls to engage in "science" practices as children. Generally, adults—parents, teachers, child-care givers—encourage boys into active play with tools such as building blocks, or Lego sets. These practices develop spatial ability and risk-taking skills that are beneficial in learning science. Girls are encouraged to play within domestic roles in passive ways—such as caring for dolls, or cleaning house—these practices are not primary dispositions in learning science (Scantlebury & Baker, 2007).

THE POTENTIAL OF COGENERATIVE DIALOGUES

Jen began to re-construct her views of science, that "rigid mold," and reconsider what she considered as scientific practices. More importantly, Jen restructured the girls' practices from evidence of their disrespect and disinterest in science to examples of how *she* needed to reconstruct her knowledge of the girls to change how she taught science. Jen engaged the girls in cogenerative dialogues to learn how she could connect science to their lives, for some girls, this generated their interest in doing science. Diva noted:

> I just like it [science] because it [is] something that I'm not used to. I'm not used to wanting to dissect *kidneys* and talk about the process of *diffusion* and stuff like that. It's just crazy. (Diva, Cogenerative Dialogue, October 2004)

For Diva, and the other girls, their new interest in science was "crazy." It was not a subject they had previously enjoyed or studied; but as their identities shifted to incorporate science, they began to change their practices in the classroom. Rather than doing their hair, the girls began to manipulate the science equipment. They started becoming engaged in the lesson by asking questions. For example, Charnae offered advice on what a student should do in class.

The more questions that you ask the more that it shows you are paying attention. I mean, and then it helps you in the long run, don't sit there and not know, that ain't going to work, if you are too scared to ask questions and then, if you feel like you can't ask questions, then hopefully you will have a friend like me who will ask for you. (Charnae, Cogenerative Dialogue, October 2004)

The dialectical relationship between agency and structure allows a subject's habitus to change over time (McNay, 1999). And over time the girls' practices changed so that they asked, and answered, questions in the class. Charnae's comment also reflected a hybridity of gender schema for females—that is, incorporating atypical gender practices into her identity. Gender norms are deeply rooted and accepted in society. Although women have moved into the public sphere through paid work, there have been few changes to their private roles. Women are still expected to assume nurturing and emotional caring for others. Thus, women's roles are a hybrid of conventional perceptions of femininity, along with re-defined, negotiated dispositions that are viewed as "regulated liberties" (McNay, 1999). In her statement, Charnae shows us how she has taken what is typically labeled a masculine practice, the asking of questions, and acquired that disposition. However, her words also illustrate a *regulated liberty*, because she also voices her emotional care of others "too scared to ask questions" and her willingness to ask questions for her friends. As Charnae acquired her science identity, we observed her advocating in the class for other girls, asking their questions, encouraging their science practices by acknowledging correct answers to questions through unconscious gestures such as patting a peer on the back. A gesture of "well done." Charnae illustrated how her gender and science identities began to evolve through investment and negotiation. Caring relationships among her peers reflect the gender norms for females, and in particular, those of an *othermother*.

Othermothers are women in the African American community who care for the members' children. Othermothers extend mothering practices such as caring and nurturing to others within community who may be younger siblings, cousins, or non-biologically related children. Again, this is an area where girls cross a cultural boundary between family and school. The othermothering practices that girls develop within their families and communities can truncate their agency within school but also enhance that agency in a different cultural field. For example, girls with othermother responsibilities often are absent from school to care for young children. These actions limit their opportunities to develop science knowledge and can have consequences on their access to learning. Many schools do not consider students missing school to care for others as "excused absences." Girls with othermother responsibilities are often in conflict with schools' structures. We have met girls labeled as truants because of their consistent tardiness. However, when discussing their circumstances, the girls explain that their othermother duties of supervising young children's journey from home to elementary school means they are late to their classes. Typically, the elementary school day begins after that of the high school and the children are either too young, or travel through tough urban neighborhoods, to travel alone from home to school. And in the afternoons, othermothers leave their school to fulfill their obligations that may mean they cannot take advantage of extra help from teachers or again may be in conflict with school rules if they miss detentions or other punishments for tardiness.

In these circumstances, the school culture is in direct conflict with the girls' community culture where showing that they care for others is important. Through othermothering practices' girls may garner social capital within their extended families. And for the girls, the placement of family and community responsibilities as a priority over their regular or punctual school attendance may negatively mediate their learning experiences. School structures can also truncate girls' agency. For example, in one school, the rules required that tardy students remained seated in the foyer until the completion of first period. The daily class schedule had the same class for first period. Thus, tardy students always missed the same class. When this first class was chemistry, those students struggled to complete the required work and in many cases, ceased attending or failed the class.

Othermothering is a practice that I have observed crossing cultural boundaries from the family and community to the classroom. These practices can garner girls' social capital in the classroom, and in different circumstances, also cultural capital. I have also observed young women enacting othermothering roles towards their peers, often male, in classrooms. In these examples, teachers often ask girls to tutor males who have been absent from the class. For the teacher, this practice allows her/him to move forward with the curriculum while students who could potentially disrupt the lesson with questions are "occupied" through the personal tutoring they receive from othermothers. This practice can be viewed as a positive experience for othermothers—they can reinforce their understanding of a subject through teaching the content to others. However, the time spent working with other students means that an othermother is not interacting with someone who has deeper science knowledge—for example, the teacher.

From a positive perspective, the othermother practices girls may develop can build social capital within their communities and the classroom. For example, Charnae had negotiated her practice of asking questions with Jen and the other girls through the cogenerative discussions. Her actions of advocating and speaking for her friends illustrate her othermother role within the classroom community. This action began to build Charnae's social capital with the girls in the cogenerative dialogue group and her cultural capital. The ability to ask questions and formulate an argument is a key scientific skill. Thus, Charnae's othermothering practices of advocating for quiet girls meant she began to build her scientific practices, and thus her cultural capital within the science class.

Another example of the girls showing how they had merged their female and cultural identities with science was the *Divas* song that Diva, Sholanda, and Gabby composed and choreographed about active transport. They began with the three girls lined up at the front of the classroom, shortest to tallest in an arrangement reminiscent of the 1960s female, African-American singing group *The Supremes*. They girls started to sing with a series of "ooh's" and synchronized body movements. As they begin singing they concurrently raised their hands to the left hand side above their heads and clapped. Their hand clapping set the rhythm for the song and as they brought their hands down from above their heads, their hips swayed. The *Divas* kept the rhythm going by slapping their hips. As they continued their song, the *Divas* pivoted on their left legs and stepped forward with the right. All the while, they kept singing and their bodies moved in choreographed steps. Their well-planned, scientifically accurate and entertaining song blended their feminine and cultural identities with scientific knowledge. The girls' swaying and hand movements showed choreographed dancing and a sense of rhythm, movement, and coordination. Their singing was tuneful and melodic. They showed a "feel for the game" in the smoothness and synchronicity of the song, with the dance. Their performance also illustrated a key part of African American culture—music in conjunction with movement, verve, and an oral tradition.

SOCIAL AGENDAS, SCIENCE IDENTITIES

Through self-definition, African-American women often resist society's strong, negative cultural stereotypes. But to achieve that self-definition and self-valuation, the girls also needed safe spaces such as within female friendship, mother/daughter relationships, or formal groups (e.g., churches and communities), to develop their voices, to share their stories, and to attain respect (Collins, 1991). For many girls, school and classrooms are the sites (cultural fields) for meeting and developing social networks and agendas rather than acquiring content knowledge. The cogenerative dialogues provided the girls a forum to develop and enhance their social network with each other and Jen. More importantly, the cogenerative dialogues also provided a space for the girls to safely speak about their lives within and outside of school. And the space allowed for the girls and Jen to listen to each other. During these opportunities, girls shared that they had few opportunities to socialize with each other outside of school. Most of them lived in areas they described as "ghetto" and their parents did not allow them to visit each other after school hours because the girls would be walking out in their neighborhood, possibly at night. The cogenerative dialogues provided girls a safe space to develop the social network and capital with each other in their science community.

The role of community is important in African-American culture and, within their communities, African American females develop strong family and kinship bonds. We noted that as the girls built their science community, their othermother identities extended to each other within and outside of science. They spoke up for one another in class, praised each other when they answered science questions, engaged in science activities such as lab activities and helped each other learn science.

The social networks built through the cogenerative dialogues slowly changed the girls' identities to being competent and capable in science. However, while cogenerative dialogues assisted girls in changing their practice, they were not sufficient to assist all girls in changing their science identities. For this to occur, the girls needed the collective support from others, their peers and teachers to adopt a science identity. Our identities are also constructed from others' perceptions. Thus, through the constant and consistent positive reinforcement that they could succeed in science, the girls' identities began to shift and they began to view themselves as capable in science. For example, during a class when Jen and the students were defining objects as biotic, Jen asked if paper was biotic. Preston, called out that paper was "not biotic." Shalonda gestured with her arm towards Jen and then spoke over the top of Preston to contradict his answer.

Shalonda: Because it's dead. But the characteristics of a living thing are made from. . . . It lists the things, the characteristics of a living thing that is made in the environment, wood isn't in the environment. It did live at one time.
Preston: But that don't?

Shalonda looks at her desk for her notes and Jen returns to her seat at the computer and speaks to the class.

Jen: Okay, if it was once living I am going to tell you right now, it can be considered biotic.
Shalonda: This paper has here than living things change over time.

As Shalonda speaks, she has the paper with her notes in hand and Jen gets out of her seat and moves towards Shalonda, while talking over Preston and a couple of other boys by saying.

Jen: Hold on, hold on, Shalonda has a really good idea.
Shalonda: Living things change over time.
Jen: Right, what could be part of that change?
Shalonda: That it has turned into paper.
Jen: Or that it is dead, it dies.
Shalonda: Yes.

Shalonda had confidence in her ideas to challenge Preston's answer; she gestured with her arm and called out to gain Jen's attention. When Preston questioned her response she used data to support her claim. Shalonda did not back down but rather, sought confirmation of her answer from her notes. Also, Jen validated Shalonda's claim, not only through her verbal affirmation that Shalonda "had a really good idea" but also through her movement from the computer desk into the room to stand near Shalonda's desk while they engaged in a discussion of living things. After Shalonda's affirmation that that paper is biotic, Jen returned to her desk to make note that paper was biotic. Following Preston's answer, Jen listened to Shalonda's explanation and then several students, including Preston had overlapping speech to challenge Shalonda's statement. Jen let the students speak but when Shalonda garnered her attention through signaling and speaking with her notes in her hand, because Jen had structured Shalonda as a science student, she showed respect for her idea and used her power position as the teacher to provide Shalonda with the space to voice her argument that paper was biotic. Through this exchange we see an example of how African American women use dialogue to validate knowledge claims and also how using call and response is a pattern of discourse used in their culture.

Shalonda did not wait quietly with her raised hand for Jen to select her to answer a question. She verbally responded to Preston's answer and gestured to gain Jen's attention while she spoke up. Jen and Shalonda connected during their conversation about whether paper was biotic and being connected to others when sharing knowledge is an important facet of black feminist theory. Also, African-American culture validates the use of emotion to validate knowledge claims. Thus, the girls' cultural experiences are in contrast to science's culture, which is portrayed as rational and logical without emotion. Shalonda showed her hybrid identity in using her emotive voice to answer the question, and also justifying her knowledge claims through the use of data and evidence. The other students listened to the conversation between Shalonda and Jen; they did not interrupt either the teacher or their peer. Shalonda used her voice and claimed her right to speak about science in the class. When a student questioned her answer, knowing her answer was correct, she sought data to support her argument. Through her actions, Shalonda showed her involvement in the lesson and her interest in science.

Randee did not appear to assume individual responsibility for her learning until she got involved in cogenerative dialogues. In part, based on her prior schooling experiences, Randee did not construct herself as capable of learning science. She enacted practices in the classroom that prioritized a social agenda with boys. Further, at the beginning of the year Randee told Jen, she could not learn science. She needed personalized attention from Jen to focus her science learning, and throughout the lesson would check with Jen to see if she was successfully completing the assignments. Randee used Jen as a resource to begin her science studies. Through her involvement with cogenerative dialogues, Jen consistently assisted Randee; and Randee's classroom practice of using science as social time with boys began to change. As Randee began to construct a science identity, she chose to re-seat herself in the classroom to be closer to her "sisters" and away from the "disruptive" boys who distracted her from her studies. Moreover, she began to learn science without Jen's personalized instruction. In the next section, we discuss the teaching practices Jen initiated to engage her female students in science.

TEACHING PRACTICES THAT CAN CHANGE GIRLS' SCIENCE IDENTITIES

In her initial teaching years, Jen made a concerted effort to connect with the African American boys in her science classes (LaVan & Beers, 2005). The priority was largely due to the problems she was experiencing with resistant and disruptive males. Having had success in using cogenerative dialogues as a strategy to engage boys in science, Jen used the same approach with her female students. Over several months, Jen ran cogenerative dialogues with selected girls and through these experiences, she deepened her understanding of the girls' science identities and the girls began to re-construct their perceptions of themselves as scientists.

Jen	I know I need to help you all practice note taking for college and you also need to be able to refer to something for the test, but sometimes I'm not sure how to effectively help you learn this skill. Like when I was giving notes and talking about the notes this morning, I felt like I was speaking to myself and many of you were not paying attention. Did you feel that too?
Taylor	Well I know it's really hard to follow what you're sayin[g] when I'm writing.
Chani	I do get lost sometimes because I'm trying to listen and get it all down before you erase it.
Charnae	Yeah sometimes I be tryin' to concentrate so much on getting it [the notes] down that I miss what you say.
Jen	So what do you suggest?
Charnae	Maybe you could write the notes on the board and have us copy them before you start talking.
Jen	And you will have to remind me to slow down and tell me when you don't understand something. Okay?
Sholanda	Okay.
Charnae	That's good. We can do that.

Jen told the girls of her difficulty in multi-tasking when she was presenting science to the class, writing notes, and attempting to engage the students in a discussion on the science topics. The girls suggested to her that they could copy the notes that Jen posted on the board before the interactive lecture. They suggested that Jen ask them more questions and slow down when explaining concepts. The girls also showed their willingness to accept individual responsibility for their learning by agreeing to take a more active role in the class by asking questions and alerting Jen when they did not understand the content.

Jen implemented these changes and in the ensuing weeks, the girls increased their science participation. For example, girls referred to the notes they copied from the board earlier in the period as a resource for their answering questions, providing definitions or interjecting comments. Often, the girls would raise their hands to answer a question, look down at their notes and then provide their answer to the question. Further, in many instances, the girls translated their notes into their own understandings, asked questions and made related comments. This not only served as a resource for the individual girl's learning, as they practiced using the vocabulary words and the new concepts, but also for the other students in the class as they listened and used the girl's comments.

Cogenerative dialogues are a pedagogy|research tool that teachers have used with their students to improve teaching and learning. When foregrounding issues related to females, for example how to connect girls to science, or how to reframe pedagogical practices to improve girls' understanding of science, cogenerative dialogues could be considered a feminist pedagogy|research tool (Scantlebury & LaVan, 2006).

Gender is a primary symbolic distinction. Stereotypes of masculinity and femininity are examples of these distinctions and schools are notorious for reinforcing gender (and heterosexual) stereotypes. Feminine characteristics girls are expected to embody are caring, nurturing, emotional expressiveness and a focus on developing relationships. In contrast, society expects boys to reflect the masculine stereotype of independence, emotional detachment, rational, and autonomous. These symbolic distinctions unfold into the separation of public and private arenas. The feminine habitus is culturally appropriate for the private, domestic sphere and the masculine habitus is aligned with the public sphere. Gender inequality is a form of symbolic domination, that is, the characteristics associated with femininity are often accepted as "natural" for females. This binary dichotomy, masculine versus feminine, set up symbolic distinctions are used to justify a differential education between girls and boys.

For example, teachers use girls' othermothering skills to nurture and care for others in the class. Typically, those students are boys who have higher truancy rates that their female peers. Their absence from school often means they receive personal, one-on-one tutoring from others. Yet, when girls enact a feminine habitus of caring and listening to others, waiting for their teachers' attention rather than demanding it through culturally inappropriate practices, such as calling out, demanding attention, disrupting classes, they often receive little to none of the teacher's attention. Jen, like many of her colleagues, did not focus on her girls' learning until their practice of personal grooming began to frustrate her. Girls, often, enact a feminine habitus where they disengage with learning but do not disrupt the lesson. Thus teachers often ignore the "quiet schoolgirls" because their attentions are focused on the more (usually male) demanding students.

Another way gender inequality is manifested is evident in teachers' reactions to girls who do not reflect the stereotypical qualities of the "quiet schoolgirl." Teachers often complain about "those loud black girls," girls whose habitus reflects more of their African American culture than mainstream white culture. However, depending on the field and how teachers construct the girls, I have observed a different reaction to that habitus. Ivory, was a student-researcher, from City High. She played on the school's basketball team, and also on the local courts in her neighborhood. She was the only girl who played at the neighborhood courts. Ivory's habitus in the classroom often reflected her assertive style developed in another cultural field—the basketball court. Her science teacher did not silence Ivory or admonish her for practices, such as calling out answers or yelling for his assistance during laboratory sessions. While, other teachers may have labeled Ivory's practices as "unfeminine," her science teacher provided a safe, public space for her voice and interest in learning science. Ivory, like her male peers, garnered her teacher's attention through her assertive practices. In this field, the teacher accepted the porous cultural boundaries between his classroom and the neighborhood courts and did not react negatively to Ivory's "nonfeminine" practices.

While Ivory developed practices that aligned with the masculine science practices and had a teacher who primarily constructed her as a scientist, for other girls, we found that the all female setting of the cogenerative dialogue provided a different field for them to develop their science practices. In this setting, the girls can focus on issues pertinent to their learning and lives, the development and attainment of science practices can be foregrounded, rather than unconscious gender practices. In the all-female cogenerative dialogues conversations ranged from school issues to home and gave girls the chance to share how they learned science, build relationships, and establish a community that supported their shifting identities from flyy girls to scientists and in between. The girls developed hybrid identities that used practices and tools from the dominant culture—science, into their own practices.

By establishing cogenerative dialogues, Jen showed an ethic of care towards the girls, and that emotive connection with them assisted her goal of improving how she taught them science. The girls developed their science identities through the cogenerative dialogues because Jen constructed them as capable, college bound, science students. Jen assumed the girls could learn science; she valued their learning, and cared about their progress. Charnae noted the importance of having teachers show they cared about her academic progress, and compared the situation to having a doctor check up one's health.

> You check, like you in the doctor's office you get checked up on. You get your medicine and they'll come back "How are you feelin?" You see what I'm saying? Don't *baby me* but at least be sensitive to my situation. (Charnae, Cogenerative Dialogue, 2005)

Charnae willingly assumed individual responsibility for her learning. She did not want teachers to "baby" her but she did expect teachers to recognize that circumstances could impact her learning situation. The cogenerative dialogues provided Jen a forum to see how the girls were "feelin," and the chance to gather their advice on how to enhance their science learning.

As the girls assumed more active roles in the classroom discussions, they began to exhibit greater excitement in taking part in the classroom activity which was illustrated by the increased volume of the girls' voices, their calling out, and raising and moving their hands to get Jen's attention. Jen did not shut down the girls' enthusiastic engagement in science by demanding that they exhibit stereotypical gender-roles of a "quiet schoolgirl," that is waiting with their hands—raised until selected to answer the teacher's questions. Through her work in learning about her students' lives and culture, Jen understood that the girls' calling-out, overlapping speech, verbal enthusiasm, and emotive statements were culturally produced practices. More importantly, for the girls an ethic of care included being recognized for their individuality, use of emotive speech, and empathic capacities. Through the cogenerative dialogues, Jen showed her appreciation for the uniqueness and empathy regarding their lives. The girls' engagement with science increased over time, which then led to their increased participation in science.

Cogenerative dialogues provided Jen with insights into the girls' lives. Through regular dialogues, Jen began to appreciate the importance of social networks to the girls. She restructured the classroom to optimize social networks by encouraging the girls to sit next to each other during classroom activities and to make side comments during interactive lectures. She also created additional collaborative projects, allowing the girls to discuss results of experiments and activities within their groups, and had the groups reporting their collective conclusions to the class.

CODA

Cogenerative dialogues can be a forum for teachers to develop deeper understandings of girls' learning and social needs in science. African American girls are often the "outsiders within" science. Although present in the class, Jen observed that her female students did not engage in science and she introduced cogenerative dialogues as a pedagogy|research tool to change that situation. Although, teachers often misinterpret girls' outspokenness as disruptive behavior rather than an aspect of their cultural dispositions, through her prior research with cogenerative dialogues, Jen understood her students' cultural dispositions. As the teaching population becomes increasingly white, female, and middle-class, it is crucial that teachers become cognizant of the importance of voice, emotion, and an ethic of care for African American students, especially girls.

As the girls developed hybrid science identities through cogenerative dialogues they changed their practices, became actively involved in the class, and learned science. For example, as Randee developed a scholarly-science identity she moved herself away from the boys to sit near to the girls from the cogenerative dialogue group. However, girls may have fragile science identities and their identities as science learners may not extend outside of the class. For many girls it is important that they develop their social networks as ways to establish solidarity and support for their evolving identities as scientists. Cogenerative dialogues are one strategy that teachers may use to develop an understanding of the identity issues that high school girls struggle with in science and change the girls' position within science.

NOTES

My thanks go to Jennifer Beers and the girls from Charter High School and to Sarah-Kate LaVan for their involvement with this research.

1   *Flyy Girl* is a novel read by the students in English class about African-American girls growing up in Philadelphia. The lead character is Tracy, an attractive sexually active teenage girl. Tracey is flyy. She has high social capital because of her physical attractiveness and external adornments (that is, fashionable clothes, jewelry and hair styles). Raheema is her studious next-door neighbor, sometimes girlfriend.

REFERENCES

Anderson, E. (1999). *Code of the street: Decency, violence, and the moral life of the inner city*. New York: W.W. Norton.

Collins, P. H. (1991). *Black feminist thought: Knowledge, consciousness, and the politics of empowerment*. New York: Routledge.

LaVan, S-K., & Beers, J. (2005). The role of cogenerative dialogue in learning to teach and transforming learning environments. In K. Tobin, R. Elmesky, & G. Seiler (Eds.), *Improving urban science education: new roles for teachers, students, and researchers* (pp. 149–166). New York: Peter Lang.

McNay, L. (1999). Gender, habitus and the field: Pierre Bourdieu and the limits of reflexivity. *Theory Culture and Society, 16*, 95–117.

McNay, L. (2000). *Gender and agency*. Malden, MA: Polity.

Scantlebury, K. (2005). Meeting the needs and adapting to the capital of a Queen Mother and an Ol' Head: Gender equity in urban high school science. In K. Tobin, R. Elmesky, & G. Seiler (Eds.), *Im-*

proving urban science education: new roles for teachers, students, and researchers (pp. 201–212). New York: Rowman & Littlefield.

Scantlebury, K., & Baker, D. (2007). Gender issues in science education research: Remembering where the difference lies. In S. Abell & N. Lederman. (Eds.), *Handbook of research on science education.* (pp. 257–286). Mahwah, NJ: Lawrence Erlbaum Associates.

Scantlebury, K., & LaVan, S-K. (2006). Re-visioning cogenerative dialogues as feminist pedagogy [32 paragraphs]. *Forum Qualitative Sozialforschung / Forum: Qualitative Social Research* [On-line Journal], 7(2). Available at: http://www.qualitative-research.net/fqs-texte/2-06/06-2-41-e.htm.

Turner, J. H. (2002). *Face to face: Toward a sociological theory of interpersonal behavior.* Stanford, CA: Stanford University Press.

Tyree, O. (1993). *Flyy girl.* New York: Simon & Schuster.

KAREN TONSO, KATHRYN SCANTLEBURY,
WOLFF-MICHAEL ROTH, KENNETH TOBIN

7. GENDERED IDENTITIES

Michael: The two chapters are quite different in their orientation. In Karen's chapter, I can see a greater symmetry with respect to the contribution of culture to the production and reproduction of identity, that is, with respect to being positioned as much as positioning oneself. In Kathryn's chapter, the emphasis is on agency, that is, on positioning at the cost of theorizing and articulating the way in which human beings are subject to passivity even to a greater extent than to agency—we are recipients of (hosts to) our intentions, intentionality, and agency.

Ken: Your emphasis on passivity is important and consistent with Kathryn's allusions to identities being constructed by self and others—and similarly, that as females structure their social lives they are simultaneously structured by others. Following Michael's lead, my research squad came to the same theoretical point by following Bourdieu's suggestion that if it appears there is empirical support for A, then consider the likelihood of empirical support also for the converse—that is, for not A. Hence, those who make the case for agency, can enrich their interpretations by considering directly the evidence supporting "not agency"—or as you point out, passivity. In so doing agency and passivity can be considered along with associated contradictions as dialectical partners.

Karen: All research is necessarily partial. Though both chapters make important contributions to understanding how things go in these two sites, the scope of our research gazes differs dramatically. As Foucault might have said it, I intend to think rather subversively about engineering education as a normalizing technology. My study (a portion reported here) has a clear agenda to explicitly investigate what engineering is and what engineers might be in a specific site, as forms of engineering practice and kinds of engineers are being produced in all their variety, and to use these empirical findings as a way to critique the "normal" senses of engineering and engineer. Identity at PES is, as you note, a production of an engineer self in a cultural world, and cultural *recognition* actually trumps individual agency at PES. For a person like myself who sees all actions as mediated in context, the concept "agency" seems to provide less explanatory power than needed to understand what went on at the campus. I could go no further than mediated action.

Ken: Excuse me for a moment Karen. If agency is dialectically related to structure, then power to act is dynamic and continuously changing as structures unfold. I mention this to emphasize that dynamic structures continuously unfold at macro-, meso-, and microlevels of social life. My point is that for me, the use of agency in an interpretive framework extends beyond mediated action to include an examination of material and schematic structures and the contradictions that arise as social life is enacted in given fields. So, if identity is seen within an agency|structure framework, it not only changes as structures change, but also is written by self and others.

Karen: Well, Ken, I think that we are saying the same things—though you from an educational sociology vantage point and I from an educational anthropology one. Outside of these two realms there are some scholars who see agency as somehow unbounded, as unstructured if you will (and many among engineering educators se the world in these simplistic ways). I try to make the distinction in my work that I do not see such an agency, but do not want to get caught in a structuralist bind of seeing persons as not having any capacity to act. Thus, I tend to refer to mediated action as a way not to be misread as aligning with this unrealistic notion of agency.

Kathryn's study helps us understand in great detail how a teacher persuades her students into being people who engage science—which is surely a step in the right direction for any person living in a country where scientific decisions play such a central role in the everyday lives of citizens, but without a critique of what such a science might be or how it might work on different students. But, clearly, she is critiquing, in a sly way, business-as-usual teaching practices in science; she illustrates one way to take these girls' views seriously—to make them full participants in their own education. There is always a tension when researchers like Kathryn and I study in arenas where the undertone is about how to get from "here" (the current state of affairs) to "there" (some place closer to an ideal world). Obviously, one cannot put children's education on hold until an "ideal world" comes into being and must connect students to science now; this is the real work of classrooms—too much is at stake if we do not do as good a job as we can right now. That I am intentionally engaging the "ideal world" conversation more centrally than Kathryn indicates that our conscious decisions and our location along the trajectory from here to there differ and that's fine.

Ken: If cogenerative dialogues are used as a research method the ongoing dialogues allow for all participants to change their ontologies as a result of becoming aware of what others see as salient, what they regard as contradictions, and their priorities for making changes. Also, once participants learn to see others' perspectives and respect them there can be a basis for changing the culture enacted in the fields in which participation is occurring. The uses of cogenerative dialogues as a method can catalyze continuous changes in the fields of the cogenerative dialogue and the science classroom.

Kathryn: I agree with Karen's analysis of the conflict feminist researchers' experience as we examine ways to ensure that school science, with all its limitations, becomes accessible to students, particularly those whose identity does not include science, or whose teachers may construct as non-scientists. Concurrently we critique science's ideology and cultural practices. Karen's engagement with the "ideal world" conversation is also necessary to describe a vision of how science could change. But if we care about students' continuing engagement with science, we also need strategies to connect the subject to their lives. How do we teach to ensure that they attain the skills, or tools, needed to succeed in science and thus have a choice to successfully engage with science at the post-secondary level? Thus, the students in Karen's study have attained a level of success in science, which provide Karen a context to engage in the "ideal world" conversation.

Ken: I have a strong sense that if the engineers at PES had engaged in cogenerative dialogues many of the contradictions Karen wrote about here could have been discussed by the stakeholders and possible changes could have been considered and agreed to. The research method employed was a form of participant observation that did not involve participants from each stakeholder group as researchers. Accordingly, Martin, Pam, Pete, Carson, and Samuel did not review the results that Karen reports here. If these folks had come to a cogenerative dialogue with Karen and some of the professors then much of what is reported in the paper would have been discussed and there would have perhaps been a catalyst for significant change.

Karen: Would that it were so, Ken. Would that it were so. The piece that is not presented here, but is available elsewhere, is that when a student or professor began to have dialogues like the one you suggest, it marked them as "not engineer." Thus, the culture at PES had within it recognition routines for reifying the ways things are *supposed* to be. And, your statement that I have not discussed some of these results with these five student engineers is only partially true. Some have in fact seen most of what I write. Plus, as part of the research process, there were ongoing member checks and discussions during interviews, which is the source of comments from participants that "it wouldn't matter." It's a tough call whether cogenerative dialogues would have mattered or not. I wish I could think of an engineering faculty member with enough pedagogical skill to give it a go!

*

Michael: Kathryn, one of the implications your chapter seems to have is a development from flyy girl to scientist, where you do not question the nature of science itself. Why should girls adopt practices and tools from the dominant culture rather than changing the dominant culture to adopt some of the practices of flyy girls?

Kathryn: What aspect of the nature of science should be questioned-that of research science or how science is taught in school? For the girls, Jen is a primary agent in shaping science's cultural reproduction, when Jen begins to question her assumptions and practices and re-shapes those practices, does not this begin a change in the dominant science culture experienced by the girls? Science as a culture is a macrostructure, but the flyy girls' engagement with the culture occurs at the meso- and microlevels. I suggest that at certain times the girls' practices produce a hybrid culture that engages with science while also exhibiting flyy girl practices.

Ken: Maybe we have different ways to think about science from a sociocultural perspective. I see science as a form of social life that is enacted in fields, which can be nested within one another. If a science classroom is regarded as one of the fields in which science culture is enacted then students and their teachers will coparticipate in a culture that is science. Sometimes this may be called science-like or school science. However, it is different than canonical science and reflects the cultural resources of the teacher and students. From this perspective a flyy girl doing science will be somewhat distinctive because the science she does will be built on a foundation of flyy girl culture plus the other cultural resources she has at hand as she participates in her science class. This is what happens when creolized sciences are produced through hybridization of multiple cultural resources to produce distinctive forms of culture. Perhaps what is at odds here is the extent to which Jen allowed her students to overtly use cultural resources associated with ethnicity, life in their homes and neighborhoods, youth culture, and gender. Until Jen got involved in cogenerative dialogues she seemed reluctant to allow the students to use their cultural repertoire and shut down forms of participation she deemed inappropriate. However, that would not preclude the students using multiple cultural resources to support their success in science. To put a stop to this lengthy response, from my perspective science is dynamic and changing—but can obviously change to a much greater extent if participants in cogenerative dialogue learn to recognize creolized forms of science as legitimate and worth being able to enact.

Michael: I am thinking about culture as a *plural singular*, which means, that it consists not only of the practices and artifacts we can observe in more or less the same way over periods of time but also of the possibilities to do something new and different. We would not understand otherwise how poets create *new* words and *new* ways of writing, which nevertheless immediately are intelligible and taken up by others. In this light, I am asking what it might mean to say, as Karen does, that we have to change the culture. How do we change culture if it is thought as collectively available possibilities, some of which individual members realize in concrete ways? Thomas Kuhn suggested that paradigms—ways of doing, thinking, writing, talking, transacting, and so on—do not change. They cease to exist when proponents and practitioners "die out."

Karen: I think, with respect to my research project, it is especially important that readers pay particular attention to the ways in which different scholars use the concept "culture" differently. I do not mean what many intuit when they think of culture as a static set of practices, beliefs, and artifacts, or meaning simply one's roots in a collective affiliation that somehow comes with a student into a learning setting—the sort of "culture" implicated in the term "multicultural education," something akin to an ethnicity or geopolitical affiliation. Instead, I take it in a more generative sense, something being continuously produced by a group acting together and held in shared meanings by members of the "community" being studied. Not all learning settings are in fact producing a culture (and I do not think that K–12 science education *is* producing a scientific culture), but I argue that engineering education at this site is a site of culture production. Here, by culture, I infer that there is something about the engineering campus culture that has historic persistence, and this something is being worked on in practice during everyday life in the community, and it may or may not be malleable, depending on the sort of cultural production process at play in a particular community. Being a member of such a community, a culture of engineering in my case, means that one is thought of as a person who belongs, one is thought to have a persona, for want of a better word, that signals one is a member—and we capture this notion with the term "engineer." That is, one must develop an engineer identity, or become an engineer, but we should not be deluded into thinking that this means that there is some kind of a unitary sort of thing "engineer." What I have been about in the research project in engineering education is trying very hard to place under the research gaze a careful examination of "engineering" and "engineer," and of processes for producing engineering culture and engineers at this site, instead of simply accepting a normative rendition.

When I write that the campus engineering culture has historic persistence, I mean to imply that there are taken-for-granted understandings about the way the world is *supposed* to be, and they implicitly frame—but only frame, not guarantee—how members make sense of events. Thus, I study student engineers in the present, but can see the weight of history seep in via cultural knowledge about the *imagined* normality. And, as insiders go about their everyday activities, one can watch them ensuring the culture's continuation into the future. What makes this an obdurate culture is that those whose vision for the future differs from the past were "read" as not being "real" engineers, not being members of the community, those who had a say in what the culture becomes. Sociologists use "resistance" as a way to think about how change might be nurtured in communities, and this surely works in some arenas. But, as the engineering school studied "resistance," if you want to call it that, was interpreted to mean—in the cultural meaning-making system—not being a "real" engineer. It is emblematic of a dilemma that Doris Lessing once described (in *The Grass is Singing*), saying: "When old settlers say 'One has to understand the country,' what they mean is, 'You have to get used to our ideas . . .' They are saying, in effect, 'Learn our ideas, or otherwise get out; we don't want you.'" Nonetheless, the culture only *seems* to be static, but is in fact being continuously built and made and generated through social interactions among insiders.

At PES, things could have been different. For instance, suppose the campus developed strategies to recognize and reward students like Martin for their engineering prowess, and incorporated other strategies to ferret out and lower the grades of those who misbehaved (such as illustrated by people from the went-too-far terrain like Pete or Carson). It would not be too difficult for students to complete bi-weekly peer evaluations, instead of after-the-fact evaluations as they now do, for instance. This might tilt the playing field away from those who exploit others and might give credence to the design-engineering form of practice. This might provide a field of play that would encourage people—whose transformative views currently set them up for attack—to take a more public stance about what a transformed engineering might be and what a "new" engineer could be. And, I suspect that these versions would challenge, and in time transform, the normative sense. Who knows, such moves might provide the impetus for students like Pete to take their learning of a full range of engineering skills more seriously.

As to Thomas Kuhn, culture is not, to my way of thinking, the same as a paradigm. Nonetheless, he is free to wait until practitioners die out, but my experience suggests that it is much more likely that this will not happen soon enough for me.

Ken: From my perspective, the frustration I experience might relate to what Michael very early on described as passivity. Without being conscious of all aspects of changing times, the ideologies of conservatism were promoted through political power, including views of science that were consistent with positivism. These ideologies included ways of thinking about teaching and learning, purposes of schooling, and even ways to decide what constituted effective research in the social sciences. Those with the political power swept clean the institutions associated with government—that is personnel were changed and policies were prepared to support the ideologies of those in power. Close to home for science and engineering educators, federal institutions in the United States, such as the Department of Education and the National Science Foundation, reset priorities for providing funding for research and professional development in science and engineering education. Hence, if scholars wanted federal support for research, they had to meet the goals and priorities of federal agencies that had been reconstituted to support political ideologies and hegemonies.

As a scholar, I did not change the way I thought about science, science education or research in science education because of these policies—but, it was difficult for me to obtain external support for my research because it did not conform to many aspects of federal policy. I did not buy into many aspects of the National Standards in Science Education, I did not agree with the No-Child-Left-Behind ideology, and I did not ascribe to models of educational practice and accountability that reified teachers having control over students. I did not change to fall into line with the political ideologies that created macrostructures that greatly constrained my agency. Fortunately for me, I did not need so much external funding and I always found places to do my research and maintain my productivity. Furthermore, since I was tenured and senior, I could resist the pressure to obtain federal funding—a pressure that many junior colleagues could not afford to ignore. Hence, massive funds were available to support federally approved forms of scholarship and those that sought and obtained them had the resources to become powerful in many fields in which I operate as a professional. Hence, there was a re-emergence of forms of scholarship that were reminiscent of the 1970s, when I became an active scholar. Yet, those who championed those forms of scholarship had not necessarily changed from post-positivist to positivist—folks who had been waiting in the wings suddenly had structures to support their actions and they stepped forward to produce and reproduce forms of culture that reflected earlier times.

Karen: Of course, you are completely right, the pendulum swing is what I intended to make note of.

Ken: A critical part of this ideology of the conservatives was to value the perspectives of scientists from industry and universities—and to devalue what educators might have to say about science, science education and research in the social sciences. My experience of these scientists is that their epistemologies and ontologies had not changed as much as they were given pulpits from which to preach, creating macrostructures in the form of policies and resources that changed most of the fields in which science educators practice.

Of course I will die. But my students are numerous and sociocultural theory is on the rise. Having said that, positivism saturates the perspectives of many science educators and it too will not die out, but will remain as part of the schema that dialectically relate to the practices enacted by such science educators. Hence, science educators will experience contradictions and resistance because of the plurality of cultures that count as science education. As we enact social life (as science education), what we do can be experienced as patterns that have thin coherence, not thick coherence, as many cultural sociologists like to think. The coexistence of multiple and contradictory forms of culture is for me, part of what it means to be human and to enact social life.

Kathryn: As Ken suggests we science educators negotiate the plurality of cultures, along the culture of school science reflecting Michael's vision of culture as a plural singular. Studies such as mine with Jen can provide one pathway as to how science educators can move from the meso- to macrolevels within science education. Jen's efforts with her students also show how teachers can provide the context for new and different actions. I agree with Karen that often K–12 education does not produce a scientific culture, however, using cogenerative dialogues (i.e., cogens), Jen and her students have moved towards achieving this goal. The cogens are structures that allow teachers and students to examine critically "taken-for-granted" understandings and produce forms of culture that allow students to create scientific identities. And several of Ken's graduate students have conducted research that shows an improvement in science achievement of under-represented high school students whose teachers engaged in cogens.

Ken: I worry about the idea that science culture may not be produced in a science classroom. That assertion seems unlikely. To be clear, from my perspective, many forms of culture will likely be enacted in a science classroom. As students engage with the science as it is laid out by a teacher, in a textbook, or in a lab, I argue that what is enacted is science—creolized in the sense that cultural resources from the students' lifeworlds are used to pursue goals associated with a science curriculum. What is enacted in pursuit of those goals is properly referred to as science.

Karen: You can see that we differ on whether schools produce a science culture. I wonder if we had more time and you could tell me more explicitly what you mean by "culture," if we might get closer to understanding this difference. Nonetheless, I suspect we have captured some of the tension between sociocultural theory (where Michael, Ken, and Kathryn reside) and cultural production theory (where my research resides).

*

Michael: As I was reading the chapters, I was wondering what Karen and Kathryn might think with respect to the absence of feminist critique—right into the writing genres—in science education.

Karen: And I would ask: Wouldn't you count Margaret Eisenhart as a feminist, a cultural feminist who roundly critiques gender-difference arguments? Isn't Patricia Hill Collins a feminist, one who encourages us to think carefully about the limits of white feminist theory, who posits solutions grounded in black women's molding themselves to expectations they had no part in creating? Kathryn and I do build on some feminist scholarship, just not the portion that attends overmuch, to my way of thinking, to the epistemology of scientific knowledge. My reasons are varied.

Most feminists doing empirical research are in fact doing "women's studies," literally studying girls and women. I perform "gender studies," the ways that women and men develop multiple approaches to womanhood and to manhood in specific sites of science practice and how these are made sense of in those sites. I try to understand the cultural production of masculinities and femininities and how these relate to, or map onto, other aspects of engineering practice—engineering expertise, engineer identities, and relations of power. There aren't very many researchers in science or engineering education who do gender studies; so to engage this line of scholarship in a short chapter intended for these readers overwhelms the text. It's a battle I chose not to fight in this piece, but thanks for opening up the space in the metalogue to do so.

In addition, feminist scholars of science often depend on a gender-difference argument and this contributes to getting caught in the trap of a gender dichotomy—men are to competitive as women are to collaborative, men are to self-advancement as women are to self-effacement, men are to individualized effort as women are to teamwork, men are to poor interpersonal communications skills as women are to good, etc.—a unitary masculinity (man) is set against a unitary femininity (woman) and a dichotomy just seems to follow "naturally." Eisenhart's framework chapter in *Women's Science* provides an excellent critique of the inadequacy of gender-difference positions for understanding the multiplicities inherent in forms of being. Also, terms like man/woman, male/female, masculinity/femininity make it hard to get past notions of a dichotomous reality—they tend to become in their own way normalizing technologies that get in the way of understanding the complexity of productions of genders—in the plural on both sides of the sex demarcation, such as I documented at the engineering school. It is important to actually study what manhood's men build (are framed by), and what womanhood's women generate (are seen as), and to see how aspects usually taken to be masculine or feminine can in fact be taken up in someone's masculinity/femininity—and then be given meaning in a certain context. At PES, some men certainly performed forms of masculinity expressing a hegemonic manhood (and women expressed a hegemonic womanhood as sorority woman), but others did not. In fact, Martin performed a masculinity that incorporated behaviors usually thought of as indicative of femininity. These were out of sight in the campus venues where faculty and non-teammates made sense of him as a man. Martin made his counter-hegemonic masculinity private. That is, he purposefully hid these aspects, probably because they would have brought pejorative comments about his manhood, and this action of hiding himself reinforced—instead of disrupted—normality.

Furthermore, I always fear that the appellation "gender" itself serves as a normalizing technology for miscasting gender studies as woman studies, especially among researchers working in science education. I have fought long and hard—and unsuccessfully—against having this section of the book titled "Gendered Identities." It seems an artificial naming of the links between the two studies, a misunderstanding, if you will, between the authors' intentions and the editors' over-arching goals. Aren't all of the participants in the other studies in the volume somehow developing gendered identities as well, in spite of the fact that researchers may not have attended to this issue in their research? Why are we parceled out into a special section?

What we share is a commitment to justice for all persons in educational settings. Instead of feminist studies, what Kathryn and I are up to would more clearly be termed *social justice* studies. We seem less engaged in the realm of philosophical thought associated with epistemology—how we know and what we take to be knowledge, the scholarly space where most feminists of science reside—and more closely working in the political philosophy of education, a subset of ethics. My work is aligned with political philosophers who tend toward "radical" egalitarianism—an explicit call for transforming everyday worlds to make them places where all can learn and study. Here, I count thought that flows from John Rawls *(A Theory of Justice)*, Amy Gutmann *(Democratic Education)*, Charles Taylor *(The Politics of Recognition)*, Iris Marion Young *(Justice and the Politics of Difference)*, and Kenneth Howe *(Understanding Equal Educational Opportunity)*. This line of thought gets at schooling/educating all students and thus encompasses the interests of girls, women, African Americans, and anyone else (e.g., Martin) who is set on the margins by frames of reference that cannot see them—cannot *recognize* their interests or their rights to have a say in the way their worlds are constituted.

What I have tried throughout the PES research to explain is how it came to be that the campus culture framed a particular reform attempt in such a way that it maintained engineering as a masculinized endeavor. Both forms of engineering expertise—both sets of knowledge being deployed in systematic ways on campus—were the presumed terrain of men—albeit men with different forms of masculinity—and women were miscast as people who belonged only when they could be thought of as fitting the tropes for a normatively heterosexual woman among the social over-achievers, or as a prototypic "bad girl." This made evident that a hierarchy among women students was being produced in the campus milieu. There was no middle ground—either you were successful as a woman and couldn't be an engineer, or you were successful as an engineer and couldn't be a woman. This is profoundly unfair. But, let us not forget that I also documented a hierarchy among men on the campus and called for its transformation as well, that I am a crusader for men like Martin who are being treated unfairly, as well. Thus, many would *not* see my research aims, my scholarly underpinnings, and my research goals as "feminist" in nature. In fact, my work has met with quite pointed critique from feminists both among science educators and among women's studies scholars. On the one hand, I critique the norms of scientific power, and on the other hand I give credence to injustice impacting women and men. I cannot speak for Kathryn, but appreciate that black feminist thought and more recently critical race theory both understand the limits of feminist thought to unpack multiple, intersecting oppressions.

Kathryn: Karen highlights several important issues and although I agree with her regarding the differences between women and gender studies, I would not characterize my chapter or research interests as such. I identify as a feminist scholar researching the issues that impact girls' science learning and women's participation in science. Sex and one's gendered identity that evolves from being defined as a female or male are major constructs. There is a need for scholars to engage in gender studies, men's studies, women's studies or African-American studies. At times, these fields intersect and constantly learn from each other. But I focus on the lives of girls and women because there are different issues, (e.g., othermothering) that impact their education experiences, opportunities, and participation. However, research from other scholars has shown that teaching strategies, assessment practices and use of approaches such as cogens are beneficial to all students and thus, my research with girls may also be transformative for boys.

Michael: In his book on identity, Paul Ricœur (1992) defines ethical intention as *"aiming at the 'good life' with and for others, in just institutions"* (p. 172, original emphasis). If, as educators, we aim at the "good" life *with* and *for* others, in just institutions, what more can we do than reproduce the reproduction of inequities in the way our discipline has done for nearly five decades despite differently stated goals? If we want institutions to be just—including *science, education,* and *science education* (as field and topic)—do we not have to critique and change them? How would a truly feminist approach take on this task of critique and change?

Karen: In answer to your query about needing to critique science, education, and science education, of course my answer is "yes" and my research clearly takes this tack. But I am having considerable difficulty getting my mind around a "truly feminist approach" being the needed antidote. It seems to me that so long as anyone who works in science, science education, engineering, or engineering education sees this critique as solely the responsibility of feminists—instead of seeing this as something that *all* researchers concerned with a just society must undertake—we are sunk from the get-go. Such a suggestion strikes me as a particularly troublesome move because it displaces responsibility for the transformation.

Michael: By using the adverb "truly," I didn't intend to say that there is only one feminist approach. I do believe that as any other category, "feminist approach" is heterogeneous and non-self-identical, which means, there are as many different concrete realizations of a feminist approach as there are subjects taking "it." But go on.

Karen: To suggest that there is anything like unity among feminists is a misunderstanding of that terrain. In fact, as Rosemarie Tong's *Feminist Thought* has argued for some time, what feminists share is a desire to place the issues of women center stage, and that is where their shared understanding ends. Beyond that there are as many approaches, or lines of scholarship, as there are scholarly positions generally. Thus, the phrase "truly feminist approach" has little meaning to me and I cannot imagine what one might be or how we could formulate one. In fact, as a cultural feminist (some would say radical feminist), I have quite a hard time seeing how the agenda of liberal feminists—who accept that things are fine when women can become just like men, for instance—can provide for the needed transformations. And, more centrally, understanding the oppressions of women cannot help me with what I want to know about injustice for men, nor can it help—completely—Kathryn's project about teacher-girl student race relations, power relations, and science-knowledge relations. An explicit focus on social justice, instead of solely on girls and women, provides a more robust approach to understanding the complexity of injustice, to my way of thinking.

And, finally reflective practice in the presence of an insightful critic seems a good way to approach substantive change among educators in science. Kathryn's example suggests some approaches to changing classroom practice that could begin to encourage girls who seem all-too-likely to turn away from science to see it as meaningful in their lives. This could in time lead to educators seeing how science positions these girls, something currently left out of sight.

Kathryn: Yes we have a responsibility to critique those institutions and a feminist perspective is one approach. However, other critical theories would also provide insights on aiming for a "good" life. In 1979, Adrienne Rich wrote an essay titled "Towards a Woman Centered University," which offered a blueprint for critiquing the institution. She raises issues of personal safety-namely how can one focus on study if you concurrently need to monitor your surroundings? Then there is the issue of how professors interact with their female students—do they have the same expectations and standards for women and men? Are students provided similar opportunities, regardless of their sex and/or race? If we consider science (or engineering), Karen's study implies that we have yet to reach a "woman centered or just university." Engineering, remains a strongly masculinized field, but education is not and we should engage in the critique and enact changes. Within education, are our students experiencing the "good" life *with* and *for* others? I continually monitor my assumptions, my practices, and my responses to my students. Often when I reflect upon my class, I berate myself for failing to challenge a student, or allowing target students to evolve and dominate class discussions. I have found cogenerative dialogues a powerful tool in monitoring and exposing the inequities in my practice and as a forum for my students and I too reflect upon the importance of our individual and collective responsibilities. As scholars, we can begin to assume our responsibility by self-reflection and also as we garner more status and power, challenge inequities and strive towards working in and from a just institution.

REFERENCES

Eisenhart, M. A., & Finkel, E. (1998). *Women's science: Learning and succeeding from the margins.* Chicago: University of Chicago Press.

Howe, K. R. (1997). *Understanding equal educational opportunity: Social, justice, democracy, and schooling.* New York: Teachers College Press.

Gutmann, A. (1987). *Democratic education.* Princeton, NJ: Princeton University Press.

Lessing, D. M. (1950). *The grass is singing.* New York: Crowell.

Rich, A. (1979). Towards a woman centered university. In A. Rich, *Lies secrets and silences. Selected prose 1966-1978* (pp. 125–155). New York: W.W. Norton.

Ricœur, P. (1992). *Oneself as another.* Chicago: University of Chicago Press.

Rawls, J. (1971). *Theory of justice.* Cambridge, MA: Harvard University Press.

Taylor, C. (1995). *Philosophical arguments.* Cambridge, MA: Harvard University Press.

Tong, R. (1988). *Feminist thought.* Boulder, CO: Westview.

Young, I. M (1990). *Justice and the politics of difference.* Princeton, NJ: Princeton University Press.

Part C

IDENTITY AS DIALECTIC

INTRODUCTION

Most educational research uses either of two approaches. In the psychological approach, whereby cultural phenomena such as participation, knowing, and learning are reduced to individuals generally and to their minds specifically; constructivism is no exception here. In the sociological approach, cultural phenomena such as participation, knowing, and learning are reduced to collectives; social constructivism is no exception here. There is a range of seemingly distinct approaches to cultural phenomena that have in common a dialectical framing of cultural phenomena, which provides them with the resources for either one of two reductionist approaches. Among the dialectical framings we find Anthony Giddens structuration theory, Pierre Bourdieu's practice theory that draws on the habitus|field dialectic, William Sewell's theory of culture that draws on the agency|structure dialectic, and Alexei N. Leont'ev's cultural-historical activity theory that draws on the activity|action|operation dialectic. The three chapters in this part of the book explicitly ground their work in cultural-historical activity theory, which serves them as a resource for theorizing identity and other attendant issues. The non-reductionist approach is clearly available in the way emotional-volitional and ethico-moral dimensions are theorized as integral and constitutive moments of human activity (chapters 8 [Roth], 8 [Hwang/Roth]). Identity, too, not only is a resource for action but also is an important by-product of the outcomes of activity.

Identity can be understood as dialectic in two ways. First, because activity systems constitute the minimal unit of analysis that cannot be reduced to smaller elements—in fact, activity therefore *is* the element—outcomes of activity bear the mark of the mutually constitutive aspects that structure the activity on its inside. Identity, a by-product of activity, therefore also bears the mark of the constitutive aspects, one of which is the division of labor. Therefore, the identities of environmentalists, eighth-grade students and their teachers (chapter 8), of laboratory instructor, student, and researcher in a university physics laboratory (chapter 8), and those of teachers and their elementary students (chapter 10) bear the mark of a division of labor, which is both resource and outcome of the different forms of praxis observed. Thus, whether someone *is* a teacher or learner in a particular event has to be theorized both as resource—e.g., the institutional positions individuals have—and as outcome of transactions: who is teaching and who is learning at any given instant has to be taken as an empirical matter. Only in this instance can we explain in a positive way why some lessons succeed and others do not. It allows us to understand the mediating nature of the unfolding situations such as those in chapter 9, where the student does not succeed in completing the tasks assigned for the day despite the presence of a laboratory instructor.

Identity can be understood as dialectic in a second way, subtended in the dialectic of cultural possibilities and concrete life praxis. Thus, at any one moment, the subjects of an activity do not engage in any form of action but rather in actions that realize in concrete form the activity in which they participate. Thus, for example, the laboratory instructor and physics students in chapter 9 are participants in a laboratory course and, in and through their actions, concretely realize a laboratory course. The fact that laboratory courses generally and different groups and individuals within a course realize the course in different ways shows that an activity does not *determine* the actions that realize it. Rather, an activity provides a horizon of cultural possibilities for actions and therefore comes to be realized in different ways. In the process, there are different ways in which identities are co-produced. More so, when analyzed at a microlevel, it turns out that even the same person enacts a multiplicity of identities in the course of one afternoon in a university physics laboratory. This multiplicity of identity, again, constitutes a dialectic, as "identity" (etymologically derived from idem, the same) refers us to the *same*, and multiplicity to *plurality* and *difference*. Identity, therefore, is a multiplicity, in the sense that each individual at any moment of time concretely realizes an identity, which itself is a cultural possibility and therefore a multiplicity.

The multiplicity of activity includes emotions, which are integral moments of actions. Thus, L. S. Vygotsky realized long ago the omission of affect in traditional psychological approaches to understanding knowing and learning:

> Their [intellect and affect] separation as subjects of study is a major weakness of traditional psychology, since it makes the thought process appear as an autonomous flow of "thoughts thinking themselves," segregated from the fullness of life, from the personal needs and interests, the inclinations and impulses, of the thinker. (Vygotsky, 1986, p. 10)

Without doubt, affects are important aspects of our being in the world and mediate what we know and how we know it. Yet most learning research still does not include affect, making much learning research a legitimate target for the same charges that L. S. Vygotsky had launched against "traditional psychology." Even research conducted within the framework of cultural-historical activity theory, which undergirds the three chapters in this part C, generally does not consider affect, though it plays an integral role in *Activity, Consciousness, and Personality* (Leont'ev, 1978). However, there now appears to be an emerging sense in the research community that emotion and the derivative concepts of motivation and identity ought to be included in the integral analysis of human activities generally, and to mathematical and scientific knowing and learning specifically. Thus, Ken Tobin and a number of his doctoral students draw on the work of Randall Collins and Jonathan H. Turner, both of whom are concerned with establishing a field of the *sociology of emotion*. That is, no longer are emotions considered as mere bodily states accessible to the individual who feels them, but emotions are considered to be central to face-to-face meetings and therefore to the constitution of society. Cultural-historical activity theory, a widely used framework for studying knowing and learning, currently does not have a place for the consideration of emotion and motivation and the relation of these dimensions to identity. Both chapters 8 and 9 address these lacunae by articulating ways in which the role of emotions in activity can be observed, understood, and theorized.

WOLFF-MICHAEL ROTH

8. IDENTITY IN SCIENTIFIC LITERACY

Emotional-Volitional and Ethico-Moral Dimensions

Michelle: We went there last year to Goldstream [Park]–
Jane: I go by there almost every weekend. Twice.
Michelle: when you could actually see [salmon] spawning; and this guy pulled out a salmon and did a dissection on it. It was kind a gross. And then we went a couple of months later when all the salmon had died and they were just little. And you could see eagles. We missed the eagles. But it was smelling really bad. We've got pictures of us all covering our noses. It was like gross.
Jane: You know at Centennial [Park], you know what you could probably do? But there's more fish coming in or something, you could, do something like they have at Goldstream?
Environmentalist: Yea.
Jane: Set up something like that.
Michelle: Yeah.
Environmentalist: That would be neat. Maybe one day.

As part of an interview, the eighth-graders Michelle and Jane and an environmentalist, who had been their chaperon during a field-based environmental unit one year earlier, are talking about a creek where they had seen salmon. Jane proposes that something could be done to increase the resident fish populations in the creek that they had studied for nearly five months and the results of which they had presented during an open-house event organized by an environmentalist group dedicated to the revitalization of the local watershed. Closer inspection of the account shows that it contains not *just* information but in fact gives clues about the emotional-volitional and ethico-moral dimensions of being generally and identity specifically. Thus, the two young women talk about something that could be done about a creek, the sorry state of which they had amply documented as part of their research. Their talk thereby embodies an ethical aim, namely caring for and thereby being stewards of the land that provides habitat not only for fish but the human inhabitants of the valley. Although the smell of decaying salmon is "gross," signaling negative emotional valence, attaining the ethical aim of stewardship has positive emotional valence, for otherwise it would not be worth doing. More so, this brief conversation also provides glimpses of who these female students are, that is, about their identity. They are individuals who go to a specific park about twenty kilometers from their village to observe salmon spawning and who, as told elsewhere in the interview, participated in an environmentalist open-house event where they exhibited the results of their research on the creek. Both during this open house and during the interview one year later, the young women provide ample evidence of their knowledgeable participation in activities, where scientific knowledge and practices came in handy, and quite frequently so. The scientific literacy exhibited while interacting with others and the natural world—elaborated and evidenced later in this chapter—contrasts the reluctance these young women had previously shown to science and science education. That is, although they had done research, which in part had little resemblance with the tasks most seventh-graders do in school science laboratory, the scientific literacy observable in their actions was integral to their identity; and in these actions their identity was characterized by emotional-volitional and ethico-moral dimensions not normally observable in science classrooms.

In schools, science, as other school subjects, is taught separately from everything else, during a particular block of school time and, especially in middle and even more so in high schools, in special places called laboratories. The underlying epistemology is clear. Scientific knowledge can be separated from everything else of everyday life, taught in special rooms at a particular time of the day: its theoretical aspect can be taught in and through lectures, whereas its technical part is taught through step-by-step exercises in specially designed laboratory tasks. Never ever does there appear to be a question of how all of this relates to whole persons asked to learn science, to how these persons experience themselves, their identities. In policy statements and curriculum design, even less thought is given to the emotional-volitional and ethico-moral dimensions that are an integral aspect of all actions, including those related to the deployment of theoretical and practical knowledgeability in science. Scientists have participated in constructing the atomic bomb that victimized hundreds of thousands of Japanese civilians, have developed the birth-defect-producing thalidomide, and continue to develop scores of genetically modified crops that leave their indelible mark on the environment. Yet many scientists appear to be unconcerned with the effects that the outcomes of their work have on specific individuals, who are the victims of scientific and technological "progress," and on society as a whole. All of this shows that many scientists divest themselves of what philosophers have come to discuss under the term *answerability* or *responsibility*, which is an integral element of any act, scientific or otherwise. One of the most flagrant statements concerning the question of scientists divesting themselves of answerability was told to me by a Monsanto scientist—he only ate organic food all the while developing genetically modified, herbicide resistant seed that eventually end up as GMO food on the table of other people.

The move to divest science and scientists from the responsibility for their actions and to others is made easy in an epistemology that (a) considers only theoretical and technical knowledge and (b) considers these forms of knowledge only theoretically. Both moves are part of an abstraction that detaches identity and answerability from the richness of attested life: "I cannot include my actual self and my life (*qua* movement) in the world constituted by the constructions of theoretical consciousness in abstraction from the answerable and individual historical act" (Bakhtin, 1993, p. 8–9). The obtained theoretical world of conceptual knowledge and technical skills thereby comes to exist separate from my unique being and from the ethico-moral sense of acting. As a result, people generally and science learners specifically become indifferent and fundamentally predetermined and determinate beings. That is, the ethico-moral dimensions of agency, answerability, and identity are integral aspects of agency that only exist in our practical engagement with the world. There is another component equally important but often overlooked in school science and in the discussions of scientific literacy: the emotional-volitional dimension of action.

There are two components to the term emotional-volitional. On the one hand, the volitional aspects generally are discussed in Western scholarship in terms of intentions and intentionality said to underlie and determine goals and actions. Yet cultural-historical activity theorists point out that individual goals are only one aspect of ongoing practical activity, that is, praxis. Each action in fact presupposes some societal activity in the aegis of, and for the realization of which, the action has been produced in the first place.[1] We do not just act to achieve some goal but always do something that gets partially realized through that goal; we always act *for the purpose of* and *in order to*. What Michelle and Jane are talking about and how they interact with the environmentalist and the interviewer is a function of the situation: The talk reproduces and produces a societal activity known as "interviewing about." That is, we always act in the course of being engaged in some societal activity that contributes to its maintenance and survival, that is, to the satisfaction of needs generally. Although the two components, the general motive of an activity and the specific goals that realize it get us further in understanding human actions, a third component is necessary: the nonconscious serialized operations that concretely realize the goal into a completed act. This triad—motive, goals, and conditions—constitute the unity of every instant of praxis and therefore are irreducible to one another or to some other instant of activity (A. A. Leont'ev, 1971).

Emotions, the second component of the emotional-volitional dimension of praxis, are integral to each of the three-part volitional component (Roth, 2007). In the long term, we engage in particular activities because they promise to satisfy certain needs, both in terms of the job itself and in the means that we receive in return for our work. In the short term, we set goals to be achieved and to increase the control over our life conditions, that is, our action possibilities. Achieving goals also is associated with increases in emotional valence and we generally avoid those situations that are associated with negative emotional valence and negative outcomes—though short-term costs are acceptable if there are payoffs in the long run. Finally, present emotional states, which are associated with the body and often nonconscious and even unconscious, mediate and provide context for the unfolding of operations.

The purpose of the present chapter is to articulate an empirically grounded framework for theorizing scientific literacy in a way that does not reduce the notion to theoretical and technical knowledge at the expense of other irreducible aspects of human praxis without which we cannot understand why people do what they do. These other aspects are of emotional-volitional and ethico-moral nature. My empirical materials derive from a three-year ethnographic effort in one village in the Pacific Northwest of Canada designed to study learning as individuals from different communities of practice come to interact with one another. These communities included environmentalists, citizens, high school students, First Nations elders, environmental stewards, and scientists.

BECOMING A CITIZEN THROUGH PARTICIPATION IN ENVIRONMENTALISM

The object/motive of our work was to document science both as a resource for action and a terrain to be negotiated, as various members and groups of the community engaged in issues about the embedding watershed and the water available to residents. We wanted to understand what it meant for people to participate in multiple activity systems and how these multiple forms of participation mediated their identities. As part of this effort, I also taught a unit about the environment at a local middle school (grades 6–8 [ages 11–13]) together with several resident science teachers. The curriculum was driven by the idea that science education ought to provide students with resources to develop into responsible and responsive citizens (Roth & Désautels, 2004). What I had not realized at the time of planning the study was the central role that emotional-volitional and ethico-moral aspects of being play in knowing and learning.

Setting the Stage for Learning Science in and for the Community

The science unit on water and the environment began with a series of newspaper articles, one of which in particular caught the attention of the students. Entitled "Group is a bridge over troubled waters," the article has as topic (a) the sorry state of the ocean waters surrounding the peninsula on which the community is located and (b) the efforts by some local environmentalists to revitalize and save the local creeks and watersheds that feed the ocean. The middle school students immediately understand the issues stake, as many of their parents commercially or recreationally (a) fish the ocean waters for salmon, halibut, and rockfish, (b) set traps for crab or prawns, or (c) collect a variety of shellfish including clams, mussels, and oysters. The students already know that some shellfish beds are either no longer operational or so heavily polluted that they can no longer be harvested; those students who know about this articulate their understandings in the whole-class discussion.

In the newspaper article that we read together, the feeder creeks are presented as having been damaged: dredging turned them entirely or partially into ditches, increased water flow rates through ditches and culverts eroded the habitats not only of fish but also of birds and other animals. There is talk about sewage, high coliform counts, and the damage these have done, among others, to animal life in the creek and ocean. An environmentalist is quoted to feel that "the success of the revitalization process would only come through the *involvement of the people who live in the community*" (p. 9, my emphasis). This statement in particular gets the students going. Although they, as most people in the village community, do not know the creek they immediately feel called upon—they exhibit the ability to respond, responsibility, and therefore their ethico-moral character—wanting to clean up the creek, find out more about it, document its pollution, and so forth. More so, the students are excited not only about doing research outside the classroom and

IDENTITY IN SCIENCE LITERACY

Figure 8.1. The creeks feeding the surrounding ocean waters, where commercial and hobby fishery harvest a variety of fish, crab, prawns, and shellfish, not only are denatured—straightening, shade-tree removal—but also subject to (a) industrial—effluents from local industries shown in the background—and (b) agricultural pollution—through runoff from heavily fertilized fields and livestock.

school, which constitutes a break in the often dull routines of their everyday school life but also about being enabled to contribute to their village as a whole.

The teachers, with the help of parents, environmentalists, and graduate students take the class of students every other week for an entire afternoon outdoors to get to know the creek in its various reaches (e.g., Figure 8.1), define interesting research questions and projects, collect data, and document the creek in any way they want to. After this first visit, the teachers engage the students in another whole-class discussion during which it becomes evident that it is precisely the sorry state of the creek as documented in Figure 8.1—straightened, overgrown, and without shade and therefore unsuited as fish habitat—that involves the students in ways that their normal science classes do not. Students are empathetic and concerned. The photographs I use here are just like those Michelle shoots for her exhibit at the subsequent environmentalist open house; these photos are integral elements of the emotional aspect of the creek as the object of research and care. (Some journal editors in science education categorically ask authors to remove all images, and therefore remove those aspects that spoke to and called upon the participants in identifying with their objects both emotional-volitionally and ethico-morally.) A

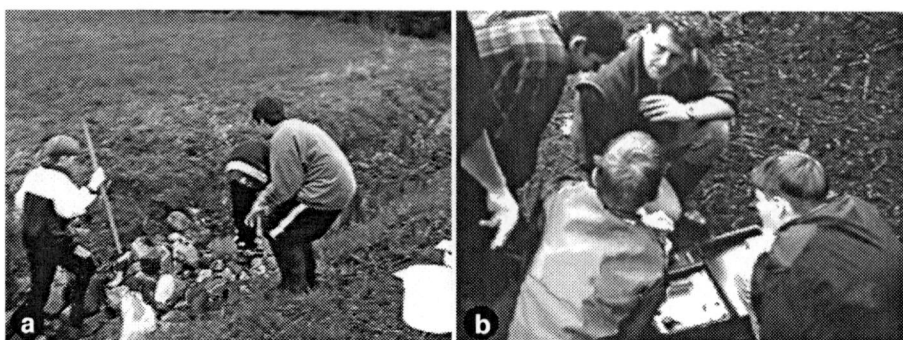

Figure 8.2. Middle school students collect a variety of data in the creek and its contributors. a. A group of students collects microorganisms just below a riffle designed to increase the dissolved oxygen levels of the water. b. Another group of students receives an introduction to scientific instruments (calorimeter, dissolved-oxygen meter), which they subsequently use to monitor the creek.

visit by the leader of the environmentalist group that has been featured in the newspaper article further deepens the commitment that students have toward the environment generally and toward the environmental health of their community and watershed specifically. Their intention is to help clean up the local creeks and to construct a better understanding of the creeks by researching it in both healthier and less healthy reaches, including different aspects of the flora and fauna within and alongside them.

Every time the class leaves the school for the creek, the student groups disperse along several kilometers of its length together with their chaperons (parents, graduate students, scientists, environmentalists, teachers), to collect the data they need to respond to their research questions and foci. For example, one group collects a variety of micro-organisms in different parts of the creek that are distinguished by the rate of flow and the presence or absence of a riffle, that is, a rock structure designed by the environmentalists to increase the oxygenation of the creek (Figure 8.2a). They intend, among others, to correlate stream speed with the distribution of the different types of microorganisms (e.g., arthropods). A biologist, who also works with and for the environmentalists, introduces students to a variety of instruments (Figure 8.2b) also used by others to collect reliable data that all can be assembled into a database and independent of the institutional position of the data-collecting agent—including school and university (college) students, stream stewards, and the environmentalists themselves.

Data and specimens are brought back to the classrooms, where students have more time to interact with others, interpret the quantitative—e.g., physical measures—and qualitative data (e.g., classification of animal and plant specimen). Once the students are working at and in the creek, and seeing the sorry state of some of its sections, they are even more eager to contribute to the common, collective cause in their own ways. The ultimate purpose of the science lesson is to enable willing

Figure 8.3. At the environmentalist open-house event, Michelle and Graeme exhibit the result of their work, produced in response to the call by the environmentalist for community participation in the preservation and revitalization of the local creeks. a. Michelle has done a qualitative study, featuring photographs and classification of environmental quality. b. Graeme (far left) talks to visitors about his quantitative study of faecal coliform in different parts of the creek as a function of land use.

students to contribute to an open-house event organized by an environmental group, whose members also participate in the teaching. In this way, students are provided with resources for engaging with environmentalists in a common cause.

Michelle and Graeme have chosen to document an aspect of the sorry state of the creek. Though working independently of one another—Michelle with three other girls—both closely work with the same environmentalist who also is a trained microbiologist. Although they work with this same facilitator, they realize their participation in environmentalism in very different ways. Michelle and her team mates do not want to engage in the measurement and correlation of physical variables but instead have in mind documenting the creek and its states in a variety of ways, including tape-recorded descriptions and reportages, photographs, audio-recordings of bird songs, leaf pressings, and so on. She does effect some qualitative determinations of the presence of fecal coliform bacteria. Later, during the open-house event and responding to a visitor's questions about the (source of the) differences between healthy creek and ditches, Michelle, standing next to her poster (Figure 8.3a) suggests:

> And the animals, you would find at the creek deer, squirrels, and more animals from the forest. But in the ditch you would find things like bugs and birds more; not the bigger animals, because they cannot live in those habitats. And there is no fish in the ditches. There's like little bugs and no fish. . . there is fish in Centennial [Park], there is cutthroat trout and stickleback. And the creek is cleaner, because it is not beside the road. And people are not dumping garbage into the creek, they dump it into the ditch, out of their cars and as they are walking by. We found much more garbage: we found pop cans, drinking things from McDonalds, French fry cases, things like that.

Graeme has enacted a very different series of projects, having enacted his participation more in the way of scientific "high culture." Thus, after reading that fecal coliform is a major polluting agent, he decides to correlate the incidence of this bacteria in different parts of one creek with his usage of the adjacent land. Speaking while standing next to his poster, which names a number of farms below which the creek features fecal coliform counts above the provincial and national norms, Graeme says, among others:

> I found at the headlands: the at the very headland farm, there was basically no coliform um right below the chicken farm, there is a three-sixty-five coliforms for a hundred mills, that's about a hundred and f– sixty above um agricultural standards.

In the context provided by our structural organization of the science lesson, students engage with a real community issue, and therefore, in evolving forms of identity very different from normal participation in school. For example, Davie, an eleven-year old, has been diagnosed, labeled, and "received treatment" as "learning disabled" within the school; there is evidence from other courses observed that he behaves in ways that were consistent with the label. Yet in the environmental unit, he not only exhibits enthusiasm and knowledgeability with respect to scientific processes and mathematics, but also teaches other teachers in the school how to conduct research, participates in the teaching of other science classes, and is a fervent exhibitor at the open-house event.

The various regular teachers with whom I teach the unit in the school note that even students traditionally not doing well become interested and engaged. I am told stories about this or that student who "is just so keen . . . I mean he wasn't a tremendously good student and yet he was the one who just put his heart and soul into every time we went out there. . . . He just kept the whole group a hundred percent motivated." That is, resident teachers observe their students to be different than in normal classroom settings; the students *are* different, which means, they exhibit different forms of identity. Michelle and Graeme also participate in this school-based environmentalism, mediated by the interactions with others from the community who agree to become involved—parents, citizens, politicians, First Nations elders, scientists, and environmentalists. Although neither Michelle nor Graeme have liked science before, their participation leads to the production and reproduction of enthusiasm. Thus, Graeme's involvement goes beyond the school as he enlists some university scientists to get access to a microbiology laboratory for conducting the tests for bacteria in the creek water samples he collects.

Concretizing Responsibility

The term *responsibility* literally means "ability to respond"; here, it is a response to the call for contributing to a common cause—understanding and doing something about the health of the local environment. The middle school students exhibit such ability, as shown by their immediate responses to come to the rescue of the ocean surrounding their village by contributing to the revitalization of the local creeks. In the course of their work of documenting the creek and its surroundings and collecting quantitative data to support their construction of relationships, the students find out a lot about their community and its history. Part of Central Saanich is located in the Hagan Creek watershed. In Central Saanich and the watershed as a whole, water has been a problem for many years. Despite being located on the West Coast, Central Saanich has a relatively dry climate (about 850 millimeters of precipitation per year) with hot dry summers and moderately wet winters. Concomitant with the climate, recent developments have exacerbated the water problem. Farmers have straightened the local creeks (Figure 8.1) thereby decreasing the amount of water retained in the soil available for filtering into and supplying the aquifer.

At the same time, the farmers draw on the creek and groundwater during the dry summer months, further increasing the pressure on the valuable resource. Other residents in the watershed have individual wells that draw on the aquifers. Their water is biologically and chemically contaminated during the dry period of the year (July to September) so that they drive five kilometers to the next gas stations to get useable water, a situation that, among others, has led to a long and acrimonious public debate. In this debate, science was both tool (for the construction of knowledge required in decision-making) and contested terrain (when people argued about whether some study constituted or did not constitute good science). Urbanization and the related increase in impervious surfaces (pavement), losses of forest cover throughout the watershed and along the stream banks (e.g., Figure 8.1), losses of wetlands and recharge areas (the areas near Figures 8.1a, b were wetlands prior to the arrival of the White settlers), and the loss of natural stream conditions further worsened the water problem.

In addition to finding out about the decreasing amounts of water available, students also find out that the water has been affected by human activity in qualitative ways as well. Storm drains and ditches conduct rainwater into Hagan Creek and its tributaries and away from these newly developed areas—along with the pollutants of suburbia, lawn chemicals and car leakage. The community of Central Saanich previously has introduced an industrial park to the watershed, which is carefully contained within a four-block boundary (at the top end of Figure 8.1a). The drains of its machine shops and biotechnology laboratories empty into "stinky ditch" (Figure 8.1a), which in turn, empties into the main creek of the watershed. To increase its potential to carry away water in a rapid manner, the creek itself has been straightened and deepened in some areas (e.g., Figure 8.1b), and much of the covering vegetation has been removed, thereby increasing erosion and pollution from the surrounding farmers' fields. These physical changes have led to increased erosion and silt load in the wet winter months, and are responsible for low water levels and high water temperatures during the dry summer months when (legal and illegal) pumping for irrigation purposes taxes the creek. These water woes of the community are periodic and repeated features in the local newspapers, community pamphlets, and fliers.

The Hagan Creek~Kennes Watershed Project, an environmental group, arose from the concerns about water quality and was "fishing for community support." The actions of the group include monitoring water quantity and quality or contributing the rewriting of community policies related to Hagan Creek, the watershed, and the quality and quantity of water. The group has created and actively promotes a stewardship program, builds riffle structures in the stream to increase cutthroat trout habitat, builds fences designed to protect the riparian areas, and monitors the number of cutthroat trout in different parts of the creek. Other activities include replanting riparian (riverine) areas for increased shading to result in a lowering of water temperature more suitable for fish. The environmentalists engage in educational activities, which includes giving presentations throughout the community or assisting me teach middle school students as part of their Hagan Creek-related investigations (e.g., Figure 8.2b).

This description shows that the environmental concerns have existed in the community for quite some time. These concerns are publicly available, both through the newspaper articles and the participation of the environmentalists in a variety of activities including town hall meetings, signs that sprout up in the watershed marking the creeks and their fish habitat, through the public talks, and so on. This situation, therefore, constitutes environmentalism as a societal activity—the director of the group receives a salary and therefore makes a living, at least partially. Environmentalism exists in the community, continuously produced and reproduced in and through every action of the environmentalists and others who did something about the sorry state of the watershed. The ethical aspect of the motive of the environmentalist activity transcends collective intentionality; this is so because environmentalism embodies the stewardship principle (for Christians the highest ethical principle) and is concerned with the long-term health of the watershed and the organisms inhabiting it, including human beings. Participating in such an activity, students not only contribute to the community but also the reproduction and production of environmentalism as a legitimate form of societal activity, and thereby change their own forms of participation—i.e., students learn.

EMOTIONAL-VOLITIONAL AND ETHICO-MORAL DIMENSIONS OF AGENCY

All approaches to the emotional-volitional and ethico-moral dimensions of the human life form hinge on *action* as the crucial theoretical notion; ethics and practical action are intimately tied. This, as the following exposition shows, reflects the fact that actions lie at the boundary between the singularity of the individual subject and its constitutive role in the plurality of the human life-form: actions mediate between individual and the collective. In addition, understanding actions and their relation to other moments of praxis is important because all (dialectical) theories of identity take them as the point of departure: we *are* what we *do*.[2] Thus, in theorizing actions we also can get a handle on theorizing ethico-moral dimensions and identity. In the present section, I begin my elaboration of a theory of identity in scientific literacy by establishing the emotional-volitional and ethico-moral dimensions of praxis; I then articulate the relationship between agency and identity in the section that follows.

Actions in Cultural-Historical Activity Theory

In dialectical social theories, agency is the dialectical complement to structure, because there cannot be agency without structure—body, environment, mind—but without agency, it makes little sense to talk about any specific structure. The concept of structure itself is dialectical in the sense that the structural aspects of an agent—animal or human—are the results of interactions with environmental structures, which are resources in and for action, but what is relevant environmental structure depends on the structure of the agent. Thus, already at the beginning of the previous century, biologists noted that a tree trunk constitutes very different structures for an insect, a bird, or a human being because of their different bodies, which leads to very different brain structures as a consequence of their experiences (Uexküll, 1928/1973).

For human beings, too, there are differences in the way individuals perceive particular settings and the things and processes that can be found therein and that constitute them. At the outset of the environmental unit, for example, dissolved oxygen and dissolved-oxygen meter do not exist as resources for Jamie, one of the middle school students engaged in environmentalism, or as object toward which his consciousness is oriented. More so, the creek and its surrounding also are fairly undifferentiated initially; during the unit, they increasingly become articulated as something having a variety of properties that initially did not exist. These properties, such as flow rate, prevalence of microorganisms by species, water quality, and so on come to exist in and through the students' participation in environmentalism. But thinking in terms of agency, which some authors also articulate in terms of *power to act*, is somewhat abstract. In fact, by its very abstract nature—it does not refer to any specific action—the notion of agency may in fact block the recognition and theorization of a particular feature that are my main concern in this chapter: the emotional and ethico-moral dimensions of agency. These are available only in the deployment of concrete action, that is, in the lived experience of praxis (Bakhtin, 1993). But considering actions in and of themselves is fraught with problems, as these are always subordinated to some larger objective—or, to express it in terms of the theory that I develop here, some object/motive (Figure 8.4).

In cultural-historical activity theory, actions are subordinated to cultural-historically developed, societal forms of contributing to the maintenance and continuation of collective life (A. N. Leont'ev, 1978).[3] The students in the environmental unit do not just do something the whole and sole purpose of which is to be graded but in fact contribute to environmentalism, which is a cultural-historical form of collective activity that is of common benefit. Writing exams, because it favors the middle class, serves partial interests; doing environmentalism serves collective interests. More so, many students are in school by force but do not buy into the motive of schooling. Students, their teachers, and everyone else participating in the teaching of the unit also participate in another societal form of activity, schooling, designed to contribute to the maintenance and survival of the species at the collective level. That is, societally mediated activities are organized forms of going about meeting basic needs of the collectivity. Farming, schooling, producing

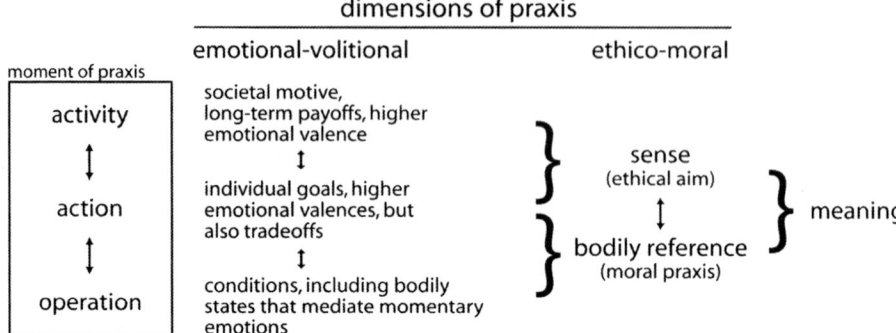

Figure 8.4. Each moment of concrete praxis realizes motive-oriented societal activity, goal-directed actions, and condition-driven operations. The different levels are associated with different aspects of the emotional-volitional and ethico-moral dimensions of knowledgeable praxis. Meaning is not an attribute but a process of sharing the common nature of being.

machines, and doing research are activities that have emerged in the course of our cultural history. On the teachers' part, for participating in schooling and thereby concretely realizing this form of activity, they receive in exchange resources (salary) that they can trade in, as part of subsequent exchange processes, for the basic necessities of their individual lives. That is, by contributing to any one of the existing activity systems, individuals can take greater control over their own life conditions by exchanging whatever they produce (goods, labor) for their personal basic necessities. *Activities* therefore are oriented toward objects/motives that have emerged at the collective level and are geared toward the survival of society.

While recognizing the subordination of action to some collective object/motive, we also need to be aware that they do not exist other than in bodily human performances. *Environmentalism* is but an abstract term that denotes a form of activity that embodies a range of action possibilities. It takes concrete actions such as measuring stream speed, taking photographs of the ditch and healthy creek, recording audiotapes to capture the spoken description of different places along the creek, and constructing and exhibiting displays at the open-house event for environmentalism to concretize the activity. The individual acts—the product of actions, existing in the seriation of nonconscious performatives—make history, as they are irretractable, once-occurrent events and therefore leave their mark on the material totality of this world. This articulation leads to a perspective in which each instant of human praxis is understood in terms of three mutually presupposing levels or orientations to be found in any performance: *activity*, *action*, and *operation* (Figure 8.4). Or rather, there is a hierarchy of societal motives, individual goals, and nonconscious operations that characterize each instant of praxis. The societal motives and individual goals constitute the *volitional* moments of agency; the operations, which are conditioned by the context, are not consciously chosen and therefore constitute a moment of passivity in the face of the power to act that they express.

Praxis: Dialectic Unit of Motives, Goals, and Operations

Activities, or rather, societally mediated motives do not get themselves realized; concrete goals are required for directing individual human subjects to realize the activities in which they participate and of which they are constitutive moments. There are many different actions that possibly realize the same form of activity—environmental monitoring, such as Michelle, Graeme, and Jamie have done, may be achieved through observing birds (Michelle), measuring water quality (Jamie), operating a water monitoring station (water technicians), or doing regular coliform counts and measurements (Graeme). My ethnographic work shows that the similar actions may realize very different activities—operating some water-monitoring station may be used to support claims that the environment is in a bad state or to highlight the existing collaboration between environmentalists and their local community. The water monitoring stations also may be used by farmers to control their water usage or by an engineer asked to evaluate water quantity and quality in a particular part of the village, where residents are not connected to the water grid but have to draw water from their own wells.

A specific action presupposes the activity that it is intended to realize; but the activity presupposes particular actions to be completed. The relationship therefore is dialectic (Figure 8.4). This dialectical relation I denote by the term *sense*. It is not that any action could *have* sense, because in a different activity, the same action would have a different sense. Actions therefore are "free floating" because many of them are possible even within one and the same activity but their sense changes with the activity they realize. The same measurement action, for example, taking dissolved-oxygen readings, realizes very different activities: in the hands of field ecologists, the instrument may lead to the construction of theoretical knowledge whereas in the hands of environmentalists, it serves to generate data in support of new demands for new community bylaws. Goals therefore constitute possibilities that need to be realized in a concrete way by concrete individuals in concrete *praxis*. Here I understand praxis in terms of once-occurrent practical action and thinking that only can be attested but never captured through however careful accounting and interpretation after the fact.

This concrete realization of a goal (plan) in and through human bodies and the tools they employ occurs through the enchainment of nonconscious operations (Figure 8.4). Operations therefore constitute sedimented, embodied, and nonconscious *ethos*: they are formed as a concrete individual interacts with others and, in the process, acquires them through explicit teaching or through unconscious mimesis. Thus, the children in Figure 8.2b are taught in an explicit manner about how to use a colorimeter, where and when to push a button and how to read the display. In the course of employing the instrument for his own purposes, Jamie in particular becomes so knowledgeable that he comes to use it without consciously thinking; that is, what initially have been conscious actions during teaching has "sunk" into the unconscious and therefore have become operations. Without having to reflect, he teaches the use of the instruments to various visitors at the open-house event where he exhibited the results of his research (Figure 8.5).

Figure 8.5. At the open-house event, Jamie shows an adult visitor how to measure the density of suspended particles in water using a colorimeter and how to interpret its display.

Operations, therefore, simultaneously are singular, because of their one-occurrent nature, and plural, because they are concretizations of collective possibilities. That is, when an individual does or says something, it is never merely subjective (singular); the doing and saying have objective quality in that they constitute possibilities for others to do and say. Thus, when Jamie knowledgeably measures water quality in terms of the amount of suspended particles using a colorimeter, inserting a test tube and pushing buttons, his doing is intelligible and he can simultaneously explain what he is using the instrument for. But although his operations come forth without being consciously controlled, they do not occur in a willy-nilly fashion. Rather, they are occasioned by the current context, including the current state in the realization of a goal and the material structure of the instrument, which affords and constrains what it can be used for. And, of course, the dialectical relation to action and activity constrains their nonconscious performance.

From this exposition it is evident that actions and operations presuppose each other: Operations concretely realize and therefore presuppose actions (goals); but actions are free-floating possibilities as long as operations have not realized them. Jamie uses the colorimeter in specific ways *in order to* measure the density of suspended particles in the creek water (goal); but the goal is realized through his adding water to the test tube, zeroing the instrument, entering the test tube into its place, and pushing the appropriate button to take the reading. Jamie does not have to consciously think about his performance, all of these moments of the measurement actions come forth, constituting the embodied form of his knowledgeability. This, of course, has consequences for his identity, which, if the trope "I am what I do" is correct, has nonconscious aspects to it because of the nonconscious nature of operations.

An action, such as measuring the suspended particle concentration, and the operations that realize it—taking zero reading, pushing buttons, filling a test-tube with a water sample, entering a sample into the instrument, taking readings—therefore also stand in a dialectical relationship. This relationship I denote by the term *reference*: the goal is the referent that frames the range of operations called upon. This term, reference, connotes the fact that some actual individual *concretely* realizes what simultaneously is but a possibility; and possibilities are, because of the collective nature of activity, inherently possibilities for others.

This shows that goals (actions) are suspended between the collective motive (activity) that mediate their sense and the concrete operations that realize them in and through a bodily performance. Each goal (action), therefore, has a double significance in the sense that it is directed not only toward the self but also toward the other. It is intelligible not only to the person acting but also to others in the culture who recognize in the action something they could have done themselves: in action others "*recognize* themselves as *mutually recognizing* one another" (Hegel, 1807/1977, p.112 [¶184]). Actions therefore mediate between individual and collective, and therefore, between possible identities and those that are concretely realized.

The relation between sense and reference also is dialectical because of the mediating function goals (actions) have between societal activity and concrete human praxis that concrete individuals realize with their material bodies. This relation I denote by the term *meaning*. Meaning therefore is not something individuals can "make" in a strict sense, as the activities in which they participate have cultural-historical origins embodying collective motives and experiences. Meaning also is not something that is merely determined collectively; to be appropriated by newcomers to a certain culture—newborns or immigrants—because it is grounded in concrete operations. That is, meaning exists in concrete, embodied human praxis, which itself is organized by cultural-historically evolved patterned activities. Rather than receiving meaning—through construction or otherwise—new tools or words *accrue* to meaning when they find their place in concrete praxis, which reproduces and produces collectively recognizable forms of activity. In collective praxis, *meaning* is the sharing of the "with" that founds what is characteristically human in being (Nancy, 2000).

From this perspective, neither the ensemble of observable actions nor the ensemble of artifacts and tools, nor a combination of these, yields culture. Culture exceeds what can be observed at any one instant, for otherwise we are unable to understand cultural change. For culture to change, new and therefore unobserved forms of actions already have to be possibilities. Because each practical (once-occurrent) action not only reproduces existing practices and resources but also and inherently produces them in new form, I am at each instance not only reproducing culture but also producing it in the form of possibilities that have not existed previously.

Emotional-Volitional Dimensions of Action

The previous subsection shows that human beings do not just perform but that there are societal motives and individual goals that orient, mediate, and condition what I do. Jamie does not just push buttons on the colorimeter for no reason at all, but, out in the field, to take concrete measurements of suspended particle density in different parts of the creek (goal), which realizes his participation in and commitment to environmentalism. During the open-house event, too, he pushes buttons

(Figure 8.5) but now with the express purpose of teaching others in the community how the instrument is used. Michelle and Graeme do not just produce words during the open-house event (Figure 8.3), but produce words—most or all of which they do not select consciously—to make a statement, ask a question, give an explanation (goal). Each of their speech actions contributes to realize environmentalism generally and, here, the open-house event specifically and concretely.

In this account, however, there is something missing. Even though goals and motives express the volitional dimensions of acting, they do not express why someone acts in the first place. Even intentions are not sufficient to explain why someone acts unless there is some regulating mechanism that directs or evaluates intentions. Only an emotional (emotive) dimension can explain why we do what we do: we generally act to increase emotional valence, often associated with the satisfaction of (more or less fundamental) needs. This emotional-volitional dimension of action—already can be found in single-celled organisms that orient themselves within some gradient (light, concentration) when it is correlated with a gradient of food—is an integral feature of our individual and collective makeup (Holzkamp, 1983).

The emotional dimension of performance is both driver and context for what we do. It is therefore integral to performance, to every instant of praxis, rather than being an external variable that can be added from the outside and after the fact. By noting its integrality I mean to highlight that it is a constitutive moment without which actual performances (and therefore cognition and consciousness) cannot be understood.

Whereas need satisfaction in animals occurs at the level of the individual, the emergence of society and its culture change the locus of need to the collective level. The paradigmatic example for articulating this shift is that of hunting at the instant of anthropogenesis. The hunting posse divides itself into two groups, hunters and bush-beaters; by doing their part in this division of labor, bush-beaters assure the satisfaction of their food needs by contributing to success of the collective hunt although they do not do the killing themselves. A little later in the division of labor of a hominid group, some individuals no longer participate in the hunt but fashion tools that hunters use to do the killing. Again, by providing hunters with well-fashioned and well-made tools, in exchange of which they receive a part of the kill, toolmakers can meet their basic needs without being involved in hunting altogether. Progressive division of labor leads to the cultural-historical evolution of forms of activity, all of which contributes to the maintenance and reproduction of the collective human life-form. Teaching, growing food for thousands, capturing several tons of fish are not necessary for the individual but are activities that contribute to the survival of the collective. That is, the collective motives embody the satisfaction of collective needs, and therefore, successful activities are associated with positive emotion valence. By participating in one or the other form of activity, individuals guarantee their need satisfaction achieved through a variety of exchange relations that convert labor into food, clothing, a home, and so on. Accordingly, there are emotional dimensions embodied in collective activity, which therefore both transcend and are constitutive of an individual's emotions.

At the level of the individual, there also are conscious emotional dimensions. Thus, human beings generally choose their goals such that what they do leads to a positive emotional valence (Turner, 2002). Sometimes intermediate goals are associated with a lowering of emotional valence—e.g., having to train hard and for an extended amount of time before garnering an award in a sports competition. But this generally is taken to happen when there are longer-term payoffs in sight.

With respect to knowledge, learners generally choose goals so that their action possibilities increase; and this increase in agency and control over one's environment has a higher emotional valence than some current situation. Expansive learning occurs precisely when individuals know that their labor leads to an expanded agency. It is not surprising, then, that the students in the environmental units I teach feel very positive about their experiences, which not only grants them control over their goals and production means for reaching them but also over the assessment of the extent to which they achieve their goal. Thus, Michelle and her peers (including Jane) are not required to do the same measurements others have done or follow the official curriculum objectives:

Jane:	No we wanted to go around and look at different sites instead of just–
Michelle:	Like 'cause there was this one group–
Jane:	instead of just stay at one site like cause they like alternated like after two weeks or something like that, then they went to a different site, [but we got to go–]
Michelle:	[but we like got to] like six different sites instead of just the two–
Jane:	to different sites all the time.

Their account—produced one year later during a conversation with an environmentalist—shows that how they have been able to go about participation has had a positive emotional valence: they had more positive feelings about it than about what they have seen others doing. Thus, "instead of just stay at one site" doing as others have done, collecting samples in one or two places along the creek, they "got to like six different sites." In the end, it has not been surprising that they characterize this science unit generally and their participation particularly by saying, "It was cool" and "It was lots of fun just to do the whole Hagan Creek project, and the whole class had a lot of fun." Not only has it been fun what she has done but also it has been apparent—in her poster, in her talk about what she had learned—that she has tremendously developed her knowledgeability concerning environmental issues generally and those that concern Hagan Creek specifically.

Goals, the volitional dimension of actions, are associated with emotions such that anticipated increases in emotional valence, associated with expansion of control and positive outcomes, are preferred over undesirable outcomes. There is, however, another emotional dimension not directly accessible to consciousness, because it is grounded in the current organic state of the body, levels of hormones, in all its parts, and in the connected psychophysical body systems (Damasio, 1994/ 2000). How I feel, even if I cannot articulate my emotional state, mediates what I do and how I do it. On anyone day, what the students do and what they say, for example, during the open-house event, is mediated by the kind of day they have: a good, bad, or average day. These unconscious and nonconscious emotional states constitute an aspect of the conditions that determine the type and quality of the operations that our bodies bring forth. Much like on the volitional level, where my intentions are given rather than intended, the emotional dimensions of operations are largely inaccessible to consciousness. In this respect, we are passive, hosts and hostages to our intentions and emotions, which mediate the operations that we nonconsciously bring forth. (This passivity is the neglected dialectical sibling of agency; and any truly dialectical theory therefore has to take it into account [e.g., Roth, in press].)

This exposition shows that actions cannot be understood independent of emotional-volitional dimensions. We do not just learn and become scientifically literate for no or some unstated reason—theoretical and practical knowledge in and for themselves, if they are not linked to some emotional valuation that comes with the enhancement of one's power to act, perhaps are not worth being learned. But the emotional dimension not only is integral to the volitional dimension but also in the ethical aims embodied in activities and the moral dimensions of each concrete performance.

Ethico-Moral Dimensions of Actions

So far I have established that we do not just act for the fun of acting—not unless the actions themselves lead to positive emotion valence, such as is the case of games. We do not just set and achieve goals independent of the societal motives particular forms of activities embody. Thus, each goal-directed action not only produces some result but also produces and reproduces a form of activity. That is, what I do has repercussions for the collective: the results of my actions are resources that enable or constrain the actions of others and myself; therefore, my actions contribute to the action possibilities available collectively. In and with the facticity of my doing, I contribute to the constitution of society generally and to all its individuals, including myself, specifically. In doing what they have done, Michelle, Graeme, and Jamie realize environmentalism, which, in and through their action, is both reproduced and produced in a new form. In participating, they contribute to maintaining a form of activity that shapes the landscape, policies, and local conditions in the village community. Their exhibits at the open-house event highlight and therefore make salient aspects of the physical environment that constitutes the concrete habitat for the people of their village. In this, environmentalism embodies a particular concern, which is of an ethical nature. Each act, because it affects the operations of others and mediates their goals, also has a moral dimension.

Ethical aims, such as the principle of (environmental) stewardship, lie at the interface between actions and activity, with which they share a volitional aspect. Michelle, Graeme, and Jamie participate in environmentalism with the aim of contributing to a larger good—understanding pollution and contributing to diminishing it. This ethical aim concerns both the environment and the human beings inhabiting it. This aim is embodied both in the motive that drives environmentalist activity and in the goals individuals frame to realize the activity (Figure 8.4). Each action that realizes environmentalism embodies this ethical principle of stewardship; its sense derives in part from the ethical aims that it seeks to achieve.

Although ethics and morality sometimes are treated indistinguishably, I follow others in reserving "the term ethics for the *aim* of an accomplished life and that of morality for the articulation of this aim in *norms* characterized at once by a claim to universality and by an effect of constraint" (Ricœur, 1990, p. 200, my translation). In this way, morality constitutes a more limited, though legitimate and indispensable *realization* of the ethical aim in concrete praxis; ethics thereby stands in a dialectical relation to morality. Articulated in this way, ethical aims and moral obligations map onto the dialectical framework characterizing each instant of praxis (Figure 8.4).

Moral norms constrain actions in the service of realizing the ethical aims toward the true life in and for just institutions. When Michelle and Graeme report on the pollution of Hagan Creek through fecal coliform bacteria that apparently have their origin in particular farms or farming practices (as Graeme's exhibit showed), they also mark particular farms in an implicitly negative way. Given that environmentalism in the Hagan Creek valley generally and students' own concretizations of environmentalism specifically embodies the steward principle. The term *pollution* not only means dirt and physical or biological impurities but also connotes moral impurity. Being identified as a source of pollution therefore marks the farmers and industries as acting in a morally abject way. The identification itself embodies morality in that there are possible consequences, because the actions have definitive effects within the community as a whole. This is not the least evident from the fact that Graeme has not been allowed back onto a chicken farm, associated with coliform levels that are more than double the national and provincial norms—his mother called it the "political . . . and social reality of doing some science." Michelle and her group mates, too, were accused of trespassing while attempting to gather data from a farm property.

To ascertain the credibility of his claims, Graeme negotiates access to and help from a university laboratory, where, under the auspices of an experienced microbiologist, he conducts the determination of the amount of fecal coliform bacteria in the samples he collected. There is therefore a considerable level of certainty that the measurements would hold up to scrutiny if they were checked by another test. At the same time, adherence to the scientific norm gave credibility and gravity to the claims gave weight to his decision to report the results; but the quality of his result does not abrogate Graeme from his responsibility to users of the facts he reports and to those (implicitly) accused as polluters. In such situations, where the moral norm that prohibits hurting others (here through accusation) and the duty of reporting pollution come into conflict, the ethical aims (here stewardship) serve as mediating devices. In the framework articulated, it is legitimate to seek recourse in the ethical aim whenever contradictory norms lead to impasses in practice. Practical wisdom leads us to invent forms of conduct that best satisfy the exceptions required by solicitude all the while betraying any relevant norms to the least possible extent.

Morality generally and the answerability for one's actions specifically have their origin in concrete praxis rather than in any private cogitation and pondering of action possibilities. Therefore, abstract considerations of patterned actions (practices) and theoretical knowledge inherently do not allow us access to the ethico-moral aspects of human life generally and those of human actions specifically. Practical activity—organized in and mediated by societally, culturally, and historically evolved motives—constitutes the appropriate unit of analysis. This unit cannot be broken further into elements that would serve as building blocks of the unit (Vygotsky, 1986). At best, analysts may identify ways in which the unit exhibits itself in one-sided form in any one identifiable aspect (e.g., subject, tools, norms, division of labor, community of practice). In being concretely realized in and through a material human body, the embodied action (enchained operations) not only materializes (becomes real) general possibilities but also becomes a singular historical act—it occurs only once and cannot ever be taken back. Human beings can argue about its sense and reinterpret it repeatedly, but its historical facticity cannot be undone. It is exactly out of the nature of action as a once-occurrent event that agents are subject to responsibility and answerability, that is, the ethico-moral dimension of action.

SCIENCE AND IDENTITY

This chapter presupposes that the notion of *identity* and the aspects of our everyday experience it denotes are in part grounded in what we do. That is, the material human body and its relative (structural) constancy in time in itself do not constitute sufficient ground for theorizing identity. In this section, I theorize identity before linking identity to the emotional-volitional and ethico-moral dimensions of action outlined in the previous section.

Over the past decade and a half, the concept of *identity* especially in its relation to the notion of *community of practice* has gained purchasing power in theoretical approaches to knowing and learning (Holland, Lachicotte, Skinner, & Cain, 1998). Although the notion is often approached as if it were unproblematic—for example, it is equated with the life history narrative of a person—there is more to it as gauged from the fact that philosophers, after elaborating attendant issues for an entire book, do not come to definite conclusions but are left with numerous aporias (e.g., Ricœur, 1990). The self continues to be a riddle. In this section, I articulate a dialectical approach to identity, which, because it is framed dialectically, continues to be a notion based on inner contradictions. Let us return to Michelle and her participation in the environmentalist open-house event. Her participation in *this* event constitutes her as an individual who does *this* rather than some other thing and therefore constitutes an aspect of who she is. Because communication not only articulates topics but also constitutes forms of actions, studying conversations enables us to understand who someone is. We return to the open-house event.

In this episode, a visitor to the open-house event asked Michelle to talk about her findings. Michelle, standing next to her exhibit, compares the healthy sections

IDENTITY IN SCIENCE LITERACY

of the creek with some of the parts that had been straightened and dredged and thereby rendered into ditches. As she talks, she points to some of the pictures.

Michelle: And then there is another ditch ((points to a photograph similar to the ones in Figure 8.1)) and this one over here ((points to another photograph)) but it some more of Centennial and over there ((points)) and then the ditches have a lot more culverts, which are dirty, and they are long and dark in the ditch
Visitor: Uh um
Michelle: but if they do have them in the creek they are usually shorter and not so dirty because there is still much water running through them. And we found out a lot of differences.
Visitor: Yea, I was going to ask you what you were finding there.
Michelle: Yea, we found so many differences that– I could not even put them all on my poster because you can't really state them all. Um there is animal differences, plant differences, everything there really and–
Visitor: And you know why? You found these differences in the–
Michelle: Yea, some of them is the water itself and the habitat around like for um for example, the water movement is rapid in the Centennial thing . . .

In this conversation, Michelle and the visitor talk about the specific findings of her research. The visitor's request reifies Michelle as the producer of the exhibit, which itself is an expression of Michelle. In this exhibit, Michelle has objectified herself. Her actions in the creek have taken the form of objects: They are objectified in the material records, photographs, and text she produced. In the conversation, Michelle is asked to account for what she has done, actions that have led to the ultimate production of the exhibit. By accounting for what she has done and found, agency is attributed to Michelle, who thereby also articulates an aspect of her Self. These are her actions and not those of another person that have produced the photographs; she accounts for the differences between the different parts of the creek, and in this, exhibits her accountability or rather answerability for what she has done. Michelle turns out to be what that person has done that can self-referentially point to herself saying "I" and "Michelle." This framing, however, does not get us out of the aporias, as actions partially are free floating, belonging, as Figure 8.4 shows, to the intersection of individual and collective. There is therefore the question about how actions come to be attributed to particular agents, especially given the fact that praxis in its entirety constitutes a unit and the individual subject is but a one-sided expression of it.

In the present episode, more is happening than the collective construction of an account of some research results that Michelle, Jane, and two other girls have produced. Michelle and the visitor are currently talking at an open-house event. In fact, the event is an environmentalist open-house event because of the particular kinds of exhibits that are found there and the kinds of interactions and conversation that occur. That is, the open house is not a thing or box into which participants are placed; rather, it is an event, and as event, it is constituted by the transactions of participants over and about exhibits.[4] These transactions constitute Michelle as an exhibiting participant, therefore an environmentalist, and the other person as a visi-

173

tor, interested (as indicated by her questions) in the findings of the various exhibitors. But the visitor and Michelle act and talk in the way they do because they know themselves to be at an environmentalist open-house event. That is, the visitor and Michelle participate *in* the event, which only exists because of their (and others') participation. That is, together, Michelle and the visitor constitute one of the many moments that make this an environmentalist open-house event. They draw on their different institutional position as resources to reproduce and produce the event as that which it is—Michelle *is* a member of the environmentalist movement (as seen from the objective fact that she exhibits), the visitor *is* a person interested in environmentalism (as seen from the objective fact of her attendance). In doing what they do the visitor and Michelle produce differential identities. Let me focus here on Michelle, though the argument is reciprocally valid for the visitor.

It is often assumed that our institutional positions are the causes for our actions. But this approach is problematic because it accounts for identity only when things work out in the way they are intended. Thus, in the present situation, Michelle is reified as an exhibitor, an expert on the question of the health of Hagan Creek, whose expertise is available in part through the inscriptions she has mustered in her exhibit (Figure 8.3a). That is, in such approaches, the fact that Michelle has studied the creek, produced a poster, and now exhibits it and accounts for the work that has produced it may be taken as cause for who she is in the conversation. But such an explanation only works because the transaction reifies her special expertise and the attribution of the exhibit to Michelle's actions. We can easily imagine conversations that would not reify Michelle as an expert concerning the creek and its environmental health. We all know science educators who treat graduate students as if they were children, philosophically unsophisticated, and unworthy of any consideration. We also know of situations where the presence of a teacher in one class leads to tremendous learning and exemplary science classes and the presence of the same teacher in another class is associated with a "bad lesson." In each case, despite evidence that a graduate student or teacher may be associated with considerable expertise in one situation, he or she comes out of other situations as less than expert. Thus, it is better to think of Michelle's expertise and identity as an expert with respect to Hagan Creek as being produced in this episode as much as it is being reproduced. Her previous experiences, including data collection, production of the poster, and the traces these actions have left in her body–mind serve as resources in the present situation. So does the institutional position of being an exhibitor and the expertise that might be attributed to it.

In this perspective, therefore, identity—the question of who someone is—is the consequence of a transaction then and there, in the thickets of the open-house event. This aspect is fleeting in the sense that Michelle, after the open-house event, returns home where she may be reified as a teenager and child. Her identity during transactions with the parents is different than that during the transactions in the open-house event, which are different again from that with her peers, on the hockey squad, or in the school. This fleeting aspect has led some to theorize identity in terms of a fractured Self (cf. Giddens, 1991).

The idea of a fractured Self, produced and reproduced differently as a function of the activities in which we participate—environmentalism, schooling, family—contradicts our everyday experience. Despite all the changes we have undergone, psychologically, socially, or biographically, we can point to a picture and say, "This is me when I was five years old." That is, in our everyday world we experience and presuppose some stable, core Self that survives all the different embodiments our Selves take as we move from activity to activity. This stability is necessary for making societal life into what it is, for if we cannot presuppose a stable referent "I" in statements such as "I promise" or "I swear," then promising and talking under oath has no meaning whatsoever. If the "I" were unstable, as the notion of a fractured Self suggests, then there would be nothing or nobody who could be held accountable and answerable for having promised or said something under oath. The stable moments of everyday praxis on which the constancy of identity is build can be discerned in the following episode.

In the following conversation one year after their science experience and the open house, Michelle and Jane are talking to a graduate student interviewing them about their engagement in environmentalism in the presence of the same environmentalist, who had been their and Graeme's chaperon during the science unit.

Interviewer: What kind of samples did you look at?
Jane: Um. Water and if we found any bugs or anything.
Michelle: And like the dirt samples and sand and figured out how moist it was, and the plants around it . . . and the tree fell on us– just about ((laughs)).
Environmentalist: Yea ((laughs)), just about. I had forgotten about that . . .
Michelle: ((Laughing)) You remember that . . . it scared me.
Environmentalist: I was– What's going to happen here? ((Laughs.)) No . . .
Michelle: I was hiding behind big, a big tree. If it falls on me, it's falling on the tree first.

In this episode, the participants reconstruct an event that had occurred while they investigated the creek and the surrounding land. Michelle begins the account of the falling tree, and subsequently retells how she was hiding behind some other tree so that she would not be hit. She talks about being scared, while at the same time laughing while recalling the event. The environmentalist contributes and thereby elaborates the event. He has been a witness and in this capacity, reifies and stabilizes the account. Later on during the same interview, Michelle returns to the episode: After they had walked away from this dangerous instant, they attempted to jump over a barbed wire, which toppled so that she "just about fell into a pothole." Here, "Michelle" *is* the result of a collective account, someone who in the middle of her (serious) investigation of the creek is imperiled by a falling tree. It provides a context to the scientific results that she has reported, something that to her and the environmentalist is an integral aspect of their experience. They have done this science in and despite of the dangers that they have been exposed to. The episode is quite singular, yet defining for Michelle and what she has done, how she has done it, and, therefore, a defining instant in her auto/biography (e.g., Roth, 2005).

This episode shows that auto/biography, as history generally, requires memory, here individual and collective. Without memory, the story about the falling tree and the roles of the different participants in the event would not exist. Nor would identity exist, because it requires accounts of a person in situation and what he or she has done. Auto/biographies are narrative forms that tell who someone is in terms of strings of events that together constitute a singular life—Michelle in the episode with the falling tree, making the decision to research the creek in six rather than in one or two locations, or presenting at the open-house event. The moment that holds all these different instantiations of Michelle and different situations together is the narrative anchored in the physical body that is constant across the situations—in contrast to the fleeting nature of Michelle as subject and Self. Not really her body is the anchor, for it, too, is changing, physically, biochemically, or hormonally. Yet there is a structural constancy that allows others to recognize Michelle in the female student during the lessons, in the environmentalist during the open-house event occurring several months later, and in the person participating in the interview one year later. Despite all the changes Michelle has undergone in the course of the sixteen months from the start to the end of her participation in my research project, including the psychological, sociological, and bodily changes, there is something that allows us to speak of the same Michelle. This constancy is achieved in part through the relative structural constancy of the body to which the experiences are ascribed and that is the locus of memory. Constancy also is achieved by means of the narrative form, which moves a constant character through the plot and thereby grounds sameness and the concept of identity.

There is a double dimension underlying narratives that allow them to take their central role in the constitution of identity. First, narratives are told in terms of characters and plots. In the account of the falling tree, Michelle and her chaperon, the environmentalist, have their different parts, play different characters. The account is part of the plot, Michelle, Jane, and the environmentalist reconstructing aspects of the environmental unit and the subsequent open-house event. Here, Michelle is the person of interest, who is threatened by a falling tree. She says that she was scared, and asks the environmentalist whether he remembers her being scared. Later on during the same interview, she comes back to recount aspects of the event again, increasing its salience: "I said, I said "you know a tree is going to fall down." It was like five minutes later CRCRAASHSH!! I like jumped out of my pants. I was like so scared and I hid behind a tree." In addition to being scared and nearly having "jumped out of [her] pants," Michelle now also is a person who has nearly predicted the possibility of a tree falling. That is, part of her (literary and literal) character is the dimension of being and feeling scared; other persons might have provided very different accounts, and therefore articulated very different identities—for example, that of a person who is not scared despite the obvious danger. Here, the character developed is that of a scared person who nevertheless knows what to do, namely hide behind another tree or trees, which would prevent the falling tree from hitting Michelle.

Second, because this is a narrative with a clear plot and a character, whose properties—e.g., being scared in such situations—are developed in the process, it is intelligible. That is, in its very singularity, the event, involving a particular character in a particular plot, transcends itself in becoming a kind of person in a kind of event. The narrative, in its very form, exposes a possibility of being that is available to others as well. It is plural although tied to the singularity of the event experienced together by *these* individuals in *this* situation. More so, because of the plural nature of the character–plot configuration—plural because it constitutes a narrative possibility for the entire collective (community, culture)—and because of the inherent intelligibility of language as such, the auto/biographical account constructed during the interview is intelligible to others, that is, as a general possibility of being. The collective nature of the conversation points us to the simultaneous autobiographical and biographical nature of the account and to nature of all auto/biographical narratives: they are singular plural as they place *particular* individuals and situations in the slots provided by *general* characters and plots.

The memory of specific measurement actions—the "real" science and scientific literacy as some science educators and scientists might say—intimately is tied to the situations in their entirety, thereby giving body to what would have been a decontextualized account. It also gives shape to the "who" of the actions and, therefore, to the identity of the participants. In the following episode, Jane and Michelle together with the environmentalist reconstruct one of their fieldwork studies.

Jane: Well we did, um, one measuring thing at like Centennial or something like that
Michelle: Oh, yea.
Jane: where we like
Michelle: measured the depth of the water.
Jane: measured it from a like a tree or something like that. I can't remember.
Environmentalist: What was that about?
Michelle: The moisture and also the depth of the water.
Environmentalist: Oh yea, yea.
Jane: Oh yea, and we were seeing
Michelle: I remember the depth part.
Jane: how much plant life or whatever was like in a certain amount of area.
Environmentalist: Yea, I remember that. It's all coming back to me now.
Interviewer: Yea. You guys have a good memory for this. It's last year that you did that, right?
Michelle: Well, I remember that part though because I just about fell in . . .
Jane: Laura fell in.
Michelle: Yea, Laura fell in and that was bad.
Jane: She was funny.

It is immediately evident that the account is constructed collectively, as different individuals contribute to constituting just what has happened and the roles individuals have played. More so, as Michelle articulates in response to the interviewer's question, she remembers having measured moisture levels (near the creek) and the depth of the water "because [she] just about fell in." Again, it is part of the plot, which constitutes her as a character, that she "just about fell in." Falling into the creek as part of conducting their scientific research is a constitutive aspect of her identity: she has done it despite the apparent dangers involved. The fact that it is a general possibility of constituting the identity of a person is evident at the very end of this excerpt, in which Jane tells the interviewer that another student actually has fallen into the creek—a situation Michelle characterized as "bad," whereas Jane described Laura to have been "funny."

Such experiences, which here are narratively reconstructed, actually happen to someone in flesh and body, during a once-occurrent event. This aspect of life as constituted by events that occur only once, events that in their original form are lived and attested to, allows us to anchor narratives with general characters and plots to concrete individuals. The concrete human body, its flesh, mediates between the collective and the individual, *ethos* and *pathos*, and collective action possibilities and concrete operations. The endangered flesh during the once-occurrent event, because it is structurally perpetuated, becomes the anchoring point of auto/biographical narratives. These therefore simultaneously are singular, pertaining to *this* individual (Michelle, Jane, Laura), and plural, pertaining to human beings in general. Because they constitute general possibilities, characters, plots, events, and identities thereby are intelligible to others, who may well have been or will find themselves to be in a similar position—Jane fell in, Michelle just about fell in.

EMOTIONAL-VOLITIONAL AND ETHICO-MORAL DIMENSIONS OF IDENTITY

In the previous sections I show (a) how emotional-volitional and ethico-moral dimensions are constitutive moments of actions and (b) how actions centrally figure in the constitution of identity. That is, identity inherently has emotional-volitional and ethico-moral character, because of its anchorage in actions and agency. Even without being directly addressed as topics of conversation, emotional volitional and ethico-moral dimensions of agency appear in accounts of what people have done, and therefore, in the constitution of their character in life history plots recounted in auto/biographical form. Throughout the interviews held one year later, the participants in the environmental unit talk about how much fun they have had in direct association with accounts of what they had learned. Similarly, the volitional and ethico-moral dimensions of actions are all over the retrospective accounts of what students have done and how they have done it.

In the following excerpt, Michelle and Jane respond to the interviewer's question concerning their learning. Contrasting her account of not knowing anything about the creek, its presence, the ocean, pollution, and so on, Michelle responds extensively, assisted by Jane and confirmed by the environmentalist, articulating the sorry state in which she has found the creek to be in.

Interviewer: So now what do you think is going to be like that information that you learned, what do you think people need to know?
Michelle: How dirty the creeks are, and how much they– 'cause they run into the ocean and stuff, and all that dirt is going to get swept into the ocean and it's on people's property and it's nice to go for a walk along the creek and to see it all filmy, it's gross, and how dirty it is. And people need to know how dirty it is and how really it affects all of us at some time.
Jane: It was kind a hard to fix that though.
Environmentalist: Yea.

Michelle: It's kind a like, well you can't live here though. It's like you've been living here for fifteen years. Okay, now you've got to pack up and move. Bye-bye.

Michelle not merely articulates the dirt in and film on the water, but also uses a term that expresses the emotional aspect of pollution: "It's gross." The term clearly has negative emotional valence, pointing to the abject quality and even moral indecency of those who cause it. As a consequence of pollution, Michelle cites the possibility to have to move, leaving a place where one has taken root. Being uprooted is not a value-free determination of an event, but laden with all the connotation of having to—as contrasted to wanting to—cut one's ties to a familiar place. Pollution, the result of the actions of others, "affects all of us at some time." Being affected negatively also is laden with negative emotional valence. Their own power to act is limited in this case, as Jane points out, with specifying the particular actions she might have been able to enact.

The ethico-moral dimensions also are available in this excerpt, as afflicting others (through pollution) occurs against the ethical principle of stewardship. This principle requires us to appreciate and guard two gifts to humanity: the environment with all its natural resources and our own human nature. Affecting others negatively through environmental pollution is morally abject and inconsistent with the principle of environmental justice, according to which the costs and impact of pollutive actions and activity cannot be redistributed to and covered by others—the patients of the agency of others. The (ethical) stewardship principle has led Michelle and Jane to articulate their finding—how dirty the creeks are—at the environmentalist open-house event; their actions that have realized this aim were therefore morally justifiable.

Being in the position of contributing to realizing the ethical aims of stewardship for the Hagan Creek watershed also requires consideration of particular norms. Thus, students have found themselves in situations where they have had to weigh the need to collect data along the creeks to support their claim of the varied nature of pollution against the norm (embodied in law) of trespassing and therefore of asking whether they could access the creek by crossing farm property. This issue of the need to access a creek at many places to construct reliable knowledge and the constraint posed by restricted access to private property is salient in the following episode. It begins with Jane's explication of why their results are limited: they have not been allowed to sample Hagan Creek at its lower reaches, which would have required them to cross either private property or an aboriginal reservation.

Jane: Well, we never actually got to really see where the creek ends up because we weren't allowed to go down there. ((General laughter.))
Environmentalist: No that's private property.
Jane: Well, we did go to that one lady's property.
Environmentalist: Which one?
Jane: Um, it was . . .
Michelle: Oh, yea.
Jane: It was like near the first site, I think, err–
Michelle: And we just kind a wandered on, oops.

Environmentalist: Yea.
Jane: Well she knew we were there— Oh that was with [teacher] 'cause she knew we were there. She came out. She thought we were like—
Michelle: She was like, what are you doing there? We were like, uh, you gave us permission. Oh yea, go ahead.

The environmentalist provides the reason why Jane, Michelle, and others have not been allowed to go to the mouth of the creek: it means crossing private property. But Michelle relativizes this explanation by reminding him that they have entered private property at least in one instance. Here, the need to sample for realizing the ethical aim overrode the norm related to the right of a landowner to limit access to her property. This account does not allow us to recover exactly why the landowner finally allows students and their chaperon to sample from her property, but there may have been some earlier communication with the teacher. In Michelle's account, there is evidence that some permission has been given, which, from her position, has legitimized their presence on the property.

The emotional-volitional and ethico-moral dimensions of identity always are distributed across all aspects of praxis, and therefore, all aspects of events that only occur once. In the students' narrative accounts, acting all the while running potential risk—and therefore, being characterized as risk takers—are present everywhere. Thus, Michelle talks about wanting to sample in culverts and meeting the resistance on the part of the environmentalist chaperon, who finally accedes and gives the girls permission to sample in these places. (The environmentalist *is* the chaperon *precisely* because he says that the young women are not allowed to do this or that.) One girl in the group has not wanted to enter the culverts, leading Michelle to characterize her as "too much of a chicken." Here, the refusal to enter and not go into the culvert becomes evidence for a trait of character. As exhibited also in the following account, the chaperon's (initial) interdiction itself constitutes an enactment of morality, as he was responsible and accountable for what happens. Although he adheres to the stewardship principle and to the curriculum generally, he also has to act such that what he will have done is according to the relevant norms. In the following, Michelle and the environmentalist articulate the legal dimensions of being on a fieldtrip during school time and the norms that such events are subject to.

Jane: We went on the other side of the bridge than the other people.
Michelle: Yea, um, but we, he wouldn't let us walk across the logs and stuff like that, for insurance.
Environmentalist: Yeah, well we don't like people falling. Well you know why, you know these days you're not even allowed to take people in these creeks. You need a certified lifeguard. That's a new rule that they had, like–
Michelle: Well that just sucks.
Environmentalist: I know. You guys wouldn't have been able to do what you did last year, this year, unless I was a certified lifeguard.
Michelle: Really?
Jane: Oh well, we all wore lifejackets.

Environmentalist: Yea, that's right.
Michelle: Oh like, help me I'm in ankle-deep water and I'm drowning.

Jane initiates the account by saying that they crossed a bridge to work on the side of the creek where others did not do their research. Michelle chimes in to explain that they have taken the bridge because "he," the environmentalist, has not allowed them to cross the creek via some fallen logs. Again, he articulates the reigning norms that constrain what they collectively can do: a lifeguard certificate is required before students can go through or over a creek. Michelle in turn articulates in an emotionally laden way the arrival of the new rules: The expression "it sucks" clearly expresses negative emotional valence. In fact, the environmentalist says that the girls *actually* have acted in ways that during the present year he would no longer have been able to permit them. The girls then relativize the norms with respect to the actual situation in which they have found themselves: First, they have worn lifejackets—a fact that the environmentalist confirms—and second, the creek only has been ankle deep, so that there has not been an imminent danger in the first place that would have mediated their action. Their actions of crossing the creek to realize the ethical aim of stewardship through the construction of knowledge would still have been defensible on moral (normative) grounds, as the context has been such that the danger the norms are intended to limit did not exist. Here, the girls have acted in morally defensible ways, acting according to the interdictions that their chaperon has stated, all the while recognizing that the current condition (ankle-deep water, lifejacket) would not have made immoral crossing the creek via the fallen logs despite the interdiction.

CONSEQUENCES FOR A THEORY AND PRAXIS OF SCIENTIFIC LITERACY

Thinking about scientific knowledge, skills, and literacy inherently involves a "fundamental split between the content or sense of a given act/activity and the historical actuality of its being" (Bakhtin, 1993, p. 2). Because the emotional-volitional and ethico-moral dimensions of action are constitutive of concrete practical action, theoretical statements of scientific knowledge, skills, and literacy inherently omit these dimensions. These have to be brought in from the outside in the form of rational (legal) norms said to guide behavior.

Theoretical and practical knowledge as stated on paper on the one hand and practical wisdom exhibited in once-occurring everyday praxis constitute "two worlds that have absolutely no communication with each other and are mutually impervious" (p. 2). School science, which focuses on the inculcation of concepts, theories, and skills articulated in policy documents, therefore inherently is blind to all those dimensions that are constitutive of everyday out-of-school knowledgeability, both theoretical (explicit knowing that) and practical (tacit knowing how). What science students learn in schools therefore is inherently useless—if I may express it in such a hyperbole—because it lacks the emotional-volitional and ethico-moral dimensions that characterize everyday knowledge. As has been pointed out, most students engage in defensive learning, they do and learn to avoid punishment, rather than in expansive learning, that is, learning for the sake of expanding their room to maneuver in the everyday world outside schools (Holzkamp, 1993). For this reason few students in regular classroom science evolve science-related identities, because their actions do not have emotional-volitional and ethico-moral dimensions other than avoiding punishment (low grades, suspension, expulsion) and expanding their action possibilities through increased symbolic capital (grades). Emotional-volitional and ethico-moral dimensions pervade and are constitutive of everyday activity, and therefore, everything we do—and this pervasiveness is evident throughout this chapter. As a consequence of the particular experiences that the students featured in this chapter had was the constitution of identities through the particular actions they had taken. Identity therefore has to be *in scientific literacy* rather than something that can be added to it from the outside.

The upshot of this is that few students develop science-related dimensions of identity unless they participate in forms of activity that embody emotional-volitional and ethico-moral dimensions. Unless students intend to do something for the greater and *common* good, they do not enact the volitional and ethical dimensions characteristic of everyday life; and unless they have the opportunity to evaluate action alternatives in terms of the differential emotional valence associated with outcomes and constrained by (moral) norms, they are not in a position to evolve identities characterized by morality. Practical wisdom[5], a central dimension of identity (Ricœur, 1990), is tied to concrete praxis in the same way that the emotional-volitional and ethico-moral dimensions are. Practical wisdom consists in inventing just behavior suited to the case: "the man of wise judgment determines at the same time the rule and the case, by grasping the situation in its full singularity" (p. 206, my translation). Unless students participate in concrete, once-occurrent practical life outside schools, where each action has ethico-moral consequences for others and for themselves, they cannot develop the form of judicious application of scientific concepts, theories, and literacy that the situation requires. It does not surprise me then that scientists disavow of the ethical nature of their work, because they attempt to divest themselves of having to answer for their actions, which they have held to be outside of all ethics and morality. How else can we understand why scientists develop atomic bombs, thalidomide, and genetic monster plants and animals? How else can we explain the behavior of scientists, who work for companies that lay proprietary claims to the genes and seeds of naturally occurring species or produce seeds that yield sterile grain so that third-world farmers cannot replant parts of their harvests?

My personal hunch is that students, who are provided with opportunities to learn in situations similar to the one presented here, where their work contributes to a greater good and therefore realizes ethical aims, develop scientifically literate identities very different from those that we observe today. More so, as shown throughout this chapter, there are emotional dimensions integral to the practical wisdom exhibited in, and developed by means of, the participation in environmental activity. These students do not have to be motivated, as the motive itself embodies ethical aims and as setting and achieving goals is related to higher emotional valence. In everyday educational praxis, I would expect such an approach to lead to individuals who, if they do not pursue scientific careers, will appreciate science much more so than past and current generations of students and citizens. Inherently, such an approach to educational praxis means that we have to radically rethink the activity system(s) in which we, educators and educational researchers, are part of. At the instant, there are no provisions—practical or theoretical—other than those articulated here to understand scientific literacy in terms of the actions people take and the emotional-volitional and ethico-moral dimensions that are *constitutive* rather than additive features.

NOTES

The preparation of this article was made possible in part by grants from the Canadian Natural Sciences and Engineering Council and Social Sciences and Humanities Council.

1. I explicitly use the adjective *societal* ("gesellschaftlich"), which is the one traditionally used in the German and Russian texts that draw on cultural-historical activity theory. For what I believe to be political reasons, English translations historically use the adjective *social*, although Marx uses its German equivalent "sozial" in very different ways than he uses the equivalent of societal. The latter term clearly is political, as it thematizes the mediating role of society rather than simply of the social situation.
2. In this chapter, I use the term "moment" in the way it is used in dialectical theories to denote any aspect of an irreducible unit. Thus, the subject and object are moments of an activity system, which constitutes the smallest unit in cultural-historical activity theory. I use the term "instant" whenever I refer to time.
3. In the languages of the cultures where activity theory has been developed, *Activity* translates the special German and Russian terms *Tätigkeit* and dejatel'nost' (activity), which contrast mere *Aktivität* (activity), which denotes any form of doing something. Activity as used here therefore has the special sense of being a societally specific form of engagement that contributes to the maintenance and survival of the human species.
4. I use the term transaction, because it better characterizes the irreducibility of the situation than the term interaction. The latter notion can be thought of what happens when two independent entities come to act taking each other into account: action lies between (Lat. *inter*) the agents. The former notion implies the irreducibility of action to an agent, because the action goes across (Lat. *trans*), linking agent and patient. This makes it possible to theorize actions in terms of the subject|object dialectic—as this is done in cultural-historical activity theory—where the object can be another person. A *trans*itive verb is a good paradigm for this perspective.
5. Practical wisdom or phronesis aims at truth in the service of action. Aristotle uses the term to articulate a form of knowledge that is neither theoretical (episteme) nor technical (techne). Following Aristotle, philosophers such as Paul Ricœur suggest that only an ethico-moral person carries out appropriate actions—those that do not harm others and therefore take into account responsibility and answerability.

REFERENCES

Bakhtin, M. M. (1993). *Toward a philosophy of the act*. Austin: University of Texas Press.
Damasio, A. R. (2000). *Descartes' error: Emotion, reason, and the human brain*. New York: HarperCollins. (First published in 1994)
Giddens, A. (1991). *Modernity and self-identity: Self and society in the late modern age*. Stanford, CA: Stanford University Press.
Hegel, G.W.F. (1977). *Phenomenology of spirit* (A. V. Miller, Trans.). Oxford: Oxford University Press. (First published in 1806)

Holland, D., Lachicotte, W., Skinner, D., & Cain, C. (1998). *Identity and agency in cultural worlds.* Cambridge, MA: Harvard University Press.

Holzkamp, K. (1983). *Grundlegung der Psychologie.* Frankfurt/M.: Campus.

Holzkamp, K. (1993). *Lernen: Subjektwissenschaftliche Grundlegung.* Frankfurt/M.: Campus.

Leont'ev, A. A. (1971): *Sprache, Sprechen, Sprechtätigkeit* [Speech, speaking, speech activity] (C. Heeschen, Trans.). Stuttgart: Kohlhammer.

Leont'ev, A. N. (1978). *Activity, consciousness and personality.* Englewood Cliffs, NY: Prentice Hall.

Nancy, J.-L. (2000). *Being singular plural.* Stanford, CA: Stanford University Press.

Ricœur, P. (1990). *Soi-même comme un autre.* Paris: Seuil. (English translation published by University of Chicago Press, 1992)

Roth, W.-M. (Ed.). (2005). *Auto/biography and auto/ethnography: Praxis of research method.* Rotterdam: SensePublishers.

Roth, W.-M. (2007). Motive, emotion, and identity at work: A contribution to third-generation cultural-historical activity theory. *Mind, Culture and Activity, 14* (1).

Roth, W.-M. (in press). Agency and passivity: Prolegomenon to scientific literacy as ethico-moral praxis. In A. Rodriguez (Ed.), *Multiple faces of agency* (pp. •••–•••). Rotterdam: SensePublishers.

Roth, W.-M., & Désautels, J. (2004). Educating for citizenship: Reappraising the role of science education *Canadian Journal for Science, Mathematics, and Technology Education, 4,* 149–168.

Turner, J. H. (2002). *Face to face: Toward a sociological theory of interpersonal behavior.* Stanford, CA: Stanford University Press.

Uexküll, J. von (1973). *Theoretische biologie* [Theoretical biology]. Frankfurt: Suhrkamp. (First published in 1928)

Vygotsky, L. S. (1986). *Thought and language.* Cambridge, MA: MIT Press.

SUNGWON HWANG, WOLFF-MICHAEL ROTH

9. DIS/CONTINUITY OF IDENTITY

"Hot Cognition" in Crossing Boundaries

Educational discourses are centered either on knowing (construction) or on written text, treating everything as text. Both forms of discourse are independent of who we are, embodied creatures with feelings, emotions, and desires. Reading any research article about knowing and learning will give the impression that human beings are like robots, driven by stored (conceptual) knowledge that some central processor—homunculus in the ear—applies to the situation. In virtually all research articles, humans are bloodless creatures that are either *determined* by the stuff and structures in their bloodless mind or *determined* by the environment, as modeled in the correlations that use class, race, parental educational level, and so forth as predictors of what someone can do and achieve. Yet we all have experienced ourselves on good days, when we happily go to work or which we spend with our families, and bad days, when everything appears to go wrong; and not to forget, there are the average days, apparently characterized by emotional indifference. That is, emotions are central to what we do, or more so, our every action is constituted in part by our current emotional state and the valence that anticipated outcomes of our actions have for us. But any (discursive, practical) action is subtended by its relation to the ongoing societally motivated cultural-historically specific activity—schooling, researching, farming—on the one hand, and by our bodies that materialize these activities through their actions, on the other. This interdependence of the cultural-historical nature of activity and the actions that realize them has consequences for those who move from one culture to another, because our bodies continue for a while to produce actions that embody the ethos of the first culture while participating in the second, where the actions may be inappropriate. This then has reverse consequences for the emotions constitutive of the actions. And because both actions and emotions are integral aspects of identity (see Roth, chapter 8), moving from one country to another can come with profound disorientations, which precisely raise questions about who we are.

The twentieth century has seen increasing levels of migration: people have been able to seek refuge in new countries, migrant laborers have left their home countries to seek work elsewhere, and students have gone to study abroad. In each case, migration comes to stand for uprooting, change of cultural context, and disorientation with respect to what is right and correct to do. That is, most people we know who immigrated to a new country, for some reason and with more or less limited competencies in the new language and culture, remember feelings of physical uneasiness that initially pervades the newly immigrated. Because we are whole persons, this sense pervades how we know, what we learn, and how we relate to people: it fundamentally mediates how we experience the world and ourselves, that is, our identities.

SungWon: I am thinking about my first days in the foreign country (Canada) to which I moved for my postdoctoral research. It emerged with feelings of blurring all things around me. Whenever I talked to others, I felt as if my voice echoed without reaching them and their words echoed without reaching me. As if standing in front of a great wall, my mouth and my body stiffened. Things did not come along as the way I anticipated and I often times did not know what had been wrong or what I should have done to act appropriately. As if sucked into a whirlpool of distorted time, I experienced continuous time delays—I used to find myself not doing things in time, things that others seemed to do with such an ease. This must not be what I had known who I was, but I seemed to be seen differently. How can I explain this. . . . Once I say something, my words float around the world. Once I am silent, I become transparent.

The (material) body is physically in the new country with others while looking at what they do and listening to what they say; it appears to live in the same world where others do. Yet what we sense with our flesh in each action is otherness, estrangement from the world rather than a sense of inhabiting the same world.[1] The person often finds herself attempting to understand what is going on, where this physical unease, emotional disconnectedness comes from, and ultimately "who I am" in this foreign country.

Michael: I still remember when I first came to Canada, how I frequently experienced the world as if separated from it through an invisible shell. I could hear the voices of the other people in the same room, but the voices entered my consciousness as if from some outer space. I saw the speaker, yet his or her voice came from somewhere else. It was as if I was disconnected, materially and certainly socially from the world that surrounded me. This feeling of disconnectedness is physical, almost nauseous. I am there, but not there. I can hear the other persons; they are close and yet so far away. A physical sense of disorientation, as if in a dream, perhaps even a nightmare. The strange aspect about all of this is that I had come, physically moved into a new country. And now it is the reverse. When I go back to the country where I lived for almost half my life, I feel disoriented, physically, and emotionally. I cannot "read" others anymore, I have lost all bearings for knowing whether in some meeting people are for or against some issue, and it is not only intellectually disconcerting but physically, and emotionally.

The physical, emotional suffering that a person experiences in a foreign country is not separable from his understanding of the world. The emotionality is already constitutive of the concrete action of looking at, listening to, and gesturing toward the world and thereby mediates knowing how the world works and who the knower himself is. That is, at the heart of suffering identity lie the issues of emotions and the flesh that feels.

In this chapter we attend to the suffering of identity that occurs in the attempt to cross the boundaries of different cultural practices, particularly in the situation of learning science. In the present-day world, an increasing number of students decide to move into new places in the short or long term in pursuit of their learning objects. Educational institutes run programs that provide students with learning opportunities in diverse communities and cultures. Participating in new activities and engaging with unfamiliar culture often bring forth different experiences and associated sufferings concerning "who I am" with respect to others and the person her- or himself. Particularly for science students the suffering of identity is even more severe when science is taught as if it were a body of knowledge independent of the knowing person.

The purpose of this chapter is to contribute to a theory of identity that does not reduce the question of identity to the sole phenomenon of cognition separated from emotion but places the inseparable relation of emotionality to cognition ("hot cognition") at the core of understanding discontinuity or continuity of identity. We particularly take an embodied dialectic approach that explicitly conceptualizes emotionality as an irreducible aspect of actions of the human flesh and therefore constitutive of what is known and who the knower is.

SUFFERING IDENTITY: AN EMBODIED DIALECTIC APPROACH

Disembodied minds do not have emotions and therefore do not suffer; like our laptop computers, they function in the same way in whichever country they are on this earth. Human beings, however, come to have a sense of estrangement in a foreign place because they have material bodies. In the above narratives, when SungWon moves to a new country, she experiences her material body as more or less the same as she did in her native Korea. The same material body leaves the former place and continues to interact with others in the new place; this body is the anchor of her identity, its pivot, where the person SungWon can point to and say "I." At the same time, the move to Canada changes the social relations she realizes in and through her flesh with respect to others. In and through her sensing flesh, she produces resources of communication that are compatible with social relations in the new place. It is at this point that experiences of strangeness and estrangement occur. In the previous culture, her material body and the social self anchored in the flesh were in alignment because the bodily schemas at the source of the action fundamentally embody the *ethos* of her native Korean culture. The actions that the flesh produces relevantly mediate social relation with others. Now in the new culture the actions sometimes are out of place, as its inherent ethos is no longer that of the culture in the new place, and the flesh thereby comes to be displaced socially; incarnate knowing (deriving from a different habitus [Bourdieu, 1990]) is socially no longer appropriate. Moving from Korea to Canada, SungWon repositions physically, but comes to be *dis*positioned, out of place, literally and metaphorically. The resulting contradiction between the material body and the flesh brings forth the physical and emotional suffering, the feelings of disconnectedness from the world. That is, at the center of her suffering identity lies the material|social body that develops and unfolds emotionality.

Suffering in and for one's identity is not limited to the individual but simultaneously is a matter of individual and collective. An individual may feel estrangement because there is already another presupposed; but this feeling is the result of a relation that constitutes both individual and collective. That is, estrangement and suffering in and for identity presuppose a Self|Other dialectic. The actions of the flesh belong to Self and at the same time makes sense in relation to Other. (See Roth, chapter 8, for a more extended discussion of the relation of actions to nonconscious operations and societally motivated activity.) It constitutes resources the flesh produces for communication. In this sense, the material body|social flesh invokes not only the experience of estrangement but also involves possibilities for resolving the suffering in the configuration of a new identity. Even the act of feeling estrangement constitutes resources for developing a sense of collectivity to one another. Therefore, understanding cultural boundary crossing and the associated dis/continuity of identity requires an explanation of how the practice of the flesh communicates the sense of estrangement and thereby mediates the development of identity.

Over the past few decades there has been a growing interest in going beyond the framework of cold cognition and integrating emotional aspects into theories of learning and identity. Yet, few studies have attended to learners' real experiences of different cultural practices and the suffering that derives from it (e.g., Roth & Harama, 2000). Rather, we find the literature to be silent about the centrality of emotion in cognition or, if ever, the consideration of emotion as an external factor to cognition. For example, in the special session held at the 2005 annual meeting of AERA (American Educational Research Association) on boundary-crossing experiences, we find most reports talked about "successful" cases in which learners rarely had problem emotionally and cognitively. Those cases, however, can be seen as evidence that learners did not experience cultural boundaries at all although researchers might suppose boundaries from their perspectives.

Concerning the role of emotionality in cognition, recent development of neuroscience provides important evidence that emotion is integral to human cognition (Damasio, 2005). Unlike Cartesian approaches that separated body and emotion from cognition, these empirical studies report that the brain systems engaged in emotion are deeply involved in the management of social cognition and behavior. Furthermore, emotion is a process in which we come to a particular conclusion without being aware of all the immediate logical steps (Depraz, Varela, & Vermersch, 2002). The complex way cognition is intertwined with emotionality in the development of identity is appropriately conceptualized using the theoretical framework of *cultural-historical activity theory* (CHAT) that inherently involves *embodied dialectic* perspectives in several points summarized as follows (see also Roth, chapter 8).

First, the term *activity* in cultural-historical activity theory denotes an object-oriented, culturally and societally mediated whole system involving motives, tools, division of labor, and rules that mediate the interactions between the members of a community of practice. Understanding the historical unfolding of an activity requires analyses at three levels: *activity* directed toward object (motive) formulated by collective entities (communities, societies), *action* directed toward goal framed by subjects (individuals, groups), and *operation* unconsciously conditioned by the social and material context (A. N. Leont'ev, 1978). This analytic framework opens up opportunities to articulate the role of emotion in the cultural-historical development of activity. On the one hand, emotions are specific bodily states of the subject of action and therefore constitute the actional background at the level of operations. On the other hand, emotions connect actions with those who share the situation and take a role of consciously setting up the goal of action in its relation to the overall activity.

Second, subject and object constitute an irreducible unit in activity. The consciousness of human subjects develops through the activity of the subjects in the world objectively given to them in their perception, and the world becomes an object by the activity of the subject. Given that emotion is constitutive of the historical unfolding of an activity (Holzkamp, 1983)—embodied both in the motive, oriented toward meeting the collective needs—the dialectic of subject and object implies that becoming familiar with specific culture involves the development of emotionality pertaining to what the activity is oriented toward and the identity pertaining to who the actors are with respect to their Selves and Others.

Third, as the subject-centered approach to cultural-historical activity emphasizes, subjectivity already presupposes intersubjectivity. Central to production and reproduction of intersubjectivity is the role of human bodies that concretely realize the cultural possibilities in communication and thereby make them available to the other (Hwang, Roth, & Pozzer-Ardenghi, 2005). The development of emotionality can be addressed using the dialectical framework of a communicative act (A. A. Leont'ev, 1971). It is through our bodies that we "consume" the performative dimensions of acting and at the same time produce the effect on others and the material world; this makes us responsible prior to any individual or collective consciousness (Levinas, 1998). Emotionality constitutes a background over and against which an illocutionary act (intention of speaking) unfolds, and evolves with an anticipation of fulfilling a perlocutionary aspect (effect of speaking). In this sense, solidarity or estrangement is understood to emerge in relation to emotionality and develop as communicators produce resources for positive emotional valence.

In what follows, we articulate our theoretical framework, that is, an embodied dialectic approach to the dis/continuity of identity by analyzing empirical case materials from an undergraduate laboratory course at a Canadian university. We describe the context of this auto/ethnographic study and develop our theoretical claims through analyses of concrete case materials. The case materials we introduce and analyze in this chapter are part of those that appeared as boundaries in the lifeworld and thus critical moments of identity. The study is in part auto/ethnographic, because it analyses the experiences of a Japanese student through her close relationship with SungWon, who shared many experientially mediated emotions arising from the dis/continuity of identity associated with their migration.

CONTEXT OF THE AUTO/ETHNOGRAPHIC STUDY

The case materials analyzed in this study derive from a research project SungWon conducted as one strand of a large-scale project directed by Michael, proposed for the purpose of understanding students' knowing and learning as they move across and between different sociomaterial and cultural-historical settings. Our *Navigating Boundaries* project took shape on the ground of two intertwined experiences: (a) SungWon's stay at the foreign country (Canada) for her postdoctoral study and (b) the research she conducted with a female Japanese student (Mariko) who allowed SungWon to participate in her everyday academic life at a Canadian university. At the time of the study, Mariko was a third year undergraduate student majoring in physics. Seven years prior to the study, she had moved from Japan to Canada for her pursuing undergraduate studies. Physics was the second of her double major, complementing the first major in geography. After successfully completing her first major as an honors student, she decided to study physics in the pursuit of her long-standing academic interest in climatology.

SungWon: Conducting a project with Mariko provided me with opportunities not only to collect empirical data concerning the issues of boundary crossing in general but also to see myself as a particular case in a self-reflective manner. In my conversations with Mariko, I also talked about my own experiences, good and bad, when I saw a relation to her physics learning in this country foreign to both of us. I found myself looking at those instances through the lens of my experiences or, the other way around, I looked at my own experiences through the lens of Mariko's experiences.

Michael: This is very much a case of what ethnopsychoanalysts refer to as *countertransference,* which occurs, for example, when researchers conduct their studies in very different cultures (Devereux, 1967). When ethnographers do fieldwork in another culture, their data, descriptions, and theories are the result of effects that the Other has on the researcher's flesh. The foreign is cognized through our hearing, seeing, smelling, touching, and feeling, all of which are incarnate modes by means of which we are connected to the world and which are at the very foundation how we make sense, literally, and what we can know about it.

SungWon: In this research project, I eventually developed an auto/ethnographic stance, a method that explicitly takes intersubjectivity as the way to arrive at knowing both researcher and researched and thus to eschew the dichotomy opposing objectivism and subjectivism (Roth, 2005).

Michael: And here *auto/ethnography* is reflexive in a double sense. First, ethnography inherently is about ourselves, as we use (new, old, theoretical, commonsense) language to articulate the other so that our accounts are both *about* the other and a concrete expression of narrative forms and contents that are characteristic of our own culture. Second, auto/ethnography is reflexive in that being about ourselves, it is also about the other, who in the description of our actions, recognizes actions that are intelligible and performable by themselves.

SungWon: During data collection and analysis, I explicitly intertwined ethnography and autoethnography. To gain a better understanding of learning activities as are salient and relevant to the learner's everyday life, I conducted intensive ethnographic fieldwork, spending many hours with Mariko in the physics laboratory and outside. I videotaped all activities related to Mariko's physics learning including both curricular and non-curricular activities. The former had classroom lectures and laboratory experiments and the latter included problem-solving tasks (doing an assignment on her own or with her classmate) and departmental colloquia. I also collected teaching materials such as lecturer's notes, transparencies, and handouts used for teaching on the one hand and student's notes, marked assignments and examinations on the other. Every learning task was documented in my field notes on a day-by-day basis.

Through the studies over a three-year period, we appreciated the significance of emotionality in crossing the boundaries of different cultural practices. In accordance with the auto/ethnographic method, we drew on our own experiences of living in a foreign country as materials for taking a first-person perspective of the data, particularly cases that emerged and became salient to ourselves. Discursive psychology and conversation analysis supply the lenses that allow us to uncover the local organization of interaction between communicators. They are ideal tools for analyzing human activity systems because they are methods to get at the implicit ways in which people make and make sense of the situations they participate in. In our concrete analysis, we attend to various resources that material bodies produce and make available to one another during communication. Typically, these are speech and gesture; we especially draw on the change of pitch as a resource for getting emotions through their expression in human communication.

Following the ethnomethodological tradition, we approach our database without presuming any stable properties attributed to an individual person or a specific culture. For example, the fact that Mariko immigrated to Canada from an Asian country might raise nationality or language as explanatory resources for boundaries at an institutional (macroscopic) level. Rather, we posited that boundaries and associated structures, if any, come into being only if a sociomaterial body experiences physical and emotional resistances in concrete situations. That is, a boundary is not really a boundary until it is experienced and perceived as such by a person, who, as a consequence, desires to traverse and go beyond it. It is a systemic condition taken into consideration to the extent that it becomes available resources for actions. "Crossing boundaries" denotes a process by means of which people deal with boundaries that emerge in their lifeworlds rather than that exist or are presupposed from a distanced third-person perspective. Thus, "a boundary is not that at which something stops but . . . that from which something *begins its presencing*" (Heidegger, 1971, p. 154), a realization that also became central to Homi Bhabha's (1994) thinking about the location of culture. It is the presencing of something that allows us to recognize a boundary.

BODY AND EMOTIONALITY IN THE DEVELOPMENT OF IDENTITY

The emergence of boundaries to learners' lifeworlds involves suffering associated with the experience of estrangement. The sense of estrangement presupposes one's relation to the other, which already bears possibilities for its negation. Thus, estrangement does not remain the experience of an individual person. It can be overcome through a changing relation, that is, through collective praxis. Change of emotionality is a social process that brings out changes concerning who a person is with respect to others. In this section, we articulate an incarnate dialectic theory of identity that takes the individual and collective production and reproduction of emotionality as its significant constitutive part. We present two theoretical claims supported by case materials. First, we suggest that in and through the body|flesh, human beings generally and science learners particularly produce|reproduce emotional valence in communication, where it is available for others and analysts. It is only because emotions are salient that others can become empathetic, that is, feel with and for the other emotions that participants take as shared. As already recognized by the philosopher Edmund Husserl, one cannot recognize emotions (e.g., anger) in the other without first taking an explicitly external perspective on one's own affects; this is the principal condition for comprehending the fleshly manifestations of others' emotions (Franck, 1981). Empathy, therefore, derives from living *with* the other, and this *with* precedes anything that can be called identity because it is the condition for both intersubjectivity and subjectivity to emerge. Second, the empathetic terrain of communication constitutes identity in a reflexive way, so that it is therefore simultaneously individual and collective.

Flesh Produces|Reproduces Emotional Valence in Communication

Our flesh is central to emotionality, for without it there would be no emotions. In and through our flesh, we communicate emotions to others and therefore make it a collective phenomenon. More so, an emotion expressed collectively mediates individual emotions—the *mood* of our workplace mediates how we feel all the while how we feel constitutes the collective mood. The emotions others articulate publicly and those that we feel and make available publicly thoroughly mediate face-to-face interactions. On the one hand, emotions are specific bodily states of the subject that constitute the background of action. Emotions therefore mediate agency, but the subject of agency has no control over these states or the emotions: we are passive hosts to our emotions. Integral to our actions, thus, are emotions, which come to be objectified in the material resources we produce (including sound). These material resources that the acting flesh produces in communication deploy emotionality to the other and thereby constitute the material ground for making variations at the collective level. On the other hand, flesh connects actions with those who share the emotionality and to take a role of consciously setting up the goal of action with respect to the collective object. In each action, a body concretely realizes the emotional valence (affective stance) that was already presupposed at the collective level and thus was available to it as a possibility for action.

DIS/CONTINUITY OF IDENTITY

At the same time, in and through our bodies, we create new emotional valence that could not have been achieved otherwise. Therefore, the role of the bodies in the production|reproduction of emotional valence constitutes an important aspect for understanding learners' experiences of otherness and therefore boundaries and the crossing of these. In what follows, we present an example of problematic situation that Mariko experienced in one of her physics labs and analyze the interactive configuration of emotionality unfolding through the human bodies.

The episode occurred near the end of Mariko's optics laboratory concerning polarization. At that moment, she realized that she did not have sufficient time to complete all tasks scheduled for a three-hour laboratory session. While referring to laboratory manuals and often times asking the lab instructor for assistance, Mariko conducted experiments of varying an angle between two plates and measuring the intensity of a light beam passing through them. She focused on the tasks without a break and now found herself not having completed them. The conversation occurred when the lab instructor Glenn came to Mariko's laboratory bench and explained those tasks remaining to be completed.

Episode 9.1[2]
01 G: ((Glenn has his hands come closer)) what angle it comes off just completely polarized in one direction ((*He puts his hands flat together and has it tilted slightly.))
        *[Figure 9.1a]
02 M: ((Mariko nods her heads)) um
03 G: BUT-
04   (0.96)
05 G: I mean, it takes a little while to get used to that, so (???).
06   ((Mariko laughs and turns her head away from Glenn.))
07 M: Glenn ((*She lightly taps the table.)) *[Figure 9.1b]
08   I feel [₁*inferiority= *[Figure 9.1c]
               [₁((Mariko looks at Glenn.))
09   =be[₂cause I ((*Mariko keeps laughing and turns her head away toward the table.))
        *[Figure 9.1d]
10 G: [₂=Eh::↓ ((*Glenn shakes his head right and left while turning his gaze away from Mariko and to the table))

Glenn points to the equipment on the laboratory bench and explains the elliptical polarization in glass. In accordance to his utterance "in one direction" his hands

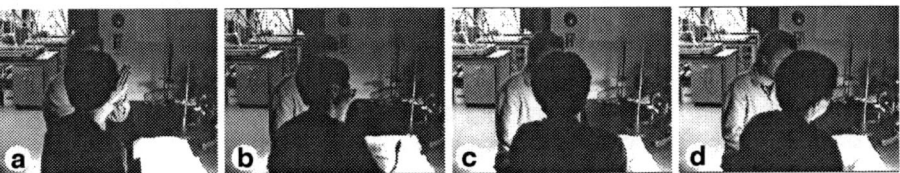

Figure 9.1. a. Glenn puts his hands flat together and Mariko looks at the gesture. b. Mariko calls on Glenn while turning the head away from him c. Mariko looks at Glenn. d. Mariko turns her head away from Glenn and at the same time Glenn shakes his head.

193

produce the shape of a tilted line (turn 01, Figure 9.1a). Mariko nods and says "um" (turn 02). Glenn utters "but" (turn 03) and then stops, which, because Mariko does not take the opportunity it provides for a turn, leaves a pause (turn 04). Glenn's continues, "it takes a little while" (turn 05). Mariko begins to laugh while turning her head away from Glenn (turn 06). She calls Glenn and slightly taps the table (turn 07, Figure 9.1b). She looks at Glenn again (Figure 9.1c) and tells him that she feels a sense of inferiority (turn 08). Mariko keeps laughing nervously (turn 09) and continues to say "because." Immediately Glenn produces the particle "Eh," but does so much louder than were his previous utterances (turn 10). This particle can be heard as an articulation of sorrow or as an interjectional interrogative particle, often inviting assent to a sentiment expressed (OED, 2006). Here, however, it is more likely an embarrassed acknowledgment of a sentiment expressed by the other that does not normally belong to the culture of dispassionate physics.

In this episode, Mariko and Glenn make their emotions available to one another through their flesh—turning, gesturing, inflecting voices, content—and thereby develop their collective understanding of the current situation: Mariko articulates her sense of inferiority and produces other signs that partially support the sense of inferiority and the embarrassment that comes with publicly acknowledging such a sense to the laboratory instructor and to the researcher present. The flesh, seat of agency and passivity, enables the participant actors' emotional engagement in the situation and at the same time enables their taking part in varying collectively available emotionality that they are responsible for. At the beginning of the episode, Glenn talks to Mariko about the motion of an elliptically polarized light beam. He uses low voice intensity and gestures to produce his communicative performances. Mariko nods and says "um," which is an interjection used to indicate hesitation, doubt, and assent. The low speech volume and brevity of the interjection suggest that she has heard him, which is the affirmative dimension of her speech act, but she also makes available—through low volume and brevity—doubts about her understanding of the content of his talk. She thereby affirms and reifies—it is only through her acknowledgment that Glenn's talk becomes an explanation—his explanatory actions both cognitively and emotionally.

Glenn continues to speak, but the next speech production is an interjection that expresses opposition; it is also uttered with an intensity that exceeds the average speech intensity of his previous utterance: "BUT." He then stops suddenly. This sudden stop therefore brings about a pause, because Mariko does not take this opportunity for taking a next turn. This indicates a change of not only the conversational topic but also the associated emotional valence. Glenn turns his eyes to the table and, in breaking the silence says, "I mean." In and with this utterance, Glenn does not introduce new content. In fact, it is a way of showing one's responsibility for a pause specifically and for the conversation generally without actually saying something. It is a way of producing time for finding a stance that allows this or other speakers to continue. The hesitation in the stance is an expression of his uncertain understanding of the current, problematic situation, involving the discomfort of being confronted with expressed emotions—e.g., embarrassment. That is, the situation is charged emotionally in a double sense: the silence by Mariko confronts Glenn, whose subsequent actions are colored by his own emotional stance, discomfort about being confronted with the emotionality (embarrassment) of another. He finally turns his face, looks at Mariko, and utters in a low, monotonous voice, "it takes a little while to get used to" (Figure 9.2a). The utterance becomes gradually faster, decreases in intensity only to become indiscernible toward the end. Glenn's presumable understanding is that it will take time for Mariko to do the remaining tasks. Yet this understanding is inseparable from the discomfort, the experienced emotionality: not only of Glenn's own but also the one he supposes to be Mariko's. The speed, pitch, and intensity of the utterance concretely realize the possible supposed emotional valence between Glenn and Mariko toward the current problematic situation in consideration of what has been anticipated.

Once materialized in and through the flesh, an emotional stance does not remain that of an individual speaker—it now is a resource available to be taken up by anyone present. That is, once materially expressed in and through the flesh, emotions are imbued with a facticity that is as materially real as any other material fact—the laboratory bench, the chairs participants sit on, and the instruments they use to measure and construct phenomena. This collective emotional stance colors the situation, which constitutes the ground for the next action. The emotional stance of one therefore comes to be embodied in the action of the other who, in acting, draws on the available resources, which now include the materially articulated emotions of the other. Glenn's own actions, now in part directed toward the new reality, enfold his own emotional valence, expressed in his hesitations (pauses) and articulations of uncertainty (low speech intensity).

In response to Glenn, Mariko begins to laugh and turns her face away from Glenn and directs her gaze toward the laboratory bench. While still laughing, she utters the instructor's name, "Glenn," and lightly taps the table with her right hand (Figure 9.1b). She returns her gaze toward him and notes, "I feel inferiority" (Figure 9.1c). The utterance, laughing, body orientation, eye gaze, and gesture, all contribute to expressing embarrassment and discomfort. The preceding transactions constitute a ground in which her actions unfold an emotional valence that is no longer her own though she necessarily contributes to its constitution. However, in and through her flesh, Mariko does not simply reproduce some presupposed negative emotional valence deriving from embarrassment. The emotion expressed changes the very ground of the collective emotional valence. Her nervous chuckle mitigates the situation by contrasting itself both to the calmness of Glenn's preceding utterances and the current utterance that has her feeling of inferiority as its content. In some sense, her chuckle anticipates the embarrassment that her utterance would confront both Glenn and herself with, a type of reality normally absent from physics, for there is nothing that ties the theoretical knowledge of physics to the emotions of individual human beings participating more or less peripherally in the discipline.

In this rather mitigated ground of emotionality, Mariko *ex*poses her *dis*position toward the current situation: "I feel inferiority." Mariko's action changes the ground on which Mariko and Glenn have to deal with the problematic situation of the unfinished and perhaps unfinishable lab. Glenn, instead of articulating more, produces the particle "eh" as if he deeply sighed. He shakes his head and turns his gaze from Mariko to the table (turn 02). The rapid decrease of pitch, together with the gesture and body orientation can be experienced as an indication that he does not agree with Marko's action. Glenn expresses resistance. His action, a concrete realization of possibilities presupposed in the interaction (he only does what he presumes to be intelligible and rational in the present situation), again changed the ground of interaction. Mariko keeps laughing and, in turn, averts her gaze. Both now have to deal with the problematic situation on the ground that their actions produced.

In this episode, the transactions continued producing changes in the salient aspects of their understanding of the current laboratory work. The first issue of their face-to-face encounter is the fact that Mariko requires more time than has been given initially. The conversational topic then changes to the feelings Mariko expresses in the situation, a sense of inferiority related to the fact that she is not able to bring to completion a laboratory task that should have taken less than the allotted three hours. In the process of articulating their understandings of the problematic situation, the two produce and reproduce emotional valence as well. Their emotions are not just some processes that rage in their bodies; rather, emotions come to be expressed materially through body position, gestures, words, prosody, and inflections. In this materiality, they constitute resources for subsequent actions in the same way that anything else present in the situation is a resource for action for all those present. Emotionality therefore is available collectively and, because participants both complete and continue each other's actions and take up and build on the resources others produce, is both an individual|collective phenomenon. It is precisely because of their public availability that human beings learn to name the effects of their bodily states using emotion words—it is only because we see our own emotional expressions through the eyes of others that we come to understand our own emotions (Franck, 1981). The configuration of collective emotionality matters for in its empathetic terrain emerge individual actors who materialize specific action possibilities and achieve who they are with respect to not only others but also themselves.

Identity Emerges from and Develops on the Empathetic Terrain of Communication

The articulation of a situation as problematic occurs in and through communication that co-expresses emotionality. Interacting individuals develop their collective understanding of the situation by empathizing with one another. Specific possibilities for action among many others emerge salient in this collective terrain of emotionality. Realized possibilities in concrete actions articulate who the actors are with respect to others and themselves as well, thereby mediating the development of identity. That is, identity emerges and develops on the empathetic terrain of communication, which simultaneously is individual and collective. In this section we exemplify how the interactive configuration of emotionality constitutes an empathetic terrain of communication and thereby mediate the development of individ-

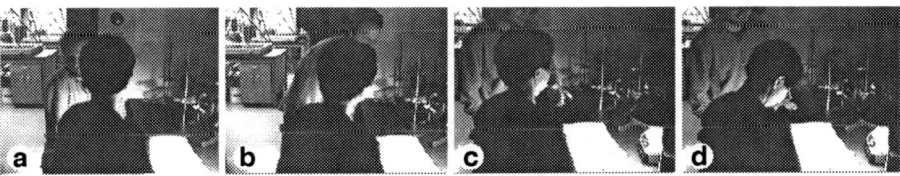

Figure 9.2. a. Glenn shrugs his shoulders. b. Glenn stands up. c. Glenn points with his left hand to Mariko while talking about the barrier. d. Mariko drops her head.

ual|collective identity. The following case materials pertain to the situation following Episode 9.1 in which Mariko and Glenn end up being confronted with their discomfort about the situation.

Episode 9.2

11 M: [₁I'm always a kind of,
        [₁((Mariko turns her head away from Glenn while laughing.))
12 G: =that's fine.
13 M: =late and I just (0.68) cannot (0.46) finish all of the lab within the time.
14   [₂(1.95)
        [₂((*Glenn shrugs his shoulders during a pause.)) *[Figure 9.2a]
15 M: What's [₃the indication, of this? ((Mariko keeps laughing.))
16 G: [₃I (???)
17 G: (1.03) ((Glenn and Marko laugh.)) no skin off my back it
18 M: ((Mariko laughs more loudly.))
19 G: ((*Glenn stands up)) it doesn't bother me. *[Figure 9.2b]
20   ((Glenn and Mariko laugh.))
21 G: just so long as you're getting something out of the labs, that's all that really matters, right?
22   (0.72)
23 M: ahuh, yeah
24 M: but still, most students,
25 G: yea, but they don't *have the language barrier you have, right? *[Figure 9.2c]
26 M: ye:[₄s (students?)
           [₄((Mariko *hangs down her head)) *[Figure 9.2d]
27 M: ((Mariko laughs.))

Mariko attempts to explain the reason for her feeling that became salient in the previous episode (turn 11). Even before her utterance comes to an end, Glenn says, "that's fine" (turn 12). Mariko articulates further that she has not completed the laboratory within the allotted time (turn 13). A pause ensues during which Glenn shrugs his shoulders (turn 14). Mariko utters "what's the indication of this" (turn 15). At the same time Glenn talks but his utterance is overlaid and mixed with Mariko's rather intense talk and laughter (turn 16). Another pause ensues. Now, both Glenn and Mariko laugh. Glenn says "no skin off my back," which may imply that it does not really matter to him (turn 17). Mariko laughs more loudly (turn 18). Glenn stands up from the table (turn 19). Glenn and Mariko continue to laugh again (turn 20). Glenn articulates that it would be fine so long as she gets something out of the laboratory (turn 21). Another pause unfolds (turn 22), now interrupted by Mariko, who utters "yea" (turn 23) and then continues "but still, most students." Here, "but" functions as a conjunctive that expresses an opposition or disagreement (turn 24). Even before the utterance ends, Glenn interrupts and articulates the situation: others do not have a language barrier that Mariko has (turn 25). Mariko says "yes," drawing it out a bit longer than usual and then lets her head drop (turn 26). She begins to laugh again as if attempting to mitigate the situation (turn 27).

In this situation, the interaction of the two proceeds with resolving the problematic situation that occurred in the previous episode and the associated emotional discomfort. By saying "I'm always" (turn 11), Mariko indicates her attempt to ascribe the responsibility for the current problematic situation to herself. As her utterance unfolds, Glenn, as if he already knows and does not want to listen to what Mariko will say, interrupts. He says, "that's fine" (turn 12), but Mariko's utterance has already gathered sufficient momentum to keep her going (turn. 13). There is a pause, and Glenn then responds with a shrugging gesture (turn 14). Mariko breaks the silence by saying, "what is the indication" (turn 13), which sounds like a rhetorical question. After another pause, Glenn explicates his stance, "no skin off my back" (turn 17). This suggests that he does not agree with the problem itself and thus implies an attempt to avoid resolving the situation by blaming Mariko. Rather, Glenn problematizes the current issue.

Mariko laughs more loudly (turn 18) and in this articulates neither explicit agreement nor explicit disagreement. If she agrees, she will have to negate the problem she just raised. If she disagrees, she will have to blame Glenn for setting the constraints. Her action provides Glenn with another opportunity to continue describing the situation from his perspective. He stands up and suggests that it does not bother him (turn 19). By standing up, Glenn indicates that there is no more issue to discuss. Glenn and Mariko begin to laugh (turn 20). Glenn now articulates what really matters: Mariko has to get something out of the lab (turn 21). He thereby indicates that his concerns are more about what Mariko learns rather than about whether she completes all the tasks.

As a result of this transaction, Mariko is identified as a student who is slow and thus whose achievement is lower than the normal. Glenn comes out of the transaction as a lab instructor who is flexible in evaluating students' achievement. Emotionality is an important aspect constituting identity in this situation: it is realized as both process and outcome of interaction. The unfolding conversation thereby constitutes an empathetic terrain, which both provided the resources on which participants draw in acting. In each turn, the two attempt to mitigate emotional discomfort. Mariko ascribes the problematic situation to herself ("because I" and "I always," turns 09, 11) rather than to Glenn, the laboratory instructor who is responsible for students' achievement in the class. Her action thereby implies her emotionality: she would not want to blame Glenn but—in contrast to the North American ethos where individuals tend to blame others for their own shortcomings and misgivings—would blame herself. Glenn's different description of the situation also involves his emotionality that would not want to blame Mariko. The collective emotional valence of the situation leads Mariko to a rather contradictory conclusion: she negates the initially raised problem and affirms the current situation (turn 23). However, Mariko immediately says, "but still" (turn 24) as if she realized this contradiction.

On this day, Mariko cannot complete the third task because she spent too much time on the first task. She does not achieve a calibration of the light meter for the first measurement and therefore has to repeat the first task. The problem partially lies with her unfamiliarity with the instrument. However, in and through communication followed, Mariko and Glenn stabilize the problem in ascribing to the structural feature that emerges as an irremovable boundary (turns 25–26) rather than in finding practical ways to enhance Mariko's familiarity with the lab instrument. Our analysis shows that this ascription, the action of drawing on the structural feature, can be seen as a consequence of interaction that unfolds, whether they consciously intend or not, in such a way of respecting the other and figuring out the problem.

Mariko appears to agree with Glenn (turn 23). Yet, the problem has not been resolved, because she is reminded that "most students" finish the same laboratory task within the allotted time. Her action constitutes an opportunity to develop the issue in a different direction than before; Glenn says, "they don't have the language barrier you have" (turn 25). The "language barrier" here functions as a discursive move that mitigates the problematic situation: the problem now becomes a structural factor. This action also involves emotionality in that it avoids blaming either participant. Other students would not have been different if they also had a language barrier. In consequence of Glenn's utterance, the language barrier emerges as an important aspect that is used situationally to explain Mariko's low achievement and therefore who Mariko is. We need to understand Mariko's identity established in this situation as a consequence of empathetic transactions that attempt to be ethical to one another in resolving a problematic situation. That is, the questions "*Who* is Mariko?" and "*Who* is Glenn?" can only be understood appropriately in consideration of the empathetic terrain on which they operate and that they produce collectively. The question of identity has to be rephrased such that the answer is: The individual|collective identities of Mariko and Glenn are the result and outcome of the communication and collective emotionality.

EMOTIONALITY, IDENTITY, AND SOLIDARITY

In this chapter, we show how emotionality is integral to the constitution of actions, and through the completed act and its results how it becomes materialized and therefore objectified in the situation as a whole. More so, emotions constitute not only contextual aspects—even thoughts, in our framework (as participative thinking [Bakhtin, 1993]), are an irreducible aspect of the activity—that mediate actions but become the very topic of face-to-face communication. Once made public and materialized in objective form, emotionality is an inherent feature of the world; it is available collectively and—consistent with cultural-historical approaches—objectively. Emotionality obtains a facticity that cannot be denied, and a facticity that becomes foundational to individual actions. But because actions are constitutive of who we are, individual|collective emotionality is constitutive of identity, which we here understand as a possibility concretely enacted by each and every person. But as a possibility, identity therefore is inherently intelligible to another, who sees in one's identity not only something that is intelligible and can be empathized with but also something that he or she could espouse. We all have our heroes that we want to resemble and whose actions we want to emulate in our behavior.

SungWon: I felt uncomfortable while sitting beside Mariko and Glenn. I empathized with Mariko's desire to be a good physics student and the deep embarrassment she ultimately expressed. I shared her feelings and, as if they were mine, I felt similarly embarrassed. Simultaneously, my discomfort seemed to be rooted in the sense that the current situation looked very irrelevant. I was asking myself why the situation had to unfold in the way it did. Mariko had told me—before Glenn arrived at her laboratory bench—that she would ask him whether she could do some extra work. However, the opportunity for asking has not been realized as we see in the presented episodes. As I analyzed the concrete data later, I came to understand that the situation developed as it did because teacher and student are in flesh and blood: beings that act empathetically and ethically to one another. Once I take emotionality as the key for understanding concrete situations of learning science and the identity brought forth from them, I concern myself with how our science classrooms and laboratories constitute terrains on which the empathetic and ethical aspect of identity contributes to expand the possibilities for teaching and learning, particularly in the case of boundary-crossing experiences. I think especially about the fact that the researcher's presence in the situation already constitutes part of the configuration of individual|collective emotionality and therefore affects not only researched but also researchers themselves.

Solidarity constitutes the recognition that we are all human beings sharing social life. We all are constitutive of life, in which being means being singular plural. Each person constitutes an Other to all other Selves, and therefore is a constitutive moment of them. Recognizing that we are constitutive of others in the same way that these others contribute to constitute us comes with a responsibility for collective life, which is the condition of our own. This responsibility is not something that has to be crafted on human lives after the fact, but constitutes the very possibility of being human (Levinas, 1998). More so, this responsibility is more ancient than any human being. Even before a baby can think and is capable of conscious reflection and deliberation, it participates in child–parent transactions. As a consequence, parents change. That is, the child's contributions to the transactions with parents are constitutive of their identities even before the child has any notion of itself as Self and of its parents as others.

CODA

Identity frequently is theorized and described as something related to individuals, what they say about themselves, and possibly their private emotions. In this chapter, we propose a radically different view of emotionality and how it mediates identity. Thus, because emotionality is a constitutive aspect of our actions, the results of these actions are marked by the emotionality of the agent—though, with Karl Marx, we hold that emotions are embodied in our completed acts and products only in an estranged form that is disconnected from our bodies. Simultaneously, however, actions are *for* others who, in the actions, recognize themselves as recognizing the other (Hegel, 1807/1977). That is, emotions are constitutive of actions and, in completed acts and products, are available as agential resources for all those co-present in the situation. The emotions therefore come to be embodied not only in the topic of the conversation but also on the very terrain on which the conversation unfolds. We are who we are not only in and for ourselves but also, and more importantly so, for others. But the emotionality of these others, constitutive as they are for our own being, also becomes an aspect of our own emotionality and action.

When someone moves from one culture into another (very different) culture, a rupture and disjunction occurs between the cultural referents for our actions all the while through our flesh we produce actions that embody the ethos and pathos of another culture. It is out of this rupture and disjunction that the experience of a dis/continuity of identity emerges. That is, whereas the bodily capabilities for certain (discursive, physical) action continue and, in their materiality, move from one activity system to another, from one culture to another, the other-oriented component of the action does not move. It is a function of a dialectical relation with the other. Pain and suffering arise precisely from the non-coincidence between the fleshly production of praxis and the sense of a misfit of action in the present (sub-) cultural setting. In the change from one culture (Korean, Japanese) to another, the divergence of incarnate practical action and its sense is more drastic than when an individual moves between quite different activity systems within the same society, where people share at least their native tongue. But our research experience has shown that this divergence can be experienced any time a person moves into a new subculture, which we have experienced, for example, while participating in the day-to-day affairs in a fish hatchery. Initially, everything has been unfamiliar and strange, not only the practical actions but also the discourses fish culturists use. The theoretical approach developed in this chapter appears to be appropriate to theorize such boundary crossing experiences as well.

NOTES

This work was supported by a grant from the Social Sciences and Humanities Research Council of Canada. We are grateful to our research participants for their contributions to this project. Our thanks go to the Victoria Chat@UVic group (Diego Ardenghi, Leanna Boyer, Yew-Jin Lee, Lilian Pozzer-Ardenghi, Giuliano Reis, and Jin Yoon) for providing a forum to discuss our ideas.

1 We follow phenomenological conventions to use *body* for referring to the physical-material body, whereas we use *flesh* to refer to the sensing body, the only one that can make sense.
2 We digitized our database originally recorded in 8mm videotapes into relevant sizes of video files. Our first analysis included transcripts of the episodes with an accuracy of 10 milliseconds at the sound and 33 milliseconds at the image levels by moving image by image. For in-depth speech analyses, we produced separate audio files of selected episodes. Using a freely available software package PRATT (available at http://www.praat.org), we analyze loudness, intensity, and pitch of utterances recorded in the sound tracks; particularly we attend to the configuration that pitch graphs provide (prosody). The episode has been transcribed following the conventions of conversation analysis (Have, 1999):

[Beginning of overlapping talk or gesture;
= Equal sign at the end of one turn and at the beginning of the next indicates latching turns, that is, there is no gap between the two speakers;
(0.96) Elapsed time in tenth of a second;
BUT Capitalized letters indicate stronger intensity of speech;
Eh:: Lengthening of a phoneme is indicated by colon;
↓ The arrow indicates a fall in intonation more clearly noticeable than normally occur;
– Dash indicates sudden stop in talk;
..,;? Punctuation marks are used to indicate characteristics of speech production, such as intonation rather than grammatical units of language;
(?) Question mark in parenthesis indicates inaudible utterance(s);
(( )) Double parentheses are used to enclose comments and descriptions.

REFERENCES

Bakhtin, M. M. (1993). *Toward a philosophy of the act*. Austin: University of Texas Press.
Bhabha, H. K. (1994). *The location of culture*. New York: Routledge.
Damasio, A. R. (2005). *Decartes' error: Emotion, reason, and the human brain*. New York: Penguin Books. (First published in 1994)
Depraz, N., Varela, F. J., & Vermersch, P. (2002). *On becoming aware: A pragmatics of experiencing*. Amsterdam: Benjamins.
Devereux, G. (1967). *From anxiety to method in the behavioral sciences*. The Hague: Mouton.
Franck, D. (1981). *Chair et corps: Sur la phénoménologie de Husserl*. Paris: Les Éditions de Minuit.
Have, P. ten (1999). *Doing conversation analysis: a practical guide*. London: Sage.
Hegel, G.W.F. (1977). *Phenomenology of spirit* (A.V. Miller, Trans.). Oxford: Oxford University Press. (First published in 1807)
Heidegger, M. (1971). *Poetry, language, thought* (A. Hofstadter, Trans.). New York: Harper & Row.
Holzkamp, K. (1983). *Grundlegung der Psychologie*. Frankfurt/M: Campus.
Hwang, S., Roth, W.-M., & Pozzer-Ardenghi, L. (2005). Understanding collaborative practice: Reading between the Lines Actions. *Outlines, 7*, 50–69.
Leont'ev, A. A. (1971): *Sprache, Sprechen, Sprechtätigkeit* [Speech, speaking, speech activity] (C. Heeschen, Trans.). Stuttgart: Kohlhammer.
Leont'ev, A. N. (1978). *Activity, consciousness and personality*. Englewood Cliffs, NJ: Prentice Hall.
Levinas, E. (1998). *Otherwise than being or beyond essence* (A. Lingis, Trans.). Pittsburgh, PA: Donquesne University Press. (First published in 1974)
OED (2006). *Oxford English Dictionary Online*, Oxford University Press, Available at: http://dictionary.oed.com/ [Date of Access: March 31, 2006].
Roth, W.-M. (2005). *Auto/biography and auto/ethnography: praxis of research method*. Rotterdam: Sense Publishers.
Roth, W.-M., & Harama, H. (2000). (Standard) English as second language: tribulations of self, *Journal of Curriculum Studies, 32*, 757–775.

MARIA VARELAS, CHRISTINE C. PAPPAS, ELI TUCKER-
RAYMOND, AMY ARSENAULT, TAMARA CIESLA, JUSTINE
KANE, SOFIA KOKKINO, JO E. SIUDA

10. IDENTITY IN ACTIVITIES

Young Children and Science

We explore identities and subjectivities of young primary-grade children across various activities in classrooms that use integrated science-literacy curricular units. Using activity theory, we study the interplay of the six essential elements of an activity (subject, object, artifacts, rules, community, and division of labor) in the context of four activities that act as curriculum genres, and consider two fundamental aspects of identity and subjectivity, namely, how children see and position themselves and how others see and position them. We consider both the role of activity as a system in the formation of students' identities and also the ways children's identities shape the activities. The four activities are: participating in a whole-class dialogic read-aloud of an information book on earthworms; performing a hands-on exploration in groups where children sort a variety of "ill-defined," "ambiguous" objects in solids, liquids, and gases; creating a classroom mural of a temperate forest ecosystem where children construct and represent forest entities using a variety of materials and organize them in a two- (or sometimes three-) dimensional space; and, sharing with an adult, their own illustrated information books that children individually composed in two units, *Matter* and *Forest*. These four activities have different goals, employ different artifacts that act as multi-modal resources, are governed by different rules, and necessitate different ways of getting the work done. Furthermore, these activities allow us to explore the ways in which the interplay of children's membership in and across various communities (e.g., those of the classroom and the school, and as writers, scientists, and artists) mediates the use of identities in meaning making. In addition, we examine the relationship between the meaning of an activity and the sense children may develop in this activity, as children engage with science, with each other, and with the teacher.

INTRODUCTION

This chapter draws on work in the *Integrated Science-Literacy Enactments* (ISLE) project that takes place in primary grade, urban classrooms. We work with teachers to develop, implement, and study science teaching and learning that is guided by several principles. We strive to offer children multi-modal opportunities to engage with both levels of scientific activity—theorizing about the world around us and collecting and processing data, or, in other words, empirical evidence, either in the form of observations or experiments. We use topics (such as rain and the water cycle, and a forest community) that children have had experiences with, or are interested in, and for which there are many children's literature information books for young children. These books that inform various hands-on explorations serve as tools of engagement in science. They allow children to interact with scientific ideas and language, in addition to visual designs, all of which become ideational resources for their inquiries. Various other texts (journaling, semantic mapping, a class mural, their own illustrated information book, data sheets) that children create themselves and with their classmates and their teacher become tools of thinking and talking science where the emphasis is on children's reasoning and sense making. This sense making is encouraged not only at the individual but also the collective level, as children work in small groups, participate in whole-class discussions with the teacher and their peers, as well as share their thinking with other adults—research assistants in their classrooms. The hands-on explorations that children do themselves (except for a couple teacher demonstrations) offer a different type of discursive space for children to think and talk science in the presence of others, and in the presence of material and symbolic artifacts. As children engage with texts, material objects, dialogue, ideas, and symbols in various classroom communities, and they are helped to put together their own understandings and ways with words with those of the scientific community, they become learners of both science and literacy.

In this chapter, we explore various integrated science-literacy activities in different classrooms: (a) a hands-on exploration that involves sorting objects into solids, liquids, and gases in a third-grade African American classroom; (b) a whole-class read-aloud of a book on earthworms in an ethnically diverse first-grade classroom; (c) a whole-class mural-making activity of a forest in a bilingual Spanish/English second-grade classroom; and (d) two dyadic conversations between the same Latino third-grade boy and a university researcher around two different information books the student had composed. We use three lenses for the conceptualization and analysis of these activities—identity, activity theory, and sense making, as we examine how young children build and use conceptions of themselves to navigate situations of learning and interaction in the context of these activities.

IDENTITY, ACTIVITY, AND SENSE MAKING

Conceptions of Identity

Identities can be tools used by people to participate in activities and negotiate meanings in communities of practice. They help us to make sense of where we fit in the "going-ons" of everyday life. In this way they are externally oriented, helping us to communicate to others who we are and what roles we play in the pursuit of co-constructed/contested goals and the meanings that result from and within this pursuit. Identities are also internally oriented. They help us make relevant to our own lives aspects of activities we engage in. They help us to choose what is meaningful for us, based on our pasts for both our present situations and for our futures.

Identities are conceptions of ourselves, but also conceptions of others about us and our conceptions of others' ways of "seeing" us as we act, behave, think, perform, feel, and position ourselves in activities. The kinds of people we (and others) think we are mediate the kinds of people we enact in activities and the kinds of meanings we make. At the same time, our participation in activities mediate who we, and others, think we are, thereby influencing our future participation and meaning making in activities. We, as well as the activities, have histories and are part of trajectories that have been formed with (or without) our involvement over various timescales. Moreover, the activities and we develop and evolve over time; however, depending on the timescale, this development may be relatively small or reflecting no change at all. Furthermore, it is also important to consider the moment-to-moment interactions between who we are and the activities we engage in by choice, or because we have to or because of a combination of the two.

Identities are not singular entities. Children and adults may have multiple identities that may be foregrounded or backgrounded in particular activities. As we engage in an activity, we negotiate (consciously or unconsciously) the identities that are played out, and, thus, certain identities get strengthened or weakened, and others are not affected that much. The interplay between the various identities we invoke allows us to put out markers to activity participants that signal our membership in certain identity domains.

A third-grade girl may be the "bossy" girl who tells everybody else what to do, an actress, an aspiring doctor, a sister of a mentally-challenged girl, a "mother figure" for her many cousins, and a "buddy" for her divorced mother. In her school classroom, depending on the activity, others can recognize her as a particular type of person. In the classroom discourse in which she may or may not participate, one of her identities may stand out more than another. Her interactions with the rest of the activity participants, what is said to her, how she is recognized, or not, in classroom discourse, what she and the other participants do, think, feel, talk during an activity reinforces, or not, a kind of person she thinks she is and a kind of person others think she is. This is a particular lens to the identity construct that Brown, Reveles, and Kelly (2005) have coined as "discursive identity," or in other words "the act of communicating identity via discursive interaction" (p. 783). In this chapter we adopt this lens and we use activity theory to further understand how discursive identities are realized in science classroom activities that share similarities, but they are also characterized by unique elements.

Although not extensively explored in this chapter, it is important to point out the distinction between subjectivities and identities. One of the differences between subjectivities and identities is their relative transparency in the moment of interaction. One can identify oneself as a teacher. It is a recognizable identity, legitimized by others with whom we interact and by the institutions in which we participate (schools, for instance). Although dynamic, institutions develop enduring identities (such as teacher and student) that exist across time. This makes these kinds of identities more visible in our everyday exchanges. Any new identities must be recognizable in terms of what has come before them. Subjectivities are often harder to recognize in the moment. For instance, as that teacher, we may not think we are positioning ourselves as a person who values only information found in "official" sources such as books, when we respond to students' responses by saying, "Did we read that?" Subjectivities happen much more quickly and are much more tacitly supported by participants than identities.

Thus, subjectivities are the ways we position ourselves and the ways others position us in our moment-to-moment interactions; they make up our subject positions, or who we are as kinds of subjects. They are the results of our positionings, the interaction of a number of kinds of identities: biological (we might be an 8 year-old girl), institutional (a third-grade student), and affinity (one of a group of girls who participates in folkloric dancing) (Gee, 2000–2001). But most importantly, they are the result of discursive (interactional) positionings that may or may not be based on biological, institutional, and affinity group kinds of identities (Davies & Harré, 1990). Subjectivities are often constituted by singular, fleeting moments or references to kinds of people or affiliations. Subjectivity is one "story" in the flow of activity. As such, our subjectivities may change over the course of an activity as a result of the kinds of discursive interactions we have. Identities, in contrast, are constituted by the accumulations and sedimentations of these singular moments. But, identities are what we use to invoke and mediate subjectivities. Thus, subjectivities and identities exist in a close dialectical relationship; identity is a mediational tool that negotiates subject positions for us, but is also *negotiated by* our subject positions.

Conceptions of Activity

Identities and subjectivities unfold and evolve within activities. Activities are historically and socially based, goal-directed, mediated co-ordinations of actions in communities of people with rules, values, and divisions of labor. That is, activities always have a motive, or a purpose that is larger than any one individual action, and people's actions within these activities are always towards one or another motive that is in part shaped by the communities they inhabit, the values and rules that such goals call for, and delineations of who gets to do what, when, and how (Cole & Engeström, 1993).

Activity systems, canonically conceived, are made of six interacting elements. Three of these elements are parts of the basic mediational triangle as composed by Vygotsky (1978)—the subject, the mediational artifacts, and the object or goal. Activity theory adds to this basic triangle by including another layer composed of the rules of the activity, the community in which it takes place, and the division of labor among activity participants.

All of these elements interact with each other. For instance, subjects are parts of communities and subjects take on the required roles that a particular division of labor entails. But communities, with rules and divisions of labor, also mediate the kinds of work subjects do. While they do not directly use activity theory, Lave and Wenger (1991) point this out in their description of the kinds of participation required in communities of practices as one apprentices to be a tailor or a midwife. What apprentices are asked to do (watch for the first few months or tend to the fire to boil water) depends on where they work (a community hospital or an indigenous midwife), who they work for (a team of nurses and doctors or their mother), and the tools they use (ultrasound machines or herbs and natural medicines).

Mediation is critical in activity theory. Artifacts mediate the reaching of the activity goals by subjects who are engaged in the activity in collaboration with others. The mediating artifacts can be material or symbolic. The materials (crayons, markers, gel pens) that children use to create a class poster of their neighborhood let them interact individually and collectively in such an activity in particular ways. To understand the children's poster we need to consider the material artifacts used as mediational devices. But we also need to consider the photographs that the children took as part of their community walk and how they talked about them in class before they made their poster. The photographs are not only material artifacts; they are symbolic devices that mediate children's actions in the world. Moreover, the dialogue that children and teacher had about them is another symbolic mediational tool that privileged some aspects of the neighborhood and suppressed others. The ideas discussed about the neighborhood are also by themselves symbolic artifacts that shape the children's product, the outcome of the poster-making activity.

The activities we explore in this chapter, which have been part of the ISLE project, offer children expanded opportunities for learning and development by explicitly and implicitly building on participants' identities as people, students, authors, artists, scientists, and classmates. These identities are embedded in and evoked by the communities of practice in which children, their peers, their teacher (as well as other interlocutors) are parts of. The activities are distinct in various ways relative to the six interlocking activity elements. In Table 10.1 we summarize the basic three elements for each of the activities. As we explore each of them below, we expand on these activity elements and also discuss the other three elements. These activities can be seen not only as separate activity structures, but also as an activity system that aims at providing children with multiple and different opportunities to work with ideas in different ways and construct a sense of the content (which is Engeström's [2001] concept of horizontal—as opposed to vertical—learning).

Sense and Meaning

As participants interact in the context of activities, everyone involved jointly constructs meaning. Roles are taken up, power relations play out, and the dynamic interaction produces the event. Participants position each other as they creatively act within institutional conventions and structures and negotiate boundaries. The participants' discursive identities are primarily communicated and developed via language, but other modalities also shape and are shaped by the participants' identities and the activities themselves. Nonlinguistic modalities, such as images, body movements, use of space, gestures, facial expressions (including gaze), sound, and

Table 10.1. The three basic activity elements for each of the four activities studied.

Basic Activity Elements	Activities Explored			
	Hands-On Exploration	Mural Making	Read-Aloud	Book Sharing
Subjects	African American 3rd graders and a white teacher	Bilingual Spanish-English speaking 2nd graders and a Latina bilingual teacher	Ethnically-diverse 1st graders and an African American teacher	A Latino 3rd grader and a Caucasian researcher
Objects/Goals	Sort objects in solids, liquids, or gases, and give reasons	Create a class mural representing ideas about the forest	For the teacher to share dialogically the book with children, creating spaces for their comments, ideas, questions, answers around the ideas presented in the book	Understand children's intentions in representation and organization of ideas in their books
Mediating Material Artifacts	Assortment of everyday objects (e.g., baggie with shaving cream, clay, drinking straw, can of soup, baggie with salt, balloon), datasheet	Construction paper, scissors, crayons, markers, colored pencils, glue, pipe cleaners, cotton balls, hole-punch, tissue paper, and a tri-fold sturdy poster board	Information book *Earthworms*	Child's own illustrated information books, *Storms* and *Chipmunks*

length of time, all contribute to the meanings we make in our lived experiences (Kress & van Leeuwen, 2001).

Linguistic and nonlinguistic modes of expression and development of meaning need to be contextualized within the interaction among the discourse partners. Vygotsky's differentiation between the meaning and the sense of a word is relevant here (despite the fact that Vygotsky focused only on language). In *Thinking and Speech*, Vygotsky (1987) made the distinction between the *meaning* of a word as it may appear in a dictionary, and the personal *sense* that a word has for a speaker as a result of the contexts in which the speaker has used and heard the word before and during the specific interaction. Sense does not only encompass intellectual content, but it is also made up of the affective overtone that a word carries with it as it is used over and over in various contexts.

A word's sense is the aggregate of all the psychological facts that arise in our consciousness as a result of the word. Sense is a dynamic, fluid, and complex formation, which has several zones that vary in their stability. Meaning is only one of these zones of the sense that the word acquires in the context of speech. In different contexts, a word's sense changes. In contrast, meaning is a comparatively fixed and stable point, one that remains constant with all the changes of the word's sense that are associated with its use in various contexts. . . . The actual meaning of a word is inconstant. In one operation the word emerges with one meaning; in another, another is acquired. Isolated in the lexicon, the word has only one meaning. However, this meaning is nothing more than a potential that can be realized in living speech, and in living speech meaning is only a cornerstone in the edifice of sense. (p. 276)

The ultimate goal of all the activities we discuss in this chapter is children's learning, broadly defined. Learning for us encompasses many aspects, including seeing oneself as a learner, questioning ideas, arguing about ways of reasoning about something, explaining, developing theories, examining evidence, externalizing the thinking that has gone into making or producing something, designing a representation of an entity. We do not associate learning exclusively with figuring out the "right" ways of thinking about an idea, a phenomenon, or concept—the ways that have been appropriated by a particular community of practice, which, in our case is the scientific practice—although this is, of course, an important and necessary aspect of learning. In this way, the sense, rather than the meaning (of particular science content), that children express and develop during these activities, is of particular interest to us, as we attempt to understand how discursive identities are played out in these activities and how children's positionings in the activities through a historical, developmental, and moment-to-moment interactional lens shape and are shaped by the activities.

SORTING OBJECTS: NEGOTIATING AMBIGUITY AND AGREEMENT

In this section, we explore a hands-on task in Jennifer's third-grade class. The activity involves two stages: first children in small groups sort everyday objects as solids, liquids, and gases; and, then, the whole class discusses the categorization of these various objects and develops whole-class groupings. Here, we focus on the first stage and particularly on one of the groups of children that were called "teams." (A key to the transcription symbols used is included in the Appendix.)

The class had read the book *What's the World Made Of?: Solids, Liquids, and Gases* (Zoehfeld, 1998), an important semiotic tool for this sorting activity that provided children with both prototypes of the three states of matter and specific characteristics of each state. At the beginning of the sorting activity, the teacher modeled for the class the sorting of a few objects (different from those the children would sort). Jennifer also explained how the datasheet (Figure 10.1) that each student should fill out was organized. Jennifer used three objects, a stuffed turtle, a bottle of spray cleaner, and an empty clear plastic cylinder as her examples. She began by showing them how they would sort the objects into three groups and place them on large sheets of paper labeled "Solids," "Liquids," and "Gases." She asked students, "What is this?" having them consider whether it was a solid, liquid or gas, as she made repeated references to the read-aloud book session and the

SOLIDS, LIQUIDS, GASES DATA SHEET

SOLIDS	LIQUIDS	GASES
Object	Object	Object
Explanation	Explanation	Explanation
Object	Object	Object
Explanation	Explanation	Explanation
Object	Object	Object
Explanation	Explanation	Explanation
Object	Object	Object
Explanation	Explanation	Explanation

Figure 10.1. Datasheet used by the students to record and explain their decisions.

three states of matter. She also asked the class to tell her the reasons for their judgments. As she went through the objects, she outlined the process of sorting by writing the steps on the board: *sort objects, draw the object, and explain* (on the datasheet). She engaged the students in conversation about the reasons they gave. She distinguished between the contents and the container of objects, and she acted on the objects to highlight parts of them. For example, she shook the bottle of spray cleaner so the children could hear the liquid inside. She tapped on the spray bottle and asked, "What do you think this is?" She also tapped on the plastic cylinder while she asked what state of matter it represented, but then she smiled at the class to indicate that there was another possible answer when they responded with solid. Daniel suggested that it belonged in the gases group because it had air inside. Jennifer was explicit about telling the children that there could be more than one possible answer and that they did not all have to agree with each other or with her. However, each person had to have an explanation on his or her datasheet.

Each of the five groups was then given a brown lunch bag with the objects they needed to sort out. Each child was also given a datasheet where he or she had to categorize each object and write out the reason for that decision. In one of the groups, four kids—three girls, Jamilia, Chantrelle, and Latessa, and a boy, Lawrence—try to work together to sort the objects and fill out the datasheet that they were given. Lawrence pulls the baggie with the salt out of the brown bag, but Jamilia takes it out of his hand, and then first identifies it with excitement and specifies that it is heavy: "Ooo, this salt // it's salt, Miss Hankes ~ it's heavy." As she says this, she tosses the baggie from one hand to the other probably feeling its "heaviness." Chantrelle takes out the cork and asks Jennifer, who has visited the group, what it is, and Jamilia takes out the baggie with the shaving cream, saying "look, look!," taking it to another group to show them, but as nobody pays attention to her, she returns to her group. The artifacts used in this activity generate quite an excitement for the children who want to touch them, manipulate them, feel them, and, as we will see below, not necessarily share them freely with each other. Also, there does not seem to be a particular way that children are dividing among themselves what needs to be done in this activity. Each one of them focuses on the item she or he chooses at the moment.

Chantrelle offers a characteristic for the baggie with the shaving cream, "I think this don't keep its shape." However, Latessa turns to another object—the helium balloon asking, "What is the balloon?" Lawrence answers, "It's a gas. It's a gas and a solid" while he is tapping the balloon. Lawrence is probably considering both the content (gas) and the boundary surface (solid) of the balloon. Latessa, following Chantrelle's example, gives a macroscopic property of her item, the balloon ("it keeps its shape") that seems to be related to the boundary surface that Latessa, like Lawrence, taps. But Jamilia turns to yet another object, the clay, saying, "[it is] hard, this is a solid." Latessa agrees with Jamilia's idea ("That's a solid") and takes the clay from Jamilia and categorizes it as a solid.

Latessa is a leader. She is the first one to put an object in one of the state of matter groups, and she is the one who feels the need for conversation about the items among their group members, "Okay let's start talking about it, y'all." For Latessa, one of the rules of this activity is that the group members discuss each object and decide how to categorize it. This is a rule that Jennifer, the teacher, had explicitly stated as she was modeling this activity to the children ("Your first task as a team is to decide if you should put the objects on the solids paper, the gases paper, or the liquids paper. That's your first task, to take all the items that you have and to decide where they belong. . . . Then explain why it goes in that category." After letting her group "fool around" with the activity artifacts for a while, Latessa steps in and encourages the group to come together and talk about the items they have to sort.

Also, Latessa, as she picks up the tube of paint, continues in the same turn, "Ooo, I know what this is. This is a ~ this is a liquid because it don't keep its shape." She models for her peers what she encourages them to do, namely, to keep going with the items, and share their thinking about the group the item belongs to, and the reasons behind it. Latessa is considered one of the "smarter" students in the class. Other students tend to go to her for help. She is also the "good student" who always knows what to do and frequently raises her hands with answers. She usually has insightful comments and she does not misbehave or act silly, though sometimes she talks a lot.

In contrast, Jamilia is seen in a different way, as the following excerpt shows:

1	Jamilia:	[She smells the sponge, as she responds to Latessa's remarks about the paint.] Duh, paint don't keep its shape. [She puts the sponge down in solids category.]
2	Latessa:	I know.
3	Jamilia:	All right [in a softer tone]. [Showing to Latessa the clay that sits in the solids group] Clay don't keep its shape.

211

Jamilia speaks with an "attitude" ("duh, paint don't keep its shape") (unit 1). It is this tone of voice that usually gets her in trouble with her classmates and the teacher. Often, other students will get frustrated with her comments and conflicts arise. But, Latessa does not engage in any conflict with her this time (unit 2) and Jamilia backs off (unit 3). Whether an object keeps its shape or not continues to dominate the discussion as the only way of determining the state of matter that an item is in. So far, though, only Latessa has provided a comment linking explicitly the liquid state of matter and the macroscopic property of not keeping its shape when she was commenting on the tube of paint. This explicitness may have led Jamilia to discuss again the clay that has already been placed in the solids group. Although Jamilia does not indicate that she is confused, her comments seem to imply a contradiction—on the one hand, she thinks that clay does not keep its shape (unit 3); and on the other hand, clay sits in the solids group. She attempts to engage Latessa in thinking about this. We expand below on Jamilia's sense of clay not keeping its shape.

4 Chantrelle: [Holding up the baggie of salt] This could keep its shape.
5 Jamilia: You can tell this [the straw] don't keep its shape. [She bends the straw and drops it in solids.]
6 Latessa: It could bend and stuff.
7 Chantrelle: This [the baggie of salt] could keep its shape.

As Chantrelle focuses on another item, the group does not take up her remarks about the salt keeping its shape (unit 4). Her contributions seem to be ignored by the rest of the group members. Chantrelle is a very soft-spoken child. When she is speaking in class it is often very difficult for anyone to hear her. Her classmates tend to want to speak for her, but the teacher does not let them, preferring to wait for Chantrelle to speak more loudly for herself. However, in this boisterous classroom, if a child is not loud enough, she or he cannot be heard, and as a result, Chantrelle is usually "run over" by her classmates, despite the fact that she has good ideas and she is well liked. Thus, her "pale" voice presence contributes to her classmates' behavior towards her. In this activity, where no explicit division of labor has been enacted in this team, its members switch from one item to another in a rather chaotic way. Consequently, Chantrelle struggles to be heard as Jamilia and Latessa seem to be talking only to each other.

In unit 5, Jamilia uses the "not keeping its shape" reasoning to add the drinking straw to the solids category. Her remark is not scientifically correct—solids keep their shape, and liquids and gases do not. However, when we say that solids keep their shape, we assume that this is true only when there is no external influence on the object. The straw changes its shape because Jamilia bends it (similarly to the clay that changes shapes when somebody works with it). Therefore, the observation that straw does not keep its shape is partially correct. At the same time, Jamilia somehow believes that the straw is a solid and puts it in the solid category.

8 Latessa: [Picking up the can of soup and tapping it with her pencil] Ooo, this is a // inside this // this is a solid.

9	Lawrence:	Where the clay at?
10	Latessa:	Inside it's a liquid	
11	Chantrelle:	[She points to the can.] It's a solid.
12	Latessa:	[Ignoring Chantrelle] But outside it's not a liquid. [She adds the can as a solid on her datasheet]. C'mon we got to talk about it.

Chantrelle's second comment about the salt (unit 7) is still ignored. Latessa focuses now on the can of soup and seems to have a conversation with herself, thinking aloud about the two parts of the soup can: its contents (the "inside," unit 10) that is liquid, and the container (the "outside," unit 12) that is a solid. She is the first to make explicit the distinction between the inside and the outside of an object. As noted earlier, Lawrence has considered both the inside and the outside of an object (the balloon), but he has not differentiated between the two stating "it's a gas and a solid." Chantrelle speaks what she thinks, but Latessa does not seem to hear her (unit 11), even though she eventually agrees with Chantrelle. Latessa also tries to rally the team to a discussion, "C'mon we got to talk about it" (unit 12); however, no one takes her up on it, even though this is one of the rules of the activity. Jamilia now considers a sponge.

13	Jamilia:	[Holding the sponge] This ain't got no shape. [Puts down the sponge and picks up shaving cream.] Ooo shaving cream!
14	Latessa:	[Picks up sponge in response to Jamilia.] It keeps [squeezes sponge] ~ when you drop it, it still keeps its shape. [She drops the sponge twice and places it with solids.]
15	Jamilia:	Y'all supposed to be doing this. [She begins sorting objects into groups, somehow implying that Lawrence's behavior was problematic.]
16	Chantrelle:	He [Lawrence] ain't doing nothing [wrong].

Jamilia moves to a different item—the sponge—and puzzled about it, she picks it up and then passes it down, remarking that it has no shape (unit 13). We do not know why she thinks that but Latessa's action, following Jamilia's comment, may give us an insight. As Latessa squeezes the sponge, its shape changes. That may be Jamilia's sense of the sponge not having any shape. But, Latessa focuses on another characteristic of the sponge. Although she squeezes the sponge, she changes her mind and considers its shape when it is dropped (she drops it twice, noticing that it keeps its shape, and then she places it in the solids category [unit 14]). It is important to see Latessa's action to go back to the sponge (even though Jamilia had left it to turn to the shaving cream) in the light of who Latessa has been in class and in this activity. As the leader, the organizer of the group, who believes that having an extensive discussion on the items is an important rule of the activity, she appropriates this rule in her talk and actions explaining why the sponge is a solid, and by physically placing it with the solids group.

After Latessa places the sponge in the solids group, Jamilia reminds everybody that they should all be sorting the objects (unit 15)—a comment that Chantrelle perceives as threatening to Lawrence and she comes to his "rescuc" (unit 16). The other children know Jamilia as a tattletale—she "tells on" other students, reports their misdoings to the teacher, and corrects what she perceives to be "bad" behavior. Chantrelle tries to prevent her from reporting on Lawrence by explicitly signaling to Jamilia that he is not doing anything wrong. After that, Jamilia begins to move objects around. She takes the soup can, water bottle, paint tube, shaving cream, and hand sanitizer, and places all of them in the liquids group, focusing apparently on the contents of the objects rather than the containers. She places salt in the gases group. Lawrence and Latessa join her in the sorting. Lawrence walks around the table to where Jamilia is standing and asks where the clay is. He finds it (this is off camera) and carries it back to his seat. He is rolling the clay in his hands and across the surface of the desk. He continues to off camera for the most part, but it appears that he is just exploring what he can do with it. He says it is a liquid and then quickly corrects himself to say that it is a solid.

17 Lawrence: This [the clay] a uh liquid. It's a solid. [Shows Chantrelle the clay and holds it out so she can squeeze it.] Look, it's hard.
18 Chantrelle: [Chantrelle squeezes the clay.]
19 Lawrence: [He places the clay in solids and then focuses on another object]. This [straw] a gas [as he puts it in the gases group]. [Then he touches the baggie of salt, which is in the same group, but leaves it there.]
20 Chantrelle: [Picks up the baggie of salt] This a solid.
21 Lawrence: [Taking each object as he names it a gas and placing it in the gases group] This [empty balloon] is a gas. This [baggie of air] a gas. This [the helium balloon] a gas right here.
22 Jamilia: The shaving cream [she picks up the baggie with the shaving cream from the liquids category and squeezes it so that it spreads throughout.]
23 Latessa: The [helium] balloon is gas and it keeps its shape [she taps the balloon several times].
24 Jamilia: Oooh, it's [shaving cream] soft.
25 Chantrelle: This [salt] a solid ~ and it keeps its shape.

Lawrence goes back to the clay, an object that already has been placed with the solids group by Latessa (unit 17). He probably feels that he has not been committed to this decision earlier and now he wants to think it through for himself. Lawrence has grown to become interested in science. He has become much more verbal in science class, and he is usually ready to contribute, but his reasoning is often different from scientifically accepted ways. He seems to be in his own "little part of the world" at times and he often comes back to ideas after the class has discussed them. Other children think of him as knowledgeable because he speaks with authority and has scientific-sounding reasons that they may not understand, but that they may be inclined to believe because it sounds good.

Although Chantrelle has not been actively sorting objects, Lawrence brings her in by allowing her to manipulate the clay (unit 17). It is not clear how Chantrelle's action makes Lawrence decide to put the clay in the solids group. Perhaps Chantrelle's squeezing of the clay helps Lawrence think that clay is hard and is therefore a solid. For Lawrence, it is also clear that objects that have a gas in them should be in the gases group, as he focuses on the contents of the balloons and the baggie filled with air (unit 21). Latessa continues to think about both parts of the

IDENTITY IN ACTIVITIES

helium balloon, the inside that has gas and the outside (that she taps) that "keeps its shape," (unit 23). And Chantrelle still tries to get her point across about salt being solid because "it keeps its shape," (unit 25) but nobody hears her now either. Jamilia (in unit 24) turns her attention to the baggie with the shaving cream that she has earlier placed in the liquids group. She is taken by the feel of it—"it's soft."

26	Jamilia:	[To Lawrence, trying to take the rubber band from him] That's a solid.
27	Lawrence:	[He does not let Jamilia have the rubber band.]
28	Latessa:	[To Chantrelle] No, it [the salt] don't. [Touches the baggie of salt.] If it's out of the bag, it don't keep its shape. I don't get it.
29	Jamilia:	[Looking at Latessa and pointing to salt] It's made out of gas. You can tell. [She makes this comment with complete confidence.]
30	Lawrence:	[Walks over to Jamilia's desk and tries to take shaving cream from her.] This is a gas I mean um |
31	Jamilia	[Pulls the baggie away.] It's liquid.
32	Lawrence:	That [shaving cream] ain't no liquid!

The artifacts used in this activity invite these children to engage in an occasional "tug of war" over them. Jamilia and Lawrence have several such exchanges. Each is very protective of the object she or he handles at that time. When Latessa touches the baggie of salt (unit 28) she eventually acknowledges what Chantrelle has been saying all along, namely, that salt keeps its shape. However, Latessa objects to Chantrelle's position—"If it's out of the bag, it don't keep its shape," even though she admits she is confused—"I don't get it" (unit 28). Chantrelle has the reputation of a good student, so Latessa does not dismiss her thinking altogether when she finally hears it. Jamilia, with her authoritative tone, totally dismisses Chantrelle's point and instead announces that the baggie of salt is "made out of gas," (unit 26). Nobody challenges Jamilia at this point, as Lawrence's attention is caught up by the shaving cream that he tries to take away from Jamilia (unit 27). Lawrence thinks the baggie with the shaving cream is a gas, but Jamilia continues to think it's a liquid, to which Lawrence categorically objects. As Jennifer, the teacher, briefly joins the group, the conversation about the shaving cream continues.

33	Teacher:	Okay, you guys had a question. Then I'm going to that team over there. What was your question?
34	Jamilia:	[Squeezes shaving cream.] Miss Hankes, I can feel ~ it feels real nice.
35	Latessa:	[Points to shaving cream.] What is that?
36	Teacher:	Shaving cream. What do you guys think?
37	Jamilia:	It's soft.
38	Lawrence:	I think it's // I think it's um a solid because you can squeeze it [making a squeezing motion with his hands, as he is trying to squeeze the bag].
39	Jamilia:	[She pulls away the bag of shaving cream from Lawrence's hands.]
40	Teacher:	You think it's a solid. What does somebody else think?
41	Jamilia:	You can't squeeze a solid [putting the baggie with the shaving cream in the liquids group].

Lawrence seems to associate squeezing with solids (unit 38). When Chantrelle squeezed the clay he put it in the solids group (units 18 and 19). He uses the same approach for the shaving cream. But Jamilia totally opposes Lawrence's idea (unit 41). She actually makes her comment with considerable sarcasm, seeming to think that the idea that a solid can be squeezed is absurd and places the baggie with the shaving cream in the liquids group where she had placed it earlier during their group conversation. We need to remember that at the beginning of the activity, Jamilia categorized clay as a solid because it was "hard." The group will not discuss this any more.

Focusing on one of the groups in Jennifer's class offered us an opportunity to see how identity, positioning, who a child thinks he or she is, and who others think the child is, intermingle in an activity that involves several artifacts and a team that adopted a non-specified division of labor. The four children's discursive identities varied, but they reveal who the children have been and have been thought of by their peers in their classroom community. The activity, with its lack of structure in terms of how and by whom the objects are sorted, allowed children to realize their sense of what needs to be done and how. The process of sorting, and reasoning about particular choices, was not linear. Children went back and forth between objects and reasons, making sense, their own meaning, of what solids, liquids, and gases look like, what they can and cannot do, and what characteristics they have and do not have. Part of the struggle was that the four children did not share a common framework, a common knowledge. But that was exactly the goal of this activity, namely, to develop this common sense of the states of matter and their characteristics. As children were creating this sense, they were touching, feeling, and manipulating the items that they were taking them from each other. They were also communicating to each other, referring to ideas and objects by pointing, gesturing, moving, as well as identifying them by language. Thus, a glimpse of multimodal sense making comes through this multimodal activity, as different children's identities mediated their actions, engagement, and ways of being.

READ-ALOUDS: DISCURSIVE SPACES FOR QUESTIONS AND ANSWERS

First graders in Sharon's class settle down for the read-aloud of the book *Earthworms* (Llewellyn, 2000). As part of the Forest unit, the class has already read and discussed three other information books, *In the Forest* (First Discovery, 2002), *A Forest Community* (Massie, 2000), and *Animals under the Ground* (Fowler, 1997). For this read-aloud activity, Sharon sits in a chair and begins by having the book facing towards her in her lap. Children, sitting on the floor facing her, are very attentive and look ready for a read aloud on worms—an exciting topic for them.

1 Teacher: Do any of you know why they are called earthworms? Alam.
2 Alam: They help the earth.
3 Teacher: How?
4 Alam: By um smoothing the soil and to help the plants grow and without the animals eating them they don't get killed.
5 Teacher: Okay. Demario.

IDENTITY IN ACTIVITIES

6 Demario: Um well this is what earthworms do. Well, when it rains they climb out under the soil.
7 Teacher: Do you know why?
8 Demario: Um because they get squished.
9 Cs: (*** ***)
10 Teacher: I can't hear. Vittoria.
11 Vittoria: Earthworms they come out // um they come out of the ground.
12 Teacher: Um, John
13 John: Earthworms come out of the ground when it rains because um // they come out when it rains because um they like it wet. When it isn't raining outside, they stay in the ground because the dirt is damp. But then when it starts raining outside they get up in holes and they get out of the ground and then they go all over the place. Not all over the place, they just stay near usually.
14 Teacher: Okay. Jamar, what do you know about earthworms?
15 Jamar: They move a lot.
16 Amato: Ooo! Ooo! [He is sitting up on his knees with his hand raised high in the air.]
17 Teacher: Now Amato.
18 Amato: Um they come out because when they come and when it is raining their house will be their new house.
19 Teacher: How does their house get destroyed?
20 Olivia: They don't have a house.
21 Teacher: No you need to listen, he might be right. What destroys it?
22 Amato: [He does not respond.]
23 Teacher: What destroys it? What were you talking about? Do you remember?
24 Amato: Rain. [He has a big smile on his face.]
25 Teacher: Oh, okay! Denzel.
26 Denzel: Rain can't destroy their houses because their houses is mud and rain makes it (***).
27 Teacher: Okay. Olivia.
28 Olivia: They are called earthworms because they live in the earth except they are just underground. [Claps hands in front of her and speaks with a soft voice.]
29 Teacher: Okay let's see what goes on. So we all agree that earthworms live underground?
30 Cs: Yes.
31 Amato: This is going to be a good one.
32 Teacher: [Holds up the book and begins to read.]

Before Sharon, the teacher, starts the read-aloud, she engages the children in sharing what they already know about earthworms. In this way she sets the tone for the rest of the read-aloud. One of the rules of this activity (that the teacher and the students have been working hard to enact in these two integrated science literacy units) is that children think about the ideas presented in the book and share their thinking with the rest of the class. We will see below how this is enacted in this particular read-aloud. So far, Sharon has called on several children to contribute—Alam, Demario, Vittoria, John, Jamar, Amato, Olivia, and Denzel—eight children out of the 30 children in her class. What is also salient during this beginning of the read-aloud is that children, sometimes prompted by Sharon as she asks them how and why (i.e., units, 3, 7, 19, 21, 23)—offer their reasoning behind the various ideas they bring up. This is another implicit rule of this activity and the unit as a

217

whole. Sharon repeatedly encourages children to explain, to think about why and how something happens, which is an important feature of scientific practice. What we also see is a collective meaning making that unfolds in the classroom. With the exception of Jamar's comment ("they move a lot," unit 15), all the other participating children talk about how the worms live in the soil, what kind of "house" they have, and how the rain interacts with their home. Furthermore, in unit 21, we see Sharon protecting Amato's participation in the classroom discourse. In unit 20, Olivia objected to Amato's idea about earthworms having "houses" probably knowing that the word "house" denotes a building, a structure that she knows worms cannot build. However, the teacher validates Amato's idea as potentially correct possibly interpreting his word "house" in the sense of a "home" where the worms live—the soil that other children have been talking about. It is worth noting that Amato has had a strained relationship with the teacher. He has been going to the principal's office several times throughout the year. Amato is a younger and very immature first grader, who has a hard time sitting in his chair and listening to directions. He is often doing something of his own interest when he is supposed to be working on his journal or other independent work. He tends to lure his close seatmates into any off-task excitement that he can. He enjoys silent reading very much and listens to the read-alouds more than he did other "uninteresting" work. He has a difficult time writing his thoughts down on paper; however his oral contributions are richer and meaningful. Right from the start, Amato is the only child to explicitly reveal his expectation of this particular read-aloud ("This is going to be a good one" [unit 31]). His excitement about this read-aloud and its discussion shows through his body movements and facial expressions (units 16 and 24).

33 Teacher: ALL SORTS OF {EARTH} DIG UP {A} PATCH OF EARTH IN A GARDEN OR PARK. LOOK CLOSELY AND YOU WILL PROBABLY SEE A WRIGGLY WORM. WORMS THAT LIVE IN THE SOIL ARE CALLED EARTHWORMS. THERE ARE SEVERAL KINDS OF EARTHWORMS. MOST EARTHWORMS ARE REDDISH BROWN AND ABOUT AS LONG AS YOUR HAND.
34 Cs: [Hold up their hands to estimate the length of worms.]
35 C1: That is big.
36 C2: That is long.
37 Teacher: THERE ARE MANY DIFFERENT KINDS OF WORMS AROUND THE WORLD. WORMS COME IN MANY COLORS AND SIZES. SOME ARE SO TINY THAT THEY ARE HARD TO SEE. OTHERS, SUCH AS THE GIANT EARTHWORM OF AUSTRALIA, ARE AS LONG AS A CAR.
38 Cs: Awesome!
39 Teacher: [Reading the caption under a picture] THIS GIANT WORM FROM AUSTRALIA [points to the Australian worm so the kids can see it] IS AS LONG AS A CAR // THIS GIANT WORM FROM AUSTRALIA IS AS LONG AS A CAR! IT IS MUCH LARGER THAN A GARDEN WORM. So there is a little bitty garden worm right there. [Points to the small worm depicted next to the long Australian one.]
40 Cs: Laughs.

41 Teacher: THE BRANDLING WORM HAS A {STRIPE} A {STRIPPING} SKIN {THAT} MAKES IT EASY TO RECOGNIZE. [Points to the brandling worm so the kids can see it.]
42 Amato: [Pointing to the book] Miss Gill, the biggest one is the king.
43 Teacher: Okay. [Turns book back towards herself to read.]
44 Alam: If there is a king.
45 Teacher: Okay, who wants to hear about finding earthworms?
46 Cs: [All raise their hands.]
47 Teacher: Okay then we need to settle down. [Turns the book to face the kids so they can see the words and pictures.] FINDING EARTHWORMS. EARTHWORMS LIVE IN {DENSE} SOIL WHERE THERE ARE BITS OF DEAD PLANTS TO EAT. MOST OF THEM LIVE NEAR THE TOP OF THE SOIL. EARTHWORMS ARE EASY TO FIND. DIG IN A GARDEN WITH A SPADE OR FORK. YOU ARE SURE TO FIND SOME IN THE SOIL. [Shows children the picture of a magnified image of an earthworm.]
48 Cs: Euuew, slimy.
49 Amato: I use a stick for the part where |
50 Teacher: EARTHWORMS LIVE IN WOODS, MEADOWS, AND GARDENS. THEY LIVE UNDER {THE} LAWNS AND AMONG {DRIED} LEAVES UNDER BUSHES AND TREES. WORMS ALSO LOVE COMPOST HEAPS, WHERE PEOPLE PILE VEGETABLE PEELINGS AND OTHER FOOD WASTE. [Points to the picture of a worm on green grass.] THERE ARE VERY FEW WORMS IN SANDY SOIL. THIS IS BECAUSE RAINWATER DRAINS AWAY QUICKLY AND THE SOIL BECOMES TOO DRY FOR THEM. So what kind of soil does the earthworm like?
51 C1: Wet soil.
52 Teacher: Okay let's find out. Now we will learn about the worm's body. Are you ready? [Turns the book towards herself to read.]
53 Cs: Yes.
54 John: I know some body parts. I know the middle part is the saddle.
55 Amato: Euuew.
56 Teacher: Uhhuh
57 Alam: Don't tell us all of the good |
58 Teacher: What else do you know?
59 John: Um you can tell which one is the um head because the head is usually um farther away from the tail. And the tail is closer. And the head |
60 Teacher: Closer to what?
61 Amato: [Is sitting up on his knees, playing with the computer chair and talking to Carlos.]
62 John: The head is closer to the saddle.
63 Teacher: Oh okay.
64 John: The tail.
65 Teacher: [In a very calm voice] Amato you don't want to go see [principal's name] so you need to sit and listen, all right? Demario.
66 Demario: (*** ***)

As the teacher begins reading the text, Amato and other children offer their thinking. Amato makes two relevant contributions. His first remark (unit 42), "the biggest one is the king," possibly associates length with the superiority of a king.

219

He also seems to be disgusted by worms and thus he shares that he uses a stick to dig for worms (unit 49). Sharon accepts his first contribution, but seems to ignore the second one by cutting him off as she returns to the book. Subsequently Amato misbehaves and Sharon calmly disciplines him (unit 65). In the meantime, another boy, John, is acting as the knowledgeable child he has the reputation for. John loves science and enjoys it in and out of the classroom. He participates in science classes that are offered through many venues in the area. He often shares about all the interesting science kits that his parents have bought him, ranging from how to make a circuit to developing his own slushy machine. His verbal skills, when talking about science, are also well beyond those of typical first graders. He is well liked and respected by his peers. They often look to him for science information and use it in conversations and in their journals. Although he is very messy and disorganized when it came to his drawing and writing, when he is asked what he has written down or what something he has drawn depicts, he can tell without hesitation. As the read-aloud starts to unfold, John portrays himself as a contributor of facts and terminology ("saddle" [unit 54]) and how to identify a worm's head [units 59, 62]).

Sharon continues to read the book that now presents ideas about a worm's body and its characteristics, and what a worm eats. She now gets to the following part of the text on how worms breathe:

67 Teacher: EARTHWORMS BREATHE THROUGH THEIR SKINS. THEY TAKE IN AIR THAT IS TRAPPED IN THE SOIL. ON WET DAYS, WORMS AIR POCKETS FILL WITH WATER, AND WORMS HAVE TO GO {UP} TO THE SURFACE. IF THEY DIDN'T, THEY WOULD DROWN UNDERGROUND. That is what Amato was talking about. [She points to Amato across the room.] A WORM'S BODY IS DAMP BUT NOT SLIMY. THE FRONT END IS MORE POINTED THAN THE BACK. So let's look at the front. [Points to the front of the worm on the previous page.] There is the front.
68 Amato: The mouth is pointy.
69 Teacher: And there is the back. [Points to the back of the worm.]
70 C1: Euuew, they look slimy.
71 C2: They look cute.
72 C3: Euuew, disgusting.
73 Teacher: [Pauses and turns the book to face her.] FEEDING. EARTHWORMS FEED ON THE ROTTING PARTS OF DEAD PLANTS. THEY HAVE NO TEETH OR JAWS, SO THE FOOD THEY EAT HAS TO BE VERY SOFT. SOMETIMES THEY NIBBLE FOOD WITH THEIR TINY LIPS, BUT USUALLY THEY SUCK IT UP.
74 Amato: Euuew!

As Amato continues to express his disgust about aspects of worms, other children do, too. The teacher acknowledges—via an intertextual connection (Pappas, Varelas, Barry, & Rife, 2003)—the contribution he made in prior discourse at the beginning of the session (unit 18). As Sharon points to Amato in the back of the classroom and gives him credit for this idea, he is now being positioned differently.

He is a "good" student, as opposed to his earlier identity as one who may be vulnerable to being sent to the principal's office. He reiterates information from the book ("the mouth is pointy"), which Sharon does not take up. Nevertheless, his efforts underscore his stance as a student who can now "sit and listen."

One of the implicit rules of the read-aloud activity is that it is to be dialogically orchestrated so that students have spaces to offer their own comments. Thus, such an approach represents a delicate "dance" between the teacher and students as they smoothly take their steps together. It allows for teaching as improvisation, where teachers address students' ideas that evolve in a moment-by-moment fashion. Within the community of our project, we have explicitly talked about these principles of dialogical teaching, the reasons behind them, the various ways they may be realized in different classrooms, and the challenges that teachers face in enacting them. Thus, Sharon is making decisions in absolute time as to when, what, and how to respond to students when they offer comments, which obviously affects how Amato and other students may be positioned during the read-aloud activity.

Sharon continues to read more about what the worms eat:

75 Teacher: IN THE DAYTIME, WORMS USUALLY STAY UNDER THE SOIL AND FEED ON THE ROOTS OF DEAD PLANTS. AT NIGHT, WHEN IT IS DARK AND DAMP, THEY CRAWL UP TO THE SURFACE AND SEARCH FOR DEAD LEAVES. THEY DRAG THE LEAVES UNDER THE GROUND. [Points to the text under the picture.] Now under this picture it says A PILE OF DEAD LEAVES IS A FAVORITE FEEDING PLACE FOR WORMS. SOMETIMES THEY STORE THE LEAVES IN THEIR TUNNELS UNTIL THEY ROT. THIS MAKES THE SOIL HEALTHY. So where do worms live?
76 Cs: Under ground. [Vittoria is pointing her finger down.]
77 Teacher: Under ground where?
78 Cs: In the soil, in the tunnels.
79 C1: Tunnels, tunnels, tunnels.
80 Alam: So they are really helpful?
81 Teacher: They are!
82 Teacher: Olivia.
83 Olivia: Um if earthworms weren't around, then we wouldn't be alive.
84 Teacher: Why?
85 Olivia: Um because they um make the soil healthy and if the soil wasn't healthy we couldn't grow food.
86 Teacher: Um. MOVING. WORMS HAVE STRONG MUSCLES TO HELP THEM MOVE. THERE ARE RING SHAPED MUSCLES INSIDE EACH SEGMENT. THESE MAKE THE BODY SHRINK OR SPREAD OUT. OTHER MUSCLES RUN |
87 John: [Raises his hand and at the same time begins to talk.] Ms. Gill.
88 Teacher: ALONG THEIR BODY AND MAKE IT GROW SHORT OR LONG. A WORM'S DAMP BODY HELPS IT MOVE EASILY // THEIR BODY HELPS IT MOVE EASILY THROUGH THE SOIL. That is what the picture is doing. [Points to the picture of a curled worm on soil.]
89 John Can I bring a book um about worms tomorrow?

90	Teacher:	Knock yourself out! A WORM'S BRISTLES ARE VERY IMPORTANT IN HELPING IT MOVE. Where are the bristles on the worm?
91	Alam:	The back.
92	John:	All over.
93	Cs:	All over.
94	Teacher:	THE WORM DIGS ITS BRISTLES INTO THE SOIL TO ANCHOR ITSELF. That means to hold on to one place.
95	John:	Like an anchor	

In unit 75, Sharon interrupts her reading of the book to ask a factual comprehension question—"So where do worms live?" She gets answers from several children, including Alam who queries Sharon as to whether they should conclude that worms are helpful. In this way, Alam seems to seek confirmation of his earlier statement ("they help the earth" [unit 2]) that he shared at the beginning of this activity before the teacher started to read the book. It seems that the book's idea that dead leaves "make the soil healthy" (p. 13) evoked his remark. Furthermore, his remark may have encouraged Olivia's comment, "if earthworms weren't around then we wouldn't be alive." As Sharon explicitly asks her to give a rationale, Olivia provides a meaningful explanation that links healthy soil, growing food, and being alive (unit 85). Thus, here Olivia positions herself as an individual who can use information from the book and her current understandings to construct a justification for her statements. And, this is who Olivia is usually in the classroom. Although she is a very soft-spoken, reserved girl, when she participates in the discussion, she often interjects interesting comments. As for Alam, he likes to learn, but most importantly he is after doing things and thinking correctly. He has been asking for affirmation of his ideas and reasoning. Alam is very good friends with John and frequently goes to him for information, answers, and clarifications. He is persistent when he wants a question answered and he usually goes through a particular person who he considers "knowledgeable"—the teacher, John, or the student teacher. Furthermore, John continues to show who he is—the child who delves into outside the school/classroom resources and blurs the boundaries between school and non-school (unit 89). John also shows his sense making; he associates the verb to anchor with the noun ("like an anchor" [unit 95]).

A puzzling contribution by Olivia occurs during Sharon's reading of the "Tunneling" section of the book. It relates to how worms make tunnels and then how these tunnels affect the growth of plants.

96	Sharon:	WORM TUNNELS MAKE AIRY SPACES UNDER THE SOIL. THESE SPACES HELP RAIN WATER DRAIN AWAY. THE SOIL BECOMES LOOSER AND FINER, SO IT IS EASIER FOR PLANTS TO GROW [She continues reading about how worms mate.]
97	Olivia:	Are earthworms good for the plants?
98	Teacher:	What do you think? What do you think?
99	Olivia:	I don't know.
100	Teacher:	You don't really know? Listen carefully, you will find out. [She returns to reading.]

IDENTITY IN ACTIVITIES

We see here Olivia as the wonderer who asks a question as the teacher reads the book (unit 97). Olivia's question relates to how worms move, and not how worms mate which is what the teacher has turned to. Perhaps Olivia has been processing the text she has heard a minute or two ago and, thus, asks this question "out of synch." Sense making does not happen in an orderly and linear way. Dialogic read-alouds need to be the discursive spaces where questions, related or not related to ideas discussed at the moment, become addressed in various ways. In this case, Sharon asks Olivia what she thinks, but Olivia claims she doesn't know. Sharon is surprised by her answer—thus, Sharon's response of doubting her "You don't really know? (unit 100). Recall that Olivia has earlier contributed a beautiful line of reasoning, namely, "If earthworms weren't around then we wouldn't be alive ~ Um because they um make the soil healthy and if the soil wasn't healthy we couldn't grow food" (units 83 and 85). The teacher probably remembers that, and she also knows that she has just read text that talked about how the worms' movement and building tunnels helps plants to grow. How can Olivia not know the answer to her question? There are at least two possibilities. First, earlier Olivia has talked about how worms help the soil. It is not necessarily obvious to a child that this would further imply that worms help the plants. Second, Olivia may have listened, but not heard, the text that Sharon has just read about worms helping plants. She may also need an extra conversation with the teacher and the class on that in order to make sense of it. Sharon is somehow confident that the rest of the book will help her find an answer to her question ("Listen carefully, you will find out" [unit 100]). Such a teacher move implies a particular way that teacher, the children, and the book (as an important artifact of the activity) share the responsibility for the work that needs to be done. The children need to follow the book. The book has information in it that will help the children make sense of particular ideas. The teacher mediates between the book and children, making the book available, comprehensible, useable, and meaningful to them. This may be accomplished by various strategies: making public children's prior knowledge (units 1–32); pointing to a part of a book picture (units 39, 41, 47, 67, 69); asking interpretive questions (unit 60); inviting children's elaboration and explanation (units 58, 84); asking factual questions (unit 75); explaining words (unit 94); or just continuing reading the book to promote the idea that information books can serve as a resource for answers that make sense (unit 100).

The next section of the book covers how the baby worms are born. Sharon reads about young worms, and after a sentence on the idea of worms being eaten, she asks children for who they think might eat the worms.

101 Teacher: IT TAKES EIGHTEEN MONTHS FOR A YOUNG WORM TO GROW UP AND LAY EGGS OF ITS OWN. IF IT DOESN'T GET EATEN. What do you think might eat them?
102 Cs: Ooo! [Hands are raised in the air.]
103 Demario: A bird.
104 Teacher: What else?
105 Demario: A cow.
106 Teacher: What else?

107 Vittoria: A chicken or a um ~ a frog.
108 Teacher: Okay.
109 Teacher: What about some of the animals that we learned about yesterday.
110 Jasmine: A fox.
111 Teacher: I don't know about that.
112 John: A badger.
113 Teacher: Okay, a badger might eat it.
114 John: A mole, they love those for dessert.
115 Olivia: A ferret.
116 Ramon: A fish.
117 C1: I was going to say that.
118 John: Yeah, that's right.
119 Teacher: IF IT DOESN'T GET EATEN, THE WORM MAY LIVE IN THE SOIL FOR TEN YEARS OR MORE.

Here we see Sharon asking another question to the children offering them an opportunity to share their knowledge and understandings, as well as possibly make connections with other ideas they have been discussing so far in the forest unit. John's comments continue to reflect who he is—he is the "expert" scientist who (a) speaks with confidence and "colors" his statements with details ("A mole, they love those for dessert" [unit 114]); and (b) evaluates others' comments ("Yeah, that's right" [unit 118]).

Sharon finishes the book reading the Afterword, an element found in children's literature information books providing extra information on the topic (Pappas, 2006).

120 Teacher: A MEDIUM SIZED GARDEN HAS ABOUT TWENTY THOUSAND {WORMS}. A WORM'S BODY IS MADE OF ABOUT 250 SEGMENTS. A FAMOUS SCIENTIST, which you all are, NAMED CHARLES DARWIN BELIEVED THAT THE WORM IS THE WORLD'S IMPORTANT ANIMAL BECAUSE IT HELPS PLANTS GROW. Stephanie missed that // I mean Olivia missed that because she is talking. The question that you asked I just answered. WITHOUT WORMS THE SOIL WOULD BE WET, HEAVY AND HARD TO DIG. Why? Why do you think it would be wet, heavy and hard to dig if we didn't have worms to dig?
121 Vittoria: They soften it.
122 Teacher: They make it soft. How do they make it soft?
123 Vittoria: They use their bodies |
124 Teacher: They use their bodies.
125 Vittoria: They use their bodies to make it flat and soft.
126 Teacher: Oh okay. WORMS EAT ABOUT ONE-THIRD OF THEIR BODY WEIGHT EVERY DAY. {THIS IS} LIKE A PERSON EATING TWENTY LOAVES OF BREAD. EVERY CRUMB OF SOIL IN PARKS AND GARDENS HAVE PASSED THROUGH THE BODY OF A WORM. How about that? WORM CASTS ARE VERY GOOD FOR THE GARDEN. THE SOIL IN THEM IS VERY FINE AND HAS VALUABLE FOOD FROM PLANTS.
127 Amato: Ms. Gill I already know that.
128 Alam: So why don't you let us learn.

IDENTITY IN ACTIVITIES

129 Teacher: EARTHWORM FARMS SELL WORMS TO GARDENERS TO HELP THEM IMPROVE THEIR SOIL. FISH LIKE TO EAT JUICY WORMS.

Despite the variety of ideas that are brought out in the text and those the children bring up themselves, Sharon masterfully holds them in her mind. As she read about Darwin and his thinking that worms help plants grow (unit 120), she reminds Olivia that this is the answer to her earlier question (and she might have missed it because she was talking). Another rule of the read-aloud activity is that the participants are quiet enough so they can hear the text and each other. The teacher's comment to Olivia implies violation of this taken-as-shared rule. Even at the end, Sharon continues her practice of asking children questions that will allow them to construct explanations ("Why? Why do you think it would be wet, heavy and hard to dig if we didn't have worms to dig?" [unit 120]). Vittoria, with Sharon's prompting, puts two important ideas together ("They use their bodies to make it flat and soft" [unit 125]). We also see Amato sharing publicly that he already knew that worms are very helpful for the plants, a comment that Alam probably considered inappropriate (unit 128). For Alam who was an eager learner and who was continuously looking for "credible" knowledge and appropriation of his understandings, Amato's announcement that he already knew something that the teacher was presenting could have been perceived as a threat to the continuation of this activity.

Sharon finishes the Afterword element and then asks children what they thought of the book.

130 Teacher: {THAT IS} WHY FISHERMEN LIKE TO USE WORMS AS BAIT. WORMS HAVE BLOOD BUT NO HEART. And that is the end. How did you like this book?
131 Vittoria: Bad.
132 John: Actually it was good.
133 Amato: Good.
...
134 Alam: My favorite part is when their heads grow back.
135 Teacher: Okay.
136 Emily: When are you guys going to bring in the worms?
137 C2: On Monday!
138 Teacher: How many people are excited about the worms?
139 Cs: [All raise their hands.]
140 Teacher: How many people found out something interesting in this book? Okay how many people would like to tell the class about it? I didn't say write in your journals just tell the class about it. Okay we will take turns. #Amato. Do you have something that you would like to share with the class?#
141 Amato: #[He is lying down on his side rocking back and forth clearly not paying attention.]#
142 Mike: When their heads grow back and they're still alive.
143 Teacher: Stephanie. Excuse me please, you know, I see a lot of movement.
...
144 Stephanie: That there is a worm um and it is as long as a car.
145 Teacher: Okay, Olivia.

146 Olivia: Um that there are a lot of eggs in the cocoon but only some come out.
147 Teacher: Okay, Kelly.
148 Kelly: Their heads get bit off.
149 Teacher: Okay, Denzel.
150 Denzel: I didn't raise my hand.
151 Teacher: I know you are busy playing.
152 Amato: Um.
153 Teacher: Amato.

Children respond to the book in many ways. A few children respond including Amato and John who have found the book good. We have heard Amato having high expectations for the book (unit 31) and apparently the book lived up to his expectations. Alam moves the discussion towards the children's favorite part of the book. He spontaneously contributes his ("My favorite part is when their heads grow back" [unit 134]), and Sharon sticks with this direction and invites other children to share their thoughts with the class. Mike and Kelly, like Alam, pick the "beheading" of worms and the fact that their head grows back. Stephanie is fascinated by the size of a long worm, and Olivia by the big number of eggs and the fact that not all of them hatch. But, the teacher also picks Denzel who has not raised his hand. It was not unusual in Sharon's class to call on children to contribute even though they had not raised their hands. Therefore, as participants in the classroom community, the children expected to be called on. However, the read-alouds, as a distinct activity, were discursive spaces where children mostly volunteered answers, questions, comments, reactions, and wonderments. Thus, Denzel is surprised with Sharon's move to call on him and makes clear to Sharon, "I didn't raise my hand" (unit 150). But, Sharon calls on him for a reason. She has noticed him being distracted and she lets him know in her way that she knows, and that this is not acceptable. Sharon has also called on Amato earlier (unit 140) who has been visibly distracted. In contrast to Denzel, Amato does not "object" to the teacher calling on him. Perhaps Amato is used to the teacher's admonitions given his history of interactions with her, and he just does not respond. Eventually Amato makes a move to offer a contribution, the teacher nominates him, but the lesson ends there. He has shown a range of positioning in the activity this day.

MAKING A CLASS MURAL: CONNECTING, REPRESENTING, AND ARGUING

The children in Ibett's second-grade bilingual class had started working on their class mural of a forest. They were to make forest entities using various materials (construction paper, scissors, crayons, markers, colored pencils, glue, pipe cleaners, cotton balls, hole-punch, tissue paper), and arrange them on a tri-fold sturdy poster board. The goal of this activity was to create an artistic representation of the ideas about the forest ecosystem that were salient to the children. Ibett had instructed them to think about the different animals, insects, plants, and other "things" that they may encounter in the forest. Ibett asked children to think about the ideas that they had encountered in the read-alouds up until this point in the forest unit. Thus far, children have read, been read to, and have discussed, books about the forest, animals that live underground, and books on earthworms. They had also conducted a hands-on exploration with earthworms (they had observed two earthworms and recorded their observations in their journal) and they had seen and discussed a website on earthworm anatomy. Children brainstormed as a whole class while Ibett recorded their ideas on a large sheet of chart paper. Once they were finished brainstorming, Ibett assigned a number to each item on the list and to each student in the class. Children worked on the item that they were assigned to from the list. They interacted with one another during the construction of their items, sharing their thoughts and ideas of what a certain animal or plant looks like, whether the animal is big or small, and so forth.

We explore here only a small portion of the mural construction—the beginning of it—that took place during one class period. One day, as soon as some children have made artifacts for the mural, Ibett invites them to go by the poster board and start arranging on the board what they have made so far. One of the children, Enrica, pulls up a chair and sits by the poster board. Throughout the year, Enrica always contributed during the lessons. She was consistently one of the most vocal and outgoing students in the classroom. During read-alouds, Enrica regularly had her hand up and she added to the whole-class discussions with her frequent comments, personal stories, questions, and ideas. In hands-on explorations, she was as inquisitive, posing questions to her classmates, Ibett, and even to the research assistants. Although this class of children was generally friendly with one another, Enrica seemed to be ostracized and singled out by a number of her classmates. In the classroom, there were occasions where other students rolled their eyes or laughed at Enrica as she spoke during an activity. There were also a few occasions where students spoke out and challenged Enrica's ideas in an antagonizing manner. According to Ibett, Enrica was ignored in the lunchroom and some children went as far as to say that they would not drink from a specific water fountain that Enrica had drunk from. However, despite the bullying, Enrica's class participation never declined. In the mural construction activity, her physical positioning by the board was important for the ways she ended up interacting with her peers during this activity. Enrica portrayed a "teacher" identity—she challenged her classmates, she offered ways of reasoning about their choices, and she made connections between ideas. And, as we will see, she 'got on others' nerves,' but she was also 'allowed' by her peers to contribute as she did.

As a child brings a mole to put on the board, Enrica says "Oh, remember, the moles need the dirt on top. The moles leave dirt above their houses. We need to put dirt above their house." After Ibett encourages the children to "go ahead and draw the dirt," children draw a little tunnel that they tape on the mural and Enrica tries to link the two entities, the mole and the dirt ("I am going to put how it [the mole] is entering [the ground]").

Then Natalia has some worms that she wants to put on the mural. She had made them all white. So Enrica reasons aloud about the color of Natalia's worms ("Oh, these are baby worms, cause they're all white"). Enrica tries to justify Natalia's color choice by using an idea that the children have discussed in the context of a read-aloud, but have not seen when observing real earthworms. In the read-aloud of the book *Earthworms* the text states that "young earthworms are whiter than their parents" (p. 21). Ibett read this text in English, and then she strategically code-switched so that she could share this important information with the students in Spanish. While pointing to the picture of the baby worms in the book, Ibett told her students in Spanish that "when they are small they are white. When they are young you can see which ones are the babies because they are whiter than the adults." This information resulted in a brief discussion about what color the adults seem to be in the picture and what color the baby worms are.

Natalia does not respond to Enrica's comment—we do not know if this was Natalia's reason for making them white. But, for Enrica, her thinking that these may be baby worms leads her to yet another idea ("Oh, let's put them next to the cocoon, so then they look like they're coming out"). Enrica then takes one of the worms that Natalia has made and puts it next to a cocoon that is already on the mural. But she picks up one of the really big ones and so another child turns to Natalia and wonders, "They're babies. Why are you making these so big?" Natalia constructs an explanation: "it's like the camera, it just zoomed in." Thus, Natalia uses an idea from one of the read-alouds where a part of an animal or a whole animal was enlarged in an insert to show how it looks when magnified. It is likely that Natalia had not planned to make the worm big because of this reason, but that she came up with this reason as she attempted to coordinate her acceptance of Enrica's suggestion that these are baby worms (because of their color) and the other child's concern with the size of the worms. In this way, we see how the children made sense of the ideas they encountered in books and in their own experiences as they engaged with the artifacts of this specific activity and its goal and governing rules. For Enrica and the other kids, one of the implicit rules of this activity is that the choice of color and size should be justified based on what they have been learning about worms and other entities in the forest.

As Enrica kept focusing on the worms she turns to the teacher implying that something is missing from the mural and then assigns a job to another child: "Teacher, the worms eat dry leaves. Andres, you have to make dry leaves for the worms." Seeing the worms on the mural prompts Enrica to think of something else that is needed to be on it. The artifacts that the children are making in this activity become a critical mediating tool for their sense making. Eventually as the children start to put the leaves on the mural, Enrica moves one of the worms near the leaves saying, "it looks like the worm is dragging the leaf underneath the ground." She puts one worm sticking out with the leaves that are falling on the ground as if the worm is taking the leaf back in the ground. For Enrica, the mural needs to make sense. Connections need to be built between entities on the mural, processes need to be represented as well as entities, and positioning and intermingling of entities is an important part of her sense making.

Another child, Mateo has made what he calls a centipede and he has used pipe cleaners for feet, but there are only ten of them. As he puts his centipede up on the mural, the following exchange occurs:

1 Enrica: Why do we call it a centipede if it only has ten feet?

2 Mateo: You are talking too much.
3 Benita: You can talk if you have ideas. She can talk. She has ideas.
4 Enrica: Well, let's change the name, this is not a <u>centipede</u>, it only has ten legs, it's a <u>tenipede</u>.

Enrica's "sense-making" self continues to come out strong in this exchange. How could it be a centipede if it only had ten feet (unit 1)? Mateo is getting upset with her (unit 2). Mateo was one of the most vocal boys in a class—a class that was clearly dominated by the girls' voices and ideas. Mateo was bright and significantly contributed to read-aloud discussions and hands-on explorations. While he aimed at pleasing Ibett, he was also widely respected by his classmates for his humor and his honesty. Enrica is talking too much for Mateo's taste, as Enrica has a comment for everybody's contribution so far. But, Benita comes to her rescue (unit 3). While working in small groups, Benita, who was generally quiet during whole-class discussions and activities, transformed into a "no-nonsense" leader. She frequently led her group during hands-on explorations and she regularly chastised her peers when she thought they were doing something incorrectly or not worthwhile. Benita legitimizes Enrica's behavior and actions so far using the argument that everybody who has ideas should be allowed to talk. Benita does not differentiate between worthwhile and non-worthwhile ideas. For her, one of the implicit rules of this activity is that participants share their ideas. And this is a rule consistent with the classroom community that Ibett was trying to develop over the year. The rule fit the particular norms of one of the communities these children felt a part of. But, Mateo's frustration was indicative of the challenge that Ibett's classroom community had been facing, namely how to make room for all the children to contribute without feeling overpowered and overwhelmed by those "sense makers" like Enrica. Mateo's frustration could be related to his sense that his centipede is an approximation of a "real" centipede with 100 legs. How could he possibly make 100 legs? Enrica would probably settle with a label change ("tenipede" [unit 4]) for what Mateo had constructed. As Enrica is not objecting to the entity itself, but to its name, the group does not pursue the conversation further and Mateo's artifact remains on the board.

Later on, as Flor notices the mole next to the cocoon, she asks, "Do animals eat the cocoons?" to which Enrica answers, "I think so." That prompts Mateo to wonder about the position of the hedgehog that Carmen has made and put up on the mural. Before everyone started to gather around the mural, Carmen had put the hedgehog up, but then another child put the mole near it.

5 Mateo: Won't the mole eat the hedgehog?
6 Teacher: I don't know.
7 Carmen: I think I'll move mine over here. [She moved the hedgehog near the centipede.] Uhm, yummy!

Although Mateo is upset with Enrica's comments, he cannot stop from engaging in sense making. For him, like Flor, Enrica, and Natalia, the mural construction is governed by the rule that positions, sizes, colors, and orientations matter when it comes to constructing this visual representation of a forest. As the various children provide the artifacts they have made, connections are considered. Carmen seems to care about the entity, the animal (the hedgehog), that she herself has made (unit 5). She moves her hedgehog away from the mole so it will not be eaten by it, but her hedgehog can now eat the centipede and be satisfied (unit 7). Emotions are infused in the sense that the children are making about animals in a forest as they are representing their ideas on the mural.

And the conversation continues with Bernardo asking the teacher: "What are we doing, teacher?" The teacher does not know what Bernardo means with that question and her first thought is that Bernardo is very confused with this activity. She does not say anything. Then Flor answers Bernardo's question: "We're doing the forest." That answer offers Bernardo an opportunity to reveal more about his thinking ("Does the forest have the sun? So, uh, we're always doing what's underneath"). Bernardo is not confused about the activity as Ibett first thought. He is noticing that most of his peers' contributions so far have been focusing on entities under the ground or on the ground, and he knows that the forest is more than that. Bernardo then announces that they "need sun," Rocio adds "clouds too," and Marisol asks, "Can I do the grass?" Bernardo pushes the group to add above the ground entities in their forest mural. We see here how communal sense making was and how the various artifacts that were becoming part of the mural helped sense making.

Finally a few children have made a tunnel out of blue construction paper. As they are placing it under the ground in the mural, Rocio comments that "the tunnels are not blue." Enrica tries to justify the choice of color: "it looks like water is dripping into the soil, fell in the ground, into the dirt. It looks like they're drops of water in the dirt. The worms eat water." Again, Enrica tries to justify her classmates' choices. She offers a sensible reason (water dripping into the soil) for the color of the tunnel that may have been accidentally used by its creators. But, she goes further to link this with other entities on the mural—the worms ("worms eat water"). Whereas Rocio raises a concern about the color of the tunnels, Enrica not only constructs and offers a meaningful justification, but also makes a connection between two ideas represented in the mural. This was Enrica's way of making sense by engaging in this activity in her own ways—an activity that got the children involved with material artifacts that they designed themselves to represent forest entities, entities which were mostly assigned to them randomly from a list that they themselves had generated and were supplemented by other entities that the children themselves eventually identified as important to put on their class mural.

SHARING OWN INFORMATION BOOK WITH AN ADULT: EXPLICATING INTENTIONS AND REASONS

In this section, we focus on an activity with two participants, a university researcher, Eli, and a third-grade Latino student, Arturo, as Arturo shared two illustrated information books he had created. This sharing took place twice, once at the end of the Matter unit (in December) and once at the end of the Forest unit (in

April). The book-making activity, which preceded the book-sharing, was intended as a culminating learning activity for the students and, for the research project, as a summative assessment of the students' scientific ideas and representations in written language and drawings. In other words, the books were used to see how students appropriated science ideas and the genre conventions of information books. Later, the sharing of the books was used to understand children's creative intentions in their representation and organization of their ideas.

In Arturo's class, as children made the books, they engaged in varied kinds of work. Children first brainstormed topics for their books in both whole-class discussion and in their science journals. Once children picked a topic that their teacher, Neveen, suggested they "know a lot about," they wrote a rough draft. Neveen told the children to write one main idea on each page. After children had written a draft of the book, they were given a booklet made of four sheets of letter-size paper stapled and folded in the middle. At this stage, children added images to their text. As subjects in this activity, children used both ideational and material artifacts to negotiate their engagement. Ideationally, these included the concepts explored in the unit and what children knew about the genres, rules, and conventions of writing an information book. Materially, children used pens, pencils, and colored crayons to create symbols on a series of 8.5 by 5.5 inch blank pages. In this book-making activity, the books and their creation acted as the objects of the activity. In the book-sharing activity, the books became mediating artifacts. The representations that Arturo had created and the conversation around them acted as the tools Arturo used to explain his intentions and reasons in constructing the book, and thus, Arturo's explanations also mediated Eli's understanding of them.

Arturo was an eager and active participant in the book-making activity, as he was in most of the other activities of the science units, although during an interview at the beginning of the year Arturo admitted he did not see himself doing science. In his words, his "real thing" was "singing and dancing." However, Arturo was also a frequent and teacher-affirmed contributor to whole-class discussions. In small groups, Arturo often functioned as the self-appointed "manager," making sure students were on task and doing what the teacher had asked. He also frequently expressed to the adults and his peers his pleasure about his engagement in the unit activities. In other words, Arturo was seen as competent, cooperative, and comfortable in school. This carried over to his relationships with adults. More so than most of the other students, Arturo tried to engage Eli in conversation as if they were peers, addressing Eli by name and asking how Eli was doing or if Eli had seen a certain movie.

Arturo's two books were titled *Storms* and *Chipmunk*. In *Storms*, he wrote about hail and how it is formed, and he included blizzards and their effects on daily life (people have to shovel and their flights get canceled). He also wrote about the devastating effects of hurricanes and tornadoes on people and property. In *Chipmunk* he outlined chipmunks' eating habits, what chipmunks and their babies look like, how they move, their enemies, hibernation and food storage, where they live and their classification as rodents. As Eli has conversations with Arturo on his books, Eli explicitly displays his stance toward Arturo "So I want to better understand what you wrote and what you drew. I'm not gonna grade your book I just // I just want to listen to you talking about your book." Eli is constructing himself as someone who wants to understand Arturo's work. In the book-sharing activity, Arturo is constructed as a knower and Eli as a learner. As the sharing unfolds, the two activity participants continue to position themselves in the same way—Eli asks questions about Arturo's book (e.g., "Why did you decide to include what food they eat in your book?") and Arturo answers them.

As Arturo shares his book with Eli, Arturo portrays himself as an author who is concerned with realistic representations for an audience and with audience engagement and comprehension. Arturo's identity is being formed by the book-sharing activity, as it is also forming the activity. Table 10.2 displays the coding matrix we used to categorize statements by Arturo. In this table we have included three columns. The first column contains the thematic codes that we used to categorize Arturo's statements that positioned him as a particular kind of author, in which there are three major categories: (a) Accurate Representation: Arturo is concerned with representing "real things" or making his representations "look real"; (b) Audience Consideration: Arturo is concerned with whether or not his readership will understand him or his representations; and (c) Resources: the knowledge or sources of knowledge that he or his audience draw on in making sense of the book or the representations within it. The second column contains the semiotic modality under discussion, such as written language or illustrations, or if the conversation is more general, the medium, such as the book or television, that was used as the unit of representation. In this way, we are able to pay attention to the modalities Arturo was using to construct himself as a particular type of author. The second column also includes the attribute, or specific dimension of the element that is being referred to. This dimension may be as simple as the fact that it was included at all (existential) or it may refer to the color or shape of an illustration. The third column is a paraphrasing of Arturo's statements that were coded for author identity.

For instance, Arturo includes trees that have "Hawaiian branches from Hawaii," in his description of hurricanes. In this case, Arturo is concerned with accurate representations. He and Eli discuss the illustration, and Arturo calls attention to their shape, or form—that they look like real Hawaiian (palm) trees. Although hurricanes may not occur in Hawaii, the tropical climate is accurate and palm trees are prevalent in places like Florida where hurricanes do occur. At other times during the interview about his *Storms* book, Arturo is more concerned with audience reaction than with reality. In two instances (the picture of a tornado and the picture of a truck flying through the air) Arturo used color as a resource for audience comprehension. On one page he colored a tornado black, not grey so that it would not be camouflaged. On another, he did not color the door of a blue truck so that readers could see the handle and know that he had drawn a truck with a door handle.

In another part of the *Storms* book, Arturo used information from a book the class has read about an actual blizzard (Shamah, Personal Conversation, 2004). Page five of Arturo's book includes the following text: "Snow can go up to a door like in 1979. Sometimes snowstorms has 20.7 iches of snow. 60 people got kille in

Table 10.2. Examples of Arturo's coded statements during the book-sharing in the matter unit.

	Code (Thematic Content related to Arturo as an author)	Representational Element (The semiotic modality that is under discussion and the relevant attribute)	Example
STORMS – ARTURO_MATTER	Accurate Representation (AR)	Book–Non-Fiction–Existential	a) [The book should be about] real things that used to happen.
	Audience Consideration (AC)	Illustration–color	b) Black not grey tornado–doesn't want camouflaged.
	AR	Language/Illustration–real event	c) Details about the blizzard of 1979.
	AR	Illustration–shape/form	d) Hawaiian branches from Hawaii.
	AC	Illustration–existential over color	e) Didn't color door handle.
	AC	Illustration–Existential	f) I didn't want to draw someone [getting hit by lighting] to think it's creepy. [Readers would say] that's weird!
	Resources	Television–source of info	g) Describing TV shows and people's perceptions of information on them. [An extended discussion is given in the main text.]
	AR	Illustrations–Nature of…	h) It's like an illusion that makes you feel like it's real but it's not.

the snowstorm of 1979. It started on Friday nigh Jan. 12 a 2:00 a.m." (Not coincidentally, Arturo had also brought up the same information during a class read-aloud of another book about snow.) Although not during the sharing, Arturo did tell Eli on a separate occasion that he had written this page to make the book seem more real. Arturo felt that because of its alignment with reality, specific information was more interesting to readers than general information. Arturo wanted to put out a good product, something that readers would be interested in. He did not say, "Ms. Shamah said we should, 'do it *this* way,'" or "Ms. Shamah said, 'she likes it when we do *this*.'" Rather, Arturo constructs himself in relation to a hypothetical readership that would want to read his book. In the context of the book-sharing, Arturo's motivation for his first book, *Storms*, is to be an interesting author to his readers.

In the book-sharing activity, the interviewer, Eli, also set up an interaction in which he asked Arturo to explicitly consider the book he wrote and himself in rela-

VARELAS ET AL.

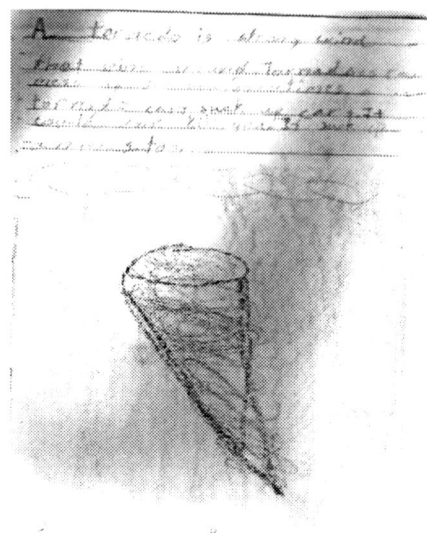

Figure 10.2. From Arturo's book entitled Storms: *"A tornado is a strong wind that spins around. Tornadoes can mess up houses. Sometimes a tornado can suck up cars. It could even kill you. It suck up animals too."*

tion to an audience. In the *Storms* book, an extended conversation ensues when Eli posed the following question:

1	Eli:	Um, do you think if another third grader // can you tell me what you think the most interesting part of your book would be for another third grader?
2	Arturo:	I'll say the tornado. [Refers to page shown in Fig. 10.2.]
3	Eli:	Why would that be the most interesting?
4	Arturo:	Cause it's like // it's like // people sometimes think it's kind of a ride, but it can kill you and sometimes on TV they	
5	Eli:	Wait, they think it's kind of what?
6	Arturo:	They think it's kind of <flying> one because [circles hand around tornado] they think they could fly in the air but they can't.
7	Eli:	Yeah. Alright.
8	Arturo:	Cause sometimes // cause sometimes shows have tornadoes and stuff and people like those kind of shows. Like Superman, stuff like that.
9	Eli:	Oh, right.
10	Arturo:	Yeah, and Hercules when he was going to <Cleveland> in the cartoon. Yeah.
11	Eli:	Okay. In the cartoon?
12	Arturo:	Yeah. In the movie "Hercules."
13	Eli:	Oh. Okay.
14	Arturo:	And some people like // and people like tornado from "The Day After Tomorrow" movie when um // when bad stuff always happened around the city. Yeah. So people like tornadoes a lot.
15	Eli:	People like tornadoes?
16	Arturo:	Yup.

17 Eli:	Okay. Do you think ah // people would enjoy your pictures more or your words more?
18 Arturo:	I'd say enjoy the pictures.
19 Eli:	Why do you say that?
20 Arturo:	Cause sometimes they look like they're almost real but you never know. It's like an illusion that's making you feel like it's real but it's not.

In this conversation, Eli uses Arturo's identity as a third grader to position Arturo as possibly knowledgeable about other children like him (in the third grade). Eli not only positions Arturo as existing within the group of third graders but also in relation to them. Arturo is both inside and outside of the group Eli mentions. The question is not, "What do *you* think is the most interesting part?"; Arturo is not being asked to consider *himself* a reader. He is the author. Arturo is able to accept his third-grade identity at the same time that he enacts his expert/author identity which functions to differentiate him from other third graders. Arturo has no problem talking about what other third graders may or may not know. He is knowledgeable about what people find exciting (tornadoes) and he uses this to engage them in his own book. He also knows what shows or movies they watch ("Superman," "Hercules," "The Day After Tomorrow" [units 8, 12, 14]) and what happens on those shows, but he is also outside of them in that he does not believe what "they" believe (unit 6). He knows tornadoes are not "a ride" as some other "people sometimes think" but that tornadoes can "kill you" (unit 4). Arturo continues to be concerned with accurate representations, especially in relation to the made up ones that appear on television and in the movies.

In sharing his second book, *Chipmunk*, Arturo is much more explicit about his concerns with getting his book right and making it "real" (unit 20). This may be in part due to the influence of his teacher's expectations that students make books that are "true" and contain "information" that will be for others to read. It may also be due to his continued participation in activities with children's information books. As a student, Arturo knew what adults expected of him and he did his best to meet those expectations. First of all, Arturo chose the topic chipmunks because it was a topic he "knew a lot about." These words echo those of Neveen's instructions and he did not hesitate to use them during the book-sharing. Second of all, Arturo wanted to make his book "look like a real one" and so he copied pictures from one of the information books on chipmunks (Whitehouse, 2004) that children had read.

Table 10.3 includes Arturo's responses to Eli's questions about his choices for the parts of his *Chipmunk* book. In places where we felt it was necessary, we included short explanations in the table itself.

Table 10.3. Examples of Arturo's statements during the book-sharing in the forest unit.

CH	Code	Representational Element	Example
	Accurate Representation (AR)	Illustration/Language relationship	a) Pictures show what the words mean.
	AR	Illustration–form	b) Wanted to make it look like a real one.

AR	Illustration–form–copying	c) I was copying (from the book *Chipmunks*).
AR–doesn't matter	Illustration–color	d) Blue clouds.
Expert	Book–existential	e) Chose chipmunks because he knows a lot.
Expert	Book–existential	f) Tells what chipmunks do.
AR–acknowledges mistake	Language–erasure	g) I screwed up.
AR	Illustration–form–copying	h) I was trying to make it look small. How I did when I was copying it. Draw like a chipmunk.
Self-depicted	Illustration–existential–self explicit representation of self in book	i) It's me [in picture].
AR/Resource	Book–existential/ISLE journals used as model for constructing page	j) Pictures on the top, lines are for writing.
AR	Illustration–existential–no dots included, but he meant them	k) Strawberry – forgot the dots. Interviewer says it looks like an acorn to him—but does not directly challenge.
Expert	Language–existential–including information	l) Information for people who might own chipmunks.
AR	Language & Illustration–existential–explains function of pictures in relation to words	m) So when you read it, you can know what it says (Strategy for comprehension). So they can know what I'm talking about [refers to same entities].
Resources	Tools–existential (what material artifacts he used).	n) I didn't have any pencils.
AR–acknowledges mistake	Language–existential (missing word)–letting Eli know what was and was not intended	o) I forgot to put "time."
AR–acknowledges mistake	Illustration–cross out.	p) I screwed up an owl.
Resources	Illustration–TV–source of information	q) If you watch TV, they tell you about owls.
AR–hedging	Illustration–size -	r) Not all of them but <u>a lot</u> of them are <u>about</u> that size.
Resources	Tool–existential	s) Couldn't find the color (of crayon). It was under the table.
Resources	Language–ISLE–source of info	t) I was thinking of what they do and I've been learning about the past few times in science.
AR–audience	Illustration–existential	u) I wanted them to know.

		["Them," in reference to audience, is generic throughout. He never says who his audience might be.]
Audience	Language–existential–gives information	v) It tells them where to search.

In this conversation, an intensified concern emerges with accurate representation, but also with audience learning. That is, in comparison to the first book he wrote, *Storms*, Arturo is more concerned with making a realistic book that will provide useful information to his audience. He is more concerned with acknowledging mistakes he had made or elements he had forgotten to include (e.g., Table 10.3, examples g, k, & p). He also makes more use of the ISLE curriculum to rationalize the choices he made. He used books the class had read to copy pictures, the format of his journal was used for his page, and he mentions generally that he uses what he has "been learning about the past few times in science" (Table 10.3, example t). For instance, in this book-sharing, Arturo discussed how he copied curricular materials as tools to make his book "look like a real one." When he was asked about the colors he used to represent a chipmunk on the cover, Arturo says: "I wanted to make it look like a real one but // I was trying my best and I was trying to look at a book to see (***) how they were doing it. So I was copying here [pointing to the chipmunk] and like these [points to the tree and chipmunk] like brown // which was dark // darker." The book Arturo refers to looking at is a children's literature book *Chipmunks* (Whitehouse, 2004) that he uses to make his own real-looking book, titled *Chipmunk*. In doing so, Arturo acknowledges the book as a resource. He identifies himself as someone who not only tried to make his illustrations real, but also tried his "best," and who used a professionally made book as a model. His motive for drawing the chipmunk a certain way was accurate representation, which is tied to his construction of the whole book as presenting accurate information for his audience, "them," to "know" about chipmunks.

Arturo also uses the relationships between text and pictures to set himself up as someone who is writing for an audience. In the following excerpt, he discusses a page in his book about chipmunk babies (shown in Fig. 10.3). Arturo has read the words, as he has done for each preceding page, and has explained the picture:

21 Eli: Okay, so how did you decide what to draw for that picture?
22 Arturo: I decided by thinking about the writing I was doing some notes that when I was gonna draw, I was gonna draw a chipmunk and the baby and that was red.
23 Eli: Okay.
24 Arturo: So they can know what I am talking about.

On this page, as on others, Arturo uses his pictures not only to re-present his words, "I decided by thinking about the writing" (unit 22) but he also uses the pictures so that his readership "can know what I am talking about" (unit 24). The pictures serve to make Arturo's words clearer. His explanation of the pictures again

Figure 10.3. From Arturo's book entitled Chipmunk *(p. 3): "Chipmunks have babies that are red. When their born they are hairless and their eyes are closed."*

constitutes Arturo as a person who is writing for an audience. In doing so, he may not be a "real" author, but he is "trying his best."

Again, while discussing the *Chipmunk* book Arturo has written, he is asked what part of the book other third graders would like the best. Arturo's answer is shorter than for the *Storms* book, perhaps because it is near the end of the sharing that was more than twice as long as the first.

25 Eli: So Arturo, if you were gonna share your book with any other third grader who wasn't in your class, what page do you think // what part of the book do you think they would like the best?
26 Arturo: I'll say // I'll say // [looking through the pages] er like // they like this part [points to last page shown in Fig. 10.4] because it tells them like // it tells them like // information where to search a lot about chipmunks.

At the time Arturo's text was written, the above website was for "Alvin and the Chipmunks." It is not known if Arturo knew that, but he did say that the website was a "real" one. As in the first book-sharing, Arturo is concerned with real information as it relates to the desires of his audience. He believes the information he shares may be of practical use to his readers and he includes certain information precisely because it has this value. Arturo is an expert and his readers have much to learn from him, but he recognizes his limitations and is able to share the burden of knowledge production with others.

The interaction within the book-sharing activity positioned Arturo as a particular kind of author. Eli and Arturo used the representations in Arturo's books, and the resources Arturo drew on in creating those representations, to mediate the book-sharing and Arturo as a kind of person. Arturo was an author concerned with audience comprehension and accurate representation in one book-sharing and he became an author concerned with accurate representation for audience learning in the other. These were similar, but slightly different identities showing that Arturo

IDENTITY IN ACTIVITIES

Figure 10.4. From Arturo's book entitled Chipmunk *(p. 13): "If you want to learn more go to www.chipmunks.com."*

has changed through his participation in activities that have expanded for him what it means to be an author/expert. In their discussion, Arturo and Eli created and sustained the conditions necessary for just such an identity. In the second book-sharing they were able to draw on a different set of ideational and material mediating artifacts (different representations, content, ideas of authorship, and resources from participating in other activities) to negotiate the activity and this constructed Arturo as a different kind of author. In turn, Arturo's particular identity of an author served the outcomes of this particular activity—to understand Arturo's book and the choices he made.

MEDIATING/ED IDENTITIES IN DYNAMIC ACTIVITIES

In the four activities we discussed in this chapter, we offered glimpses of how discursive identity is played out in activities where young people and the adults they work with engage in particular ways with various artifacts and each for their own particular purposes. Identity provided an explanation for how the children engaged in the activity and how they constructed the objects of the activity. But, identity was also constructed as the activity was unfolding, as the participants were making the activity their own. In these four activities we study how subjects/people, artifacts, and objects/purposes become central or peripheral at various points of the activity. Each child came into the activity with a particular student or person identity that has been shaped by many experiences and other activities. Who the children were and who the others thought they were explained how central or peripheral their engagement was in the activity. Children's histories shaped their role in the activities, and the activities (with their own histories) sustained the children's ways of constructing what Ivanic (1998) calls "possibilities for self-hood" and opened up new spaces for them. For Enrica, the making of the mural nurtured her "sense-maker" identity. For Enrica, the representations that were going on the mural needed to make sense; they needed an explanation. If an explanation was not offered by anyone else, she constructed it herself. As she was contributing explanations, she was also providing opportunities for other children to come up with ideas and representations. Her "sense-making" was contagious, as the activity opened up new spaces for them.

Different children also constructed the multiple goals of each of the activities differently, foregrounding some and pushing aside others. Latessa had constructed a clear goal for the sorting activity—to sort the objects and to discuss the sorting with the other group members. As she pushed her peers to reach this goal, her "leader" identity was enacted and maintained. But for the other children, the discussing part of the activity goal was more peripheral. They jumped from object to object feeling them, manipulating them, and putting them in one of the three categories without making sure that everybody had spoken and shared their ideas. The group discourse mediated by the physical objects and the ideas that children had constructed or were introduced to from the read-aloud and the teacher afforded children possibilities for various emerging self-hoods—e.g., Chantrelle as Lawrence's "buddy" and the persistent meaning-maker.

Furthermore, who the children were influenced how they engaged with the activity artifacts, both material artifacts and ideational artifacts. For Arturo, the books he had made and was sharing with Eli were about who he was as an author. As Arturo was sharing with Eli the reasons for his representations (linguistic and pictorial), he showed how he cared about realistic representations and making books that would act as learning tools for others. His books were not only material artifacts, filled with texts and images, but also sources of the ideas he was sharing with Eli, and products of his author identity.

Borrowing Bakhtin's (1981) language of centripetal and centrifugal forces, we could theorize about discursive identity in the activities we explored in two ways. On one hand, an individual's self-hood brought together the elements of an activity in a particular manner. Identity was playing a centripetal role within the activity structure. On the other hand, the multiple participants' identities were pulling the activity in multiple directions, and multiple endpoints. In this way, identity was playing a centrifugal role in the activity structure. The interplay between the centripetal and centrifugal functions of identity contributes to the fluidity of each activity and the multiplicity of practices within each activity that further nurtures the development of identity. As children used and adjusted their ways of being, talking, doing, believing, valuing, interacting within an activity, they were afforded opportunities for various ways of thinking, representing, explaining, negotiating. John's engagement in the read-aloud (both the book with its text and its pictures, and the class discussion on it) made it an activity that was about making connections, offering reasoning, sharing terminology, and going beyond the ideas presented in the book. Amato's engagement tested the rules of the activity, by breaking the rules associated with behaviors during the read-aloud. Olivia engaged with the ideas in the book trying to make them her own. Meaningful explanations were important to her during this activity as they related to the ideas brought up in the book. For Alam, the read-aloud was about affirmation of his thinking. Thus, the read-aloud was a dynamic activity, pulled and stretched, shaped and reshaped, by the various identities (dynamic entities themselves) that the children were enacting.

This complex interweaving of the different participants' self-hoods that are played out during an activity, and are influenced by it, makes the activity a "living," ever changing and evolving entity that is uniquely defined. The six classes we work with all did read-alouds, or constructed murals, or sorted objects, or engaged in book-sharings. But we should not think of each of these activities (e.g., the sorting activity) as the same activity. In different classrooms, these can be similar activities, sharing some important elements, but as wholes they are different activities.

Using an activity framework to think about our developing research on identity and activity, we see ourselves building on the analysis that we presented in this chapter (and using it as a mediating artifact) to explore activity *systems* and how identity transcends various interconnected activities and mediates across them. In the current analysis we made brief references to how some of the activities we explore are strongly linked with other activities that we did not analyze. For example the sorting activity was preceded by a read-aloud that presented children with macroscopic properties of solids, liquids, and gases. Or, for the book-sharing activity, one of the artifacts was the book that Arturo read during another activity. Furthermore, as we noted at the beginning of the chapter all these activities (and more) are part of two integrated science-literacy units that were enacted in classrooms throughout most of the school year. Studying how identity helps us understand what, when, how, why children (individually and collectively) learn science and literacy and grow and develop as science and literacy learners is an ever evolving goal for us.

APPENDIX: KEY TO SYMBOLS USED IN TRANSCRIPTS

Symbol	Meaning
//	false starts or abandoned language replaced by new language structures;
~	short pause within language unit;
~ ~	longer pause within language unit;
|	breaking off of a speaker's turn due to the next speaker's turn;
(***)	one word that is inaudible or impossible to transcribe;
(*** ***)	longer stretches of language that are inaudible and impossible to transcribe;
# #	overlapping language spoken by two or more speakers at a time;
CAPS	actual reading of a book;
{ }	teacher's miscue or modification of a text read;
[ ]	identifies what is being referred to or gestured and other nonverbal contextual information;
. . .	part of transcript has been omitted;
< >	uncertain words;
Underscore	emphasis.

NOTES

This research is part of larger project that is funded by a three-year (2004-2007) US National Science Foundation (NSF) ROLE (Research On Learning and Education) grant (REC-0411593) to M. Varelas and C. C. Pappas as Principal Investigators. The data presented, statements made, and views expressed in this article are solely the responsibilities of the authors and do not necessarily reflect the views of the National Science Foundation. We are thankful to our teacher colleagues and collaborators on this project, Anne Barry, Begoña Marnotes Cowan, Sharon Gill, Jennifer Hankes, Ibett Ortiz, and Neveen Keblawe-Shamah, who, with their practice and reflective inquiry, have played an instrumental role in conceptualizing this chapter.

REFERENCES

Bakhtin, M. M. (1981). *The dialogic imagination: Four essays by M. M. Bakhtin* (M. Holquist, Ed. M. Holquist & C. Emerson, Trans.). Austin: University of Texas Press.
Brown, B. A., Reveles, J. M., & Kelly, G. J. (2005). Scientific literacy and discursive identity: A theoretical framework for understanding science learning. *Science Education, 89*, 779–802.
Cole, M., & Engeström, Y. (1993). A cultural-historical approach to distributed cognition. In G. Salomon (Ed.), *Distributed cognitions: Psychological and educational considerations* (pp. 1–46). Cambridge: Cambridge University Press.
Davies, B., & Harré, R. (1990). Positioning: The discursive production of selves. *Journal for the Theory of Social Behavior 20*, 43–63.
Engeström, Y. (2001). Expansive learning at work: Toward an activity theoretical reconceptualization. *Journal of Education and Work, 14*, 133–156.
First Discovery. (2002). *In the forest*. New York: Scholastic.
Fowler, A. (1997). *Animals under the ground*. New York: Children's Press.
Gee, J. P. (2000–2001). Identity as an analytic lens for research in education. *Review of Research in Education, 25*, 99–125.
Ivanic, R. (1998). *Writing and identity*. Philadelphia: John Benjamins.
Kress, G., & van Leeuwen, T. (2001). *Multimodal discourse: The modes and media of contemporary discourse*. London: Arnold.
Lave, J., & Wenger, E. (1991). *Situated learning: Legitimate peripheral participation*. Cambridge: Cambridge University Press.
Llewellyn, C. (2000). *Earthworms*. New York: Franklin Watts.
Massie, E. (2000). *A forest community*. Austin, TX: Streck-Vaughn.
Pappas, C. C. (2006). The information book genre: Its role in integrated science literacy research and practice. *Reading Research Quarterly, 41*, 226–250.
Pappas, C. C., Varelas, M., Barry, A., & Rife, A. (2003). Dialogic inquiry around information texts: The role of intertextuality in constructing scientific understandings in urban primary classrooms. *Linguistics and Education, 13*, 435–482.
Vygotsky, L. S. (1986). *Thought and language*. (A. Kozulin, Ed. & Trans.). Cambridge, MA: MIT Press.
Vygotsky, L. S. (1978). *Mind in society*. Cambridge, MA: Harvard University Press.
Whitehouse, P. (2004). *Chipmunks*. Chicago: Heinemann.
Zoehfeld, K. W. (1998). *What is the world made of? Solids, liquids, and gases*. New York: Harper Collins.

WOLFF-MICHAEL ROTH, MARIA VARELAS, SUNGWON HWANG, KEN TOBIN

11. ACTIVITY, AGENCY, PASSIVITY

Michael: Personally I am interested in how activity and identity are theorized within cultural-historical and sociocultural approaches. It appears to me, Maria, that your team and I use terms very differently; and if this is the case, we have to first clarify how our terms are similar in their difference, because otherwise we will be talking past each other. The first is your use of the notion of activity, which refers to what children do. But for cultural-historical theorists, following Karl Marx, activities are historically evolved societal forms that in one way or another contribute to the survival of the collective, such as in growing food, animal husbandry, tool production (the German/Russian denote these by *Tätigkeit/deyatel'nost'* and activity theory is *Tätigkeitstheorie/teorija deyatelnosti*). You on the other hand refer to children's tasks as activity, whereas cultural-historical activity theorists of the Soviet and German school would call these tasks (the German/Russian terms are *Aktivität/aktivnost'*, the state of being/keeping busy). If we give up this difference, then the whole of activity collapses and we can no longer theorize the relationship between individual and collective as currently is done, we no longer get at the reproduction of societal inequities in schools, and we no longer get the reproduction of the working class that is described in some sociological studies, including *Learning to Labor: How Working Class Lads Get Working Class Jobs* (Willis, 1979).

Maria: We do not think of activities as what the children do, or the children's tasks. The read-aloud is an activity not because the children "do" something special than during a different task, but because it is a school curricular genre, a cultural-historical activity with particular goals, rules, artifacts, division of labor, and patterns of communication, that can shape schooling and school learning in very specific ways. It is an activity in which knowledge, ideas, facts, truths, interests, values, morals, are passed along from the more knowledgeable adults, or members of a particular sociocultural practice, to children, or novices, via texts. And depending on the activity norms, the children, via the read-aloud, may learn docility, obedience, and acceptance of ideas that others know as the truth, or they may learn to question, to wonder, to be creative in their thinking, to challenge others' ideas, to take initiative, to be assertive. Read-alouds, as an activity, may contribute to developing children's intellectual powers and making them feel proud of themselves and their thinking (what Sharon and all the other ISLE teachers were aiming at), or they may show children how to follow rules and steps, exercise very little choice and decision-making, but much mechanical and rote acceptance. In fact, these are the two extremes in the continuum of goals that Jean Anyon (1981) found for student work in schools in contrasting social class communities, with the former being in executive elite schools and the latter in working-class schools. So, read-alouds, and the other activities that we explore in our chapter, are parts of the activity system of schooling (western schooling) that inducts children in particular ways of knowing, acting, thinking, being. They are cultural practices defined by their meaning relationships to other practices, having both material components and characteristics, as well as semiotic components with cultural meanings and social values. In the curriculum that our teacher colleagues and we designed together, we enact particular notions of these activities and study how identity plays out in them. These activities are different from the one you analyze in chapter 8, Michael, and I wonder about whether and how they fit the activity profile that you painted—activities that would allow, encourage, support students to "develop science-related dimensions of identity" and that as you argue should be "forms of activity that embody emotional-volitional and ethico-moral dimensions" (p. 182).

Michael: I see you struggling with this notion of activity, having to use both for different things, which other languages do have different words, and thereby, become different concepts. Thus, I am trying to stay away from calling the things kids do in school "activity" (e.g., Aktivität/aktivnost') and thereby to reserve the term for cultural-historical formations at the collective level. What do you think about this, SungWon?

SungWon: From my perspective, "schooling" is one form of activity; it realizes the societal motive through diverse forms of institutional program. The benefit of cultural-historical activity theory is that it allows studies of the nature of change conceived in any attempt to design better ways of teaching and learning science. In my chapter I learned from the conversation between lab instructor and student that their ethical efforts toward responsibility for the other make little contribution either to bring about changes or decrease sufferings *unless* the participants deal with societal orientations of science learning that they are engaged in at the moment (university schooling in this case). Thus an overarching question that appears to me important is how school science could be designed such that it explicitly exposes the societal nature of schooling at the very moment and place of enacting the idea of science so that it allows all participants to contribute to the emergence of historically new forms of school science.

Ken: SungWon's use of institution gives me a clue to a feature of activity that is important to the framework she and Michael employ. The use of institution points to a level of social aggregation that draws attention to macro-meso interactions. It seems as if SungWon and Michael want to preserve the use of activity to the institutional level and perhaps above, if I have this right the collective they refer to also would be referenced to the institutional level.

Michael: This is precisely where I see the special affordances of activity theory: It makes us attend to multiple levels of analysis, thereby disallowing what others do in their analysis of classroom learning, namely disattend to societal structures, labor markets, social inequities while studying science learning.

Ken: In our research we decided to use *field* as the salient construct for making sense of social life. We do not use institution in our theorizing—preferring to see fields as sites for cultural production, reproduction and transformation. Fields can be discrete, overlapping, and wholly or partially nested. Hence, in a study we can zoom to focus on a collection of fields, describe individual and collective relationships accordingly, and to the extent possible take account of macrostructures. Within my research groups we have found it useful to think about activity and action in relation to the field(s) that are salient to our research. Largely based on Michael's writing we regard activity as dialectically related to action, interpreted in terms of an individual|collective relationship.

Michael: To me, the theoretical concept of activity is a way to engage in a holistic analysis in which progressively disclosed moments within the activity stand in mutually constitutive relationships. The ultimate motive is the survival of the collective; and this survival is achieved by a division of labor, which has given rise to the different systems of activity, including schooling. The motive of schooling is the reproduction of cultural practices by means of a formal institution; and the reproduction of cultural practices contributes to the survival of the collective (society). I don't even have to be conscious that this is what schooling does, and a janitor whose purpose is to exchange his labor for a salary to feed his family and give them a home, may not be conscious of the motive. In my understanding, motive is to society as goal is to the individual. *I am what I do*, materially and discursively. So in acting, I produce and reproduce identity. In *Grundrisse* (Marx, 1973), the productive process is described as one that consumes human energies but produces human subjectivities as the object world personifies itself in the human body. Thus, by laughing coyly and admitting to feel inferior, Mariko produces a particular form of schooling and therefore a particular form of identity. In many science classrooms, we actually have students resist participation in the middle-class oriented school curriculum, and thereby produce and reproduce their exclusion from the ruling classes. In resisting participation or rather, in participating in their particular ways, they contribute to the reproduction of a social class much like Willis' lads do. Because identity is tied to actions of a productive process, I don't need a special theory; it is continuously produced and reproduced whatever I do, which means, continuously for as long as I live a conscious life.

Maria: And I find SungWon's and your ideas of boundary crossing and emotional-volitional and ethico-moral dimensions important to consider around the issue of reproduction of cultural practices and identity. I think boundaries exist at both the individual and the collective levels defining the limits of different cultural practices. As people engage in cultural practices, and become part of activities, they construct/co-create social spaces that Lefebvre (1974) differentiates into perceived, conceived, and lived spaces—the lived spaces embody the real (perceived) and imagined (conceived) lifeworld of experiences, events, emotions, and choices that have ethico-moral, philosophical, and political origins. It is, I believe, in these lived spaces that participants use their identities to constitute the activities they participate in, and they reform their identities as a result of these activities. In our work, we have just started exploring the construct of "spaces" as they are created in classrooms and their relationship to identity (Kane, Hankes, Varelas, & Pappas, 2006).

Ken: We do not use *space* in our theorizing either, preferring to use *field*. From our perspective the idea of boundary or border warrants closer attention. What is a boundary? I have always thought of boundary in terms of structures. Accordingly, the material and schematic resources that bind a field, though dynamic, tend to be associated with particular forms of culture, collective and motives. So a question that might be asked is what does it mean to move to another field? We would begin to respond to such a question in terms of moving to another collective with different activities and associated motives. Hence, to find it useful in an analytic sense to claim a change of field, the participants, social bonds, enacted practices, and associated schema would differ saliently as the researcher's gaze moved during analyses of social life. For me, at any rate, fields are associated with particular cultural enactment and dynamically constituted structures. To say that the borders are weak or porous implies that new culture, produced in other fields, can be enacted as individuals and collectives pursue and meet their goals—thereby changing the field. A salient query is then—as a person enters a new field, can she pursue her goals using the culture she has as knowledge at hand (Schutz, 1967)? In so doing can she produce new hybridized forms of culture as she appropriates the culture of this field using her own cultural resources? The structures available for appropriation include those associated with the participants in this field as well as those associated with the fields in which this one is nested (i.e., macro structures that include rules and ideologies, some of which are hegemonic and the practices of powerful others).

Maria: Fields or spaces? This is an interesting point for further discussion, Ken. One of the challenges (which, at the same time, is a tremendous possibility, I think) that we are facing as educational researchers is selecting words that capture the concepts that we are developing and are working with. It is taken-as-shared that words carry meanings and take on new meanings within activities and activity systems. Thus, I wonder what gets foregrounded when we use "spaces" instead of fields. For us, "spaces" invite more thinking about boundaries than "fields" do which may foreground more openness. Also, from a mathematical perspective, a space has a set of rules/axioms associated with it. As rules and norms are indeed important in our framework (as, of course, they are in yours), "spaces" seem to push us to articulate how rules are formed, enacted, and form the nature of interactions within them. Furthermore, "spaces" in a physics perspective have several dimensions, and this idea allows us to theorize about the various dimensions that our social spaces may have. On to the notion of boundaries, I think structures shape the boundaries, but in an activity-theory view, boundaries are set dynamically by the interactions among *all* elements of the activity triangles. Certain objects/motives may stretch in particular ways the boundaries of the spaces where people interact with each other and with artifacts, and production, reproduction, or transformation of social practices may take place. It is we believe a complex and multi-faceted interplay of subjects, objects, artifacts, rules, community, and division of labor that defines the shape, permeability, size, flexibility of space boundaries.

Ken: Thanks for clarifying. It is not my goal to inflict my view of social life on you and your group. That we do not find space a useful construct at the moment reflects the other pieces of our sociocultural frame that is shaped to a large extent by Pierre Bourdieu, William Sewell, and Alfred Schutz. I see considerable strength in different groups of scholars adopting frameworks that work for them—of course while always being on the lookout for promising new ways to make sense of social life. I see many complementary aspects of the different theoretical stances adopted by the different groups involved in this metalogue.

* * *

Michael: I'm moving to another point pertaining to identity and the way your team, Maria, theorizes it. You distinguish it from subjectivity, which you define as something fleeting. Identity is something fixed in your account, sedimented from life experiences, whereas subjectivities are (personal?) stories that change in the course of individuals' participation in activity. How does this meet up with the use of identity as something fractured, multifaceted, and fleeting in the way theorists in cultural studies and sociology—including Anthony Giddens (1991)—do to capture identity? Further, when we are involved in activity, we do not reflect; most infrequently of all we reflect about who we are. I therefore like to see the whole question formulated in the way philosophers of Self approach the issue, including Paul Ricœur in *Oneself as Another*. Here, the sedimented aspect has to be maintained in narrative accounts, whereby both the accounts and the memory are aspects of collective construction, as discursive psychologists tell us (e.g., Middleton & Brown, 2005). Therefore, I am asking myself how something that is consciously constructed and perpetuated in narratives, that is, a stable self, can be invoked in fleeting activity, where the focus is altogether different, namely the object of activity, which generally is not identity itself—though in narcissistic behavior, identity actually may be the goal of the actions.

Maria: We do not address in our chapter in any systematic way similarities and differences between identities and subjectivities and your question, Michael, gives us an opportunity to theorize briefly on this. The distinctions we make between subjectivities and identities are relative. Subjectivities are relatively unstable than identities. This does not mean that identities are stable. They are not. They get constructed and re-constructed over time, in activities, with others, they have multiple dimensions, and they capture the interplay of self and others. Subjectivities are also unstable, they have shorter lifetimes than identities, and they come together (not in an additive way) to shape identities. We see subjectivities as being enacted, relatively more than identities, in the public sphere, as ways in which individuals try to find and exercise their "voice" in a shared space, positioning themselves relative to others, gaining or losing strength, aligning or clashing with others, and defining in multiple ways the particular texture of the social activity. As individuals construct and find their "voice," they form an identity that does not sediment subjectivities, but maybe metamorphizes them (to use an analogy from geology and how various-type rocks are formed) into "new" entities. In order for these ideas to make sense, of course, it is important to remember that we do not use subjectivity as a synonym of agency, but we refer to subjectivity as a mini-narrative that is composed within a social activity by an individual in the midst of her or his interactions with others.

Ken: Jonathan Turner (2002) has a similar take on identity—arguing that (heuristically) it makes sense to think of identities along a continuum ranging from core identities, to sub-identities and role identities. His role identities are associated with day-to-day enactment of social life where, for example, you may be shut down and not succeed in meeting a role. In such cases a person might blame others and accept the lack of success with little more than a shrug of the shoulders. For example, a lesson that did not go well from a teacher's perspective might be seen as an example of the kids being restless or the kids being boisterous. By blaming the kids for the lack of success of a lesson a teacher's sub-identity of being a good science teacher might not be changed. In contrast, receiving an annual review as being unsatisfactory as a science teacher might result in a search for a new job—in some other profession. Depending on how closely the teacher views self in terms of being a science teacher the core identity may or may not be changed as a result of a negative experience related to a sub-identity. Categorizing identity in this way simply reminds me that performance in some fields of activity are regarded as more salient than others and is likely to catalyze the production of emotions that could be associated with self constructions in that field. For example, based on an evaluation a person might conclude, "I am a rotten science teacher." Or, based on the same evaluation, a science teacher might conclude, "my way of teaching science will never be appreciated here. It is time for a move."

SungWon: An advantage we find in cultural-historical activity theory is that it allows us to articulate autobiographical narratives (discursive practice) as part of *collective* activity. Whereas the stability of narratives can be seen as evidence of homogeneity between activities the actors move across spatiotemporally, instability is an indication of heterogeneity. In my research project on boundary crossing, I saw Mariko's participation in different university classes and research meetings where I often heard her life stories, which she enacted with different voices, and which led to her Being seen and understood in different ways. This, together with my similar experiences, bring me to a dialectic philosophy of being-*with* conceptualizing that subjectivity appears always already with intersubjectivity (e.g., Nancy, 2000).

Ken: I am glad you mentioned what I understand as a dialectical relationship—subjectivity|intersubjectivity. As one makes sense of her social experiences she is also taking the meaning from others and testing the waters in that she wants to know about facticity—is her experience of events the same as others in the field? Hence, just as an individual does not author her identities solely, subjectivities are not crafted independently of a collective.

Michael: I use the dialectical framing because there cannot be *shared* understanding without *individual* understanding; but all individual understanding presupposes shared understanding—there is no way that we could have culture by starting with an individual, as a number of philosophers showed analyzing the faltering of Edmund Husserl's project of establishing an egology (e.g., Franck, 1981), that is, a science of the "I." The two forms of understanding are dialectically related because the very moment some first human being made an utterance to communicate something he or she presupposes that the sense of what is said is shared. And this is salient in all three chapters whenever there are two or more people talking to each other, articulating what they mean all the while presupposing that the others already have the capacity to understand the meaning. When Georg W.F. Hegel (1806/1977) writes about two self-consciousnesses that in their mutual actions "*recognize* themselves as *mutually recognizing* one another" (p. 112 [§184]), he precisely makes thematic both subjectivity and intersubjectivity.

SungWon: The emotional and ethical dimensions of intersubjectivity suggest that it is on the presupposition of empathy and solidarity with others that a stable self, narrated within one activity system could appear in or have relevance to another activity system: Who I am always is a function of the activity and whether something ascribed to me in one setting has a bearing on who I am in another setting ought to be an empirical matter. Thus, for example, we might make very different assessments as to who Mariko is if we were to follow her around while interacting with childhood friends or family in her native country. More so, who she is cannot be understood without consideration of the collective object that orients all participants in an activity—including the empathetic and ethical dimension involved in the orientation toward the motive of activity. I think that successful cases of designing school science, such as is evident in the episodes described by Maria et al. and Michael, acknowledge that developing scientific identity must have emotional and ethical dimensions at its center.

* * *

Michael: There is also the issue of sense and meaning, which is a very difficult one again because of the loss and treason these words undergo during translations and because nobody really defines how they use such terms so that even in encyclopedias of semiotics, the editor has to point out the inconsistent use of the two terms. Alexei N. Leont'ev building on Lev Vygotsky, and, following him, the psycholinguist David McNeill (2002), suggest that sense arises from the relation of individual goal (action) and collective motive (activity). Something I utter, which is part of a speech act, is produced for and therefore takes its sense from the *societal* activity, which it in fact contributes to realizing in a concrete way. That is, sense arises in the dialectical relation of goals and motives. Sense, to follow the dialectical philosopher Jean-Luc Nancy (2002), is not something we *make* but is something that is *marked, pointed to, highlighted*, and *articulated* by a variety of means, including language. Meaning, in my understanding, supersedes sense by including a second dialectical relation that exists between goal-directed actions and the unconscious operations that realize them. Words, or any other signs for that matter, therefore do not have meaning, but they accrue to always already existing meaning, as Martin Heidegger suggests, where he uses the latter in the way Nancy uses *sense*. These distinctions are important because they occur in the same plane as the distinctions that lead to different understandings of identity.

Maria: The distinction between meaning and sense is yet another complex issue. For us, the differentiation of sense and meaning is a matter of degree along various continua, namely fluidity vs. stability, ownership by the individual vs. a collective entity, amorphous vs. shaped nature. Both sense and meaning are populated by the intentions of others; they are developed and reside at the crossroads between the individual and others. But sense is relatively more owned and constructed by the individual. Meaning is relatively more "agreed upon" by others, by the disciplinary experts, by the frequent users, by the insiders of a culture, language, semiotic system. We agree, nevertheless, that words, and signs in general, do not have meanings. As Vygotsky claimed, they have meaning potentials. Their meaning potentials are shaped by the particular ways the signs are used, along with other signs, within particular sociocultural and cultural-historical systems. These meaning potentials are quite defined and stable, but they could also vary significantly across domains. The syntax/logic of a domain and the semiotic systems that govern it shape the meaning rather than the sense of a sign. Relatively speaking, the sense an individual makes of a sign is more influenced by the logic of her or his lifeworld. One of the ultimate goals of learning, and learning science in particular, is for an individual to be able to negotiate between sign sense and meaning, to successfully make the transition from sense to meaning, and, at the same time, to imbue with "sense" the meanings that she or he has been constructing within a social activity.

Ken: These seem like fine-grained distinctions. For the moment I am not so sure they have salience for me. I understand that goal|motive is best understood in terms of individual|collective and hence can see that sense arises from the tension between personal and collective objects. So, what will constitute meaning? Given where we have been in this metalogue, why not action|activity or in Schutz's terms, why not conduct|participation? (Participation, as I use it here, refers to collective conduct. Conduct is used as Schutz uses it—to include action, which is conscious and unconsciously enacted culture as well.) The strength of the latter formulation of meaning is that it retains the individual|collective as well as the conscious|unconscious aspects of participating in social life. From these vantage points I can see how such a framework could guide research and different forms of practice in science education. Of course we have already discussed meaning earlier in terms of the dialectical relationship of subjectivity|intersubjectivity.

SungWon: I think the issue of relativity is resolved once we approach semiotic topics in the paradigm of action particularly from a paradigm that frames individual and collective dialectically (e.g., Bakhtin, 1993). As is articulated through phenomenological principles, the act of touching the world comes with being touched by the world simultaneously (Derrida, 2005) and therefore the sense of an action can be said to be and at the same time not to be the individual actor's own. In this framework, learning constitutes the simultaneous change of individual and collective. Each student and teacher is a "being-in-the-world" already with others and thus any change of sense attributed to individuals comes about through the simultaneous change of the whole sociomaterial world that constitutes the presupposition of being. However, I find in many cases educational discourse rarely allows possibilities for accepting or recognizing changes in the collective, thereby reducing learning as the event occurring on an individual plane that is separate from the collective.

Michael: And this is why I think to properly understand what is going on in, for example, classes in Maria's project, we need to think of culture in a different way. It is not a box into which we *en*culturate or *ac*culturate newcomers. Rather, the children Denzel, Amato, Alam, and the others are *active* and *constitutive* members of society and culture just as much as Michelle and Jane in my chapter. It is precisely their novel production of culture that brings about the changes in culture. In a sense, therefore, teachers change in the course of teaching, because they also are constitutive members of the culture, and, as the culture changes, teachers change too. This is the point SungWon makes in articulating that there is a dialectic of individual and collective.

Maria: Absolutely! The way I conceptualize the dialectic between individual and collective is in the form of a Venn diagram where individual and collective are two sets with non-empty intersection and non-empty relative complements. There are dimensions of the individual that exist independent of a particular collective activity and others that are constituted by it; and vice versa, there are dimensions of the collective activity that are "defined" independent of the individual, and others that are constituted by the individuals. I do not focus on identities being the results of communication and collective activity as you, SungWon and Michael, write about them in your chapter, but rather I think about identities as both the shapers and the products of collective activity and emotionality. They make collective activity what it is, and they are shaped by the activity so they are not the same as before the activity (although their changes may not be noticeable until a "critical mass" is reached). What I find interesting to consider is, in terms of a particular activity, what an individual brings to it that influences it as she or he negotiates how it unfolds and evolves, and how the object/motive is accomplished given the other individuals' goals and actions. These are what I call dimensions of the individual that exist independent from a particular collective activity. Such dimensions may have emerged during, or supported by, participation/membership in institutions, fields, or spaces to some extent, but they appear to be well individually-owned at this point and relatively stable—e.g., a boy who is funny, a girl who constantly raises her hand to ask questions, a teacher who code-switches between Spanish and English all the time in her Spanish-dominant class. These are aspects of an individual's identity that he or she brings to various activities relatively independently from the nature of the activities. They may be weakened or nurtured by participation in a particular activity and they may help maintain or disrupt the form and function of the activity. Furthermore, there are dimensions/characteristics of a particular activity that are enacted relatively independent from the individuals involved in the activity at a specific time. In this way, at the same time that Arturo is a constitutive member of the book conversation activity, he also comes to experience a conversation in which Eli has relatively more control over the conversational flow, pace, and content and which Eli has enacted with several other children besides Arturo. Thus, appreciating how the individual and the collective "stand on their own," as well as are shaped by each other, allows us to conceptualize their dialectical relationship.

Ken: It seems to me that you are pointing out that agency and structure are dialectically related at both an individual and collective level.

Michael: This can be expected from cultural-historical activity theory and the way in which it is framed. Thus, agency expresses itself as production generally and the salient structures are those that are made thematic in the paradigmatic triangle representation that is used to articulate third-generation activity theory (Engeström, 1987).

* * *

Michael: Even more important to me personally right now is another issue: the asymmetry between agency and passivity, which leads to an asymmetry in the way identity comes to be theorized. Thus, if we think of identity in terms of the concept *positioning*, then clearly we see agency in the way identity emerges. But we are not only positioning ourselves, we are just as much positioned, prior to all our intentions we are and find ourselves already in a world. This is similar to the case of speaking in everyday conversations where we do not choose words so that what we say is both intentional and non-intentional, it happens to us as much as we make it happen. Identity therefore also has a tremendous passive component (Wall, 1999), just as agency, which most theorists heretofore have neglected to attend to.

Ken: Since you first mentioned this dialectical relationship we have considered its salience in our research. To us one of the key strengths of considering the dialectic is that through passivity inscriptions of many sorts occur. For example, people are gendered and racialized. In New York City where many dark skinned immigrants participate in social life, skin color can racialize people as African American, taking little or no account of the cultural resources associated with their ethnic trajectories. Thus, immigrants from Haiti and Jamaicans are treated as African American on the basis of skin color. Accordingly, others shape the structures they experience and their agencies are changed accordingly.

Maria: To echo the theme of this section, I'd say that identity is a dialectic between agency and passivity, in the sense that it is a dialectic between who we see ourselves (and others) to be, and who others see us (and themselves) to be. We are positioned by others as at the same time we are positioning ourselves. But, I wonder how much of an agency there might be in the positioning by others. Sharon had positioned Amando in a particular way, both because of how Amando and she had interacted in the past and because of the collective sense that Sharon had of that class and those students, as well as the many classes and the hundreds of students she had taught for more than two decades. Similarly, Enrica in Ibett's class was positioned by her peers in certain ways because of the ways she had chosen to portray herself so far in that community, as well as because of who others thought she was because of their past experiences with other children like Enrica. And the school, as a system, supported particular ways of "seeing" Enrica, as somebody who could potentially constrain opportunities that other children get for involvement, sharing of their thinking, and developing it further. I would also like to discuss further the idea of boundary crossing that SungWon and Michael bring out in chapter 9. I wonder how the dialectic between agency and passivity is played out as individuals encounter boundaries and engage with them (whether they cross them or not). And maybe we can go further and theorize about the role of this dialectic in the ways that boundaries get developed in the first place.

SungWon: The passivity associated with cultural-historical nature of activity arises from the human being having a social body (flesh) . . .

Michael: . . . which includes the senses, which are not theorized if we think of the *body*. The possibility of sense arises from the senses; and precisely when I intend to explore my surroundings, the possibility associated with my senses exposes me to the (material, social) world. It is passivity at the heart of agency that makes consciousness possible.

Ken: The way you use the term here seems similar to receptivity.

Michael: I think it may be more as *passibility*, which is a notion that I would like to explore to a greater extent to see what it can give us for theorizing learning and identity. I would like to see how we can develop activity theory by incorporating these terms that at present do not exist within the theory (Roth, 2007).

SungWon: The passive aspect of action and identity has been rarely recognized even among activity theorists; lots of research provides descriptions as if actors were individual minds separated from the bodies and had only an agential aspect that was expressed as high or low. When we attempted to address the embodied (material) nature of boundary-crossing—from one country to another, from everyday mundane settings to the highly ritualized settings of science laboratories—we were immediately led to take a first person perspective of ongoing activities and attend to sufferings experienced by actors including student, teacher, and researcher. One example of the dialectic of passivity and agency can be found in the case of transference and countertransference between researcher and researched, empathetically (vicariously) experienced sufferings (Hwang & Roth, 2005).

Michael: Maria, by introducing the concept of passivity, I want to go much further than you have indicated in your response. I do agree that Amando and Enrico experienced passivity, which is, in this case, just another form of agency. What I would like to do is to theorize passivity in the agent, here, Sharon and Enrico's peers. Thus, the teachers and students act in the way they do in order to realize the goals they have at the moment. But these goals are not intended; these agents receive and become hosts to intentions that they do not know the origin of.

Ken: Michael, are you saying here more than there are contradictions in the objects of an activity system? I understand the idea of being positioned, in the sense of being marginalized by others. This could be an outcome that arises without agency and associated acts of the person who is marginalized.

Michael: I am trying to communicate that passivity is at the heart of agency and therefore that there is an inner contradiction not just in the object but in the subject as well. Because of our material existence, each of us is positioned and therefore singular even prior to consciousness. This means that we are passive with respect to being positioned prior to and beyond all consciousness. Then, when we realize our goals—e.g., in speaking—our own material bodies provide resistance to the intention, which means, they have to expend energy. It is this passivity of the body that is implicit in Karl Marx's (1973) conception that *production* IS *consumption*; that is, in their productive activities, human beings expend and use up their bodies because these passively resist and therefore energy is required.

Maria: Another way to think about agency and passivity is relative to the connection, or lack of connection, between the goals of the activity participants' (ongoing) actions and the (more durable) object/motive of the collective activity system. Passivity may be the way to describe the cases when participants' goals are not quite and explicitly part of the activity—the participants may be aware of their action goals and very much intend these goals, but they may not realize how these goals build up, or not, toward the activity object. Objects may be vague, fuzzy, unformed, multi-dimensional, and thus, difficult to construct, and as a result agency may be difficult to realize.

Ken: I can see some considerable strength in the idea of identities having a component that is passive, from a first person perspective—that is, the identities are inscribed by others as we have previously discussed through agency|structure relationships.

REFERENCES

Anyon, J. (1981). Social Class and School Knowledge. *Curriculum Inquiry, 11*, 3-42.
Bakhtin, Mikhail M. (1993). *Toward a philosophy of the act*. Austin: University of Texas Press.
Bourdieu, P. (1990). *The logic of practice*. Cambridge, UK: Polity Press.
Bourdieu, P. (1992). The practice of reflexive sociology (The Paris workshop). In P. Bourdieu & L. J. D. Wacquant, *An invitation to reflexive sociology* (pp. 216–260). Chicago: University of Chicago Press.
Derrida, J. (2005). *On touching—Jean-Luc Nancy* (C. Irizarry, Trans.). Stanford, CA: Stanford University Press. (First published in 2000)
Engeström, Y. (1987). *Learning by expanding: An activity-theoretical approach to developmental research*. Helsinki: Orienta-Konsultit.
Franck, D. (1981). *Chair et corps: Sur la phénoménologie de Husserl*. Paris: Les Éditions de Minuit.
Giddens, A. (1991). *Modernity and self-identity: Self and society in the late modern age*. Stanford, CA: Stanford University Press.
Hegel, G.W.F. (1977). *Phenomenology of spirit* (A. V. Miller, Trans.). Oxford: Oxford University Press. (First published in 1806)
Hwang, S, & Roth, W.-M. (2005). Ethics in research on learning: Dialectics of praxis and praxeology. *Forum Qualitative Sozialforschung / Forum: Qualitative Social Research* [On-line Journal], *6*(1), Art. 19. Available at: http://www.qualitativeresearch.net/fqs-texte/1-05/05-1-19-e.htm.
Kane, J. M., Hankes, J., Varelas, M., & Pappas, C. C. (2006, October). *Constructing spaces for science and literacy in an urban 3rd grade classroom*. Paper presented at the annual conference of the Association for Constructivist Teaching, Lisle, IL.
Lefebvre, H. (1974). *The production of space* (E. Nicholson Smith, Trans.). Oxford: Blackwell.
Marx, K. (1973). *Grundrisse*. London: Pelican Books.
McNeill, D. (2002). Gesture and language dialectic. *Acta Linguistica Hafniensia, 34*, 7–37.
Middleton, D., & Brown, S. D. (2005). *The social psychology of experience: Studies in remembering and forgetting*. London: Sage.
Nancy, J.-L. (2002). *Hegel: The restlessness of the negative*. Stanford: Stanford University Press.
Roth, W.-M. (2007). Editorial: Theorizing passivity. *Cultural Studies of Science Education, 2*, 1–10.
Schutz, A. (1967). *The phenomenology of the social world* (G. Walsh & F. Lehnert, Trans.). Evanston, IL: Northwestern University Press.
Sewell, W. H. (1992). A theory of structure: duality, agency and transformation. *American Journal of Sociology, 98*, 1–29.
Turner, J. H. (2002). *Face to face: Toward a sociological theory of interpersonal behavior*. Stanford, CA: Stanford University Press.
Wall, T. C. (1999). *Radical passivity: Levinas, Blanchot and Agamben*. Albany: State University of New York Press.
Willis, P. (1977). *Learning to labor: How working class lads get working class jobs*. New York: Columbia University Press.

PART D

DISCURSIVE CONSTRUCTIONS OF IDENTITY

INTRODUCTION

Discourse and language are core aspects of the human condition generally and of culture in particular. Utterances not only are about objects and events, as a form of representation, but also essentially are tools for concretely realizing societally mediated motives of activity. Thus, when a carpenter in the process of building a house says to her co-worker, "Give me the framing hammer please," then she does not use language to represent something. Rather, she says something in order to get something done within the activity as a whole, that is, here the movement of a hammer from one place to another so that something else can be done: hammering a nail.

In saying "Give me the framing hammer please," the carpenter is doing more than enacting a request. First, nowadays most framing is done with nail guns; therefore, in the carpentry culture, requesting a framing hammer rather than a nail gun takes on a significance that outsiders do not easily understand. The request indicates that there might be a problem such that the job cannot be done with the nail gun; or that the hammer is used for a job that the nail gun is inappropriate for, such as pulling a nail with the claw that the framing hammer comes with. But the request tells us more: it tells us something about the knowledge of the person, as she articulates wanting a framing hammer rather than a sledgehammer or any other of the different types of hammers. In requesting a framing hammer, the person articulates special knowledge—recognizable at least by insiders—and therefore articulates an aspect of identity. In this section, chapters 13 (Brown and Kelly) and 14 (Brickhouse and Lottero-Perdue) deal with *identity* in this way, as something co-articulated in the process of (a) minority high school students talk about the physics of baseball and (b) elementary students talk about children's stories they read as part of a book club. In both chapters, involving students of different ages, identities are co-produced in the course of activities the topic of which is not at all identity. As suggested in the introduction, this aspect of identity inherently is unstable, because it is mediated in an essential way by the ongoing activity as a whole. Who someone is cannot be abstracted from the material and social conditions and resources for actions, which are irreducible aspects of the productive process and outcomes, and therefore also, the production of identity.

Identity, however, may also be the topic of talk. Returning to our carpenters, our protagonist, when asked who she is, may tell a story from a construction site where she went about a difficult job going back to using a framing hammer that many of her colleagues may find outdated. This aspect in the narrative construction of identity is apparent from the following statement of an electrician—who has been a participant in Roth's research—whose apprenticeship and training has been documented in its entirety:

After months of being a disposable set of hands on the big industrial site, working with Steve just about brings on culture shock. I've shed my fifteen-pound tool belt, and walk around with a "data geek" pouch with precision cutters and termination tools. I've been introduced to, and routinely discuss my work with, the network manager, a system analyst, and a couple of programmers. I'm a player in a team not just of electricians, but of network builders. I wear a dress shirt to work, and my jeans stay clean. Some of the guys at the other site say that data is for sissies, but I happen to like it (maybe they're just envious).

Fundamental to the analysis of such narratives as a source of identity construction has to be an account of the context *in* and *for* which the narrative is produced. That is, the narrative is not produced for itself but for an audience. Here, the electrician does not just talk about his life and the changes that occurred at the end of his apprenticeship but talks about it for the benefit of an audience (interviewer, researcher). Identity talk inherently is motivated, just as talk about any other topic; it is produced for a certain purpose, and this purpose is an irreducible aspect of the talk even if it is not articulated as such. In his chapter 12 of this Part D of the book, Yew-Jin Lee takes precisely this approach, showing us how identity emerges from interviews. More precisely, Lee articulates the different discursive repertoires available to produce such accounts. These repertoires constitute resources for producing identity precisely because of their potential to go uncontested, that is, go without further requests on the part of the interviewer to further elaborate or explain.

Central to the discursive approach is the idea that discourse is a constitutive force in social relations that establish both topic of talk and subject positions (Davies & Harré, 1990), which incorporate both conceptual repertoires and locations for persons within the structure of those who use the repertoires. Once taking a position, the speaker inevitably perceives the world from that position; the position establishes the lifeworld of the person at that instance. The person using a discourse is, in the form of identity, itself the product of discourse use. The *person* therefore continuously is constituted and reconstituted within and across situations. Positioning, the concept explicitly articulated in the chapter by Brickhouse and Lottero-Perdue, is one powerful concept in discursive social psychology for understanding personhood more generally. Because discourse is a social phenomenon, the identities produced are concrete realizations of identities possible in a particular culture.

REFERENCES

Davies, B., & Harré, R. (1990) Positioning: The discursive production of selves. *Journal for the Theory of Social Behaviour, 20*, 43–63.

YEW-JIN LEE

12. A BEAUTIFUL LIFE IN SCIENCE

Learning and Identity at Work

As veteran science educators, we all have our favorite collection of real-life success stories in science that we love to inspire others: About that child who venomously hated the subject while in school but ended up being a chemist; how that timid student evolved into an vocal campaigner for wildlife, or how that one who flunked *Genetics 101* twice chose to teach high school biology later. Less frequent but equally exciting is when we meet face to face with somebody who has led an interesting or beautiful life in science. Yet, telling others about that kind of life can be strangely problematic for we often find ourselves retelling a heroic narrative of how some poor, ignored inventor managed to overcome the odds and achieve great fame. At other times, we end up articulating variations about some die-hard fanatic who persisted in proving his or her pet hypothesis even though the establishment thought it outrageous—Thor Heyerdahl whom I met years ago at a lecture immediately comes to mind.

During my doctoral research on how people learn in their jobs at a salmon hatchery, I faced a similar situation after interviewing my key informant whom I will call Jack. I asked myself repeatedly what was the best way to recount his against-the-odds story of a boy who hated school but eventually became famous throughout the province for his knowledge of fish culture? To me it seemed that I was merely rehearsing a tired storyline in my writing albeit couched in academic language now. I wondered if I was doing him (in)justice by comparing him to Don Taso the master of the cane fields (Mintz, 1974) or Willie the bricoleur mechanic in upstate New York who could fix just about everything and anything (Harper, 1987)? And how much did Jack share in common with the high-school educated engineer-inventors at W. L. Gore & Associates who were inveterate tinkerers from a young age (Lottero-Perdue & Brickhouse, 2002)? Did I truly meet and write about the *real* Jack?

These then are some of the difficult questions regarding identities in science that I want to explore in this chapter. Although this might seem naïve or unproblematic given the prevalence of simply asking people about their biographies by journalists, talk-show hosts, and of course researchers, these aspects of a person's identity described for us are not "the real thing." Rather, just like the Jack who shared his life story so generously with us over the years, these are basically contingent and rhetorical versions of a person's identity. The Jack whom I interviewed is neither the complete nor the final version of himself that he proffered to us. Change any one of the circumstances surrounding the interview (e.g., location, turn-taking, gender of researcher, the sequence and type of questions etc.) and another aspect of Jack's identity quickly becomes salient at that point in time. The key focus now slips into asking when is a particular identity salient, through what means is this accomplished, and under what conditions do identities change (Roth, Hwang, Lee, & Goulart, 2005). Therefore, the phenomenon of identity is best considered here as an ongoing project, which I will demonstrate later through the examination of participants' discourse during research interviews.

And if it is true that issues of identity are the next "killer app" in our efforts to improve the teaching and learning of school science, then I believe that many of our conceptual tools to understand the former are inadequate for the task. For instance, many psychologists tell us that identity is something that is intimately linked with a person's developmental trajectory and normal functioning within human society. Without discovering who one actually is or can be, one would presumably encounter alienation, diminished meaning and purpose over the lifespan. Due to this inability to participate successfully in social or institutional life, one would therefore be labeled dysfunctional or a misfit. So compelling is this argument that the notion of identity capital has been proposed to characterize the internal psychological resources that are vital for the hazardous negotiation of life in late modernity.

Furthermore, demographic variables such as gender, age, family background, and socioeconomic levels are widely believed to intersect with innate identity or personality traits to produce a heady cocktail that is predictive of success in school. Based on these psychosocial theories, once a person has attained the appropriate developmental stage and has been exposed to proper classroom instruction, the burden of school failure can be pinned down to shortcomings on the part of the individual. Indeed, the different educational sorting mechanisms across the globe are all united by their affirmation of meritocracy and fairness; if one were truly capable and saw oneself as a learner, the world would be one's oyster. By the same token, the predominant modes of explaining school failure generally revolve around disadvantage and internal deficits in students' cognition and identities exemplified by the concept of learned helplessness. Hence, Jack's success story in science has an inbuilt appeal right from the start for it seems to contradict our commonsense norms about the close connection between academic pathways and mental abilities.

In this chapter, I too claim that learning science or any other subject in school has an ontological aspect surrounding it but it has little to do with the workings of the human mind and identity, at least as far as traditional psychology or sociology has conceived it. Rather, skillful performance or knowledgeability is "stretched over, not divided among, mind, body, activity and culturally organized settings" (Lave, 1988, p. 1). It does not just occur in school/formal settings but is equally commonplace in routine and mundane everyday practices—learning is a thoroughly endemic human activity. Understood in this manner, learning transforms both what we *become* and how we *act* as knowers—an expression of identity. Learning thus profoundly affects one's sense of agency and concomitant position in the social world. By extension, changing the degree and modes of participation in social practices—learning in a broad sense—presupposes changes in identity. What this simply means is that learning arises concurrently with identity rather than assuming that after a person identifies as a learner then learning can take place. This line of reasoning whereby learning and identity are not prior to each other can be traced back to the philosophy of Karl Marx and it finds its modern expression in the writings of scholars influenced by the sociocultural learning movement.

Having claimed that learning is mutually constitutive of identity, there remains the problem of empirically establishing their joint occurrence. While we have known for some time how schooling affects and molds pupils' identities and how people learn disciplinary knowledge during teaching, efforts to join these two strands together have been said to be unsystematic and unfruitful. As mentioned, most researchers have assumed an essentialist notion of identity and learning whereby these phenomena are believed to be what tangibly differentiates people at the same time as they are the very same conceptual lenses by which educational psychologists bring to their research. The most common methods to capture learning and identity as bona fide defining features of individuals are therefore paper-and-pencil surveys, psychological tests, statistical measures, participant interviews, and observations of talk and behavior.

Without attempting to deny the sophistication of these techniques, I tend to think that scholars using these methods are missing the forest for the trees by ignoring how learning and identity are actually constituted moment-by-moment during everyday practices, in classrooms and even in research contexts. Instead of decontextualized reports of learning and identity, as filtered through predetermined concepts of the analyst, what we are failing to grasp are the first-person accounts given by the participants themselves about their own learning and identity, as their unique and salient concerns. And where this information might be found is right under our noses so to speak—participants' talk. Because one is not forced to second-guess psychological states about what is happening and present (or not) within human minds, I propose to examine learning and identity in science as they are expressed through Jack's talk-in-interaction as a discursive accomplishment (Antaki & Widdicombe, 1998). Neither does the sociological argument that identity and learning are epiphenomena that are derivative of larger social forces such as class, income, and ethnicity hold much water here. Talk is really not so much communicating about learning and identity, it *does* learning and identity both for speaker and audiences—talk is work. As mentioned, we have to abandon finding out what is this or that person's identity in favor of paying attention to *when* is this or that identity relevant at this or that moment. The data that I present here insist that science educators revisit traditional notions of identity and learning in science and the methods by which to uncover their interrelationships. Some implications of these radical moves are discussed in the concluding section of this chapter.

HATCHERIES AND THE SALMONID ENHANCEMENT PROGRAM

Before I reveal how identity and learning can be seen as a (by-) product of talk, it is necessary that Shallow River Salmon Hatchery, where Jack spent the last three decades of his life, become more than just a name. It has to be a place where people in flesh and blood are found and devote the best part of their adult working lives there. Accordingly, I introduce some background information about Shallow River Hatchery and about Jack himself.

Hatcheries in British Columbia are a crucial part of the Salmonid Enhancement Program (SEP) started in 1977 by the Canadian Department of Fisheries and Oceans (DFO). Such is the centrality of the seven species of salmonids in the life and culture of the Pacific Northwest that the region is often viewed as synonymous with any river where these fish can be found. During the last century, however, many salmon runs were severely depleted due to habitat destruction, pollution, over-fishing, and dam construction. The original aim of SEP was thus a concerted (some say foolhardy) attempt to double salmon production with economic, social, cultural, and recreational benefits. Hatcheries are literally "fish factories" and are one of the most efficient means of artificial enhancement within the technological arsenal of SEP. Their raison d'être is to rear millions of juvenile salmon from the egg stage until the young fish are ready to begin their migration back to saltwater to complete their development. Shallow River was one such hatchery that specialized in manufacturing three species of Pacific salmonids—coho (*Oncorhyncus kisutch*), chinook (*O. tshawytscha*), and steelhead (*O. mykiss*). This particular hatchery stands out not only among those that I studied but also among all of the hatcheries on the Canadian Pacific coast. Scientists and support biologists from DFO consider its personnel as highly competent not only in their day-to-day jobs but also in the occasional scientific experiments they designed and conducted, although the fish culturists, whose primary job is to rear fish, were high school graduates with minimal further training. Without using proper scientific terminologies, most workers here are able to give fairly accurate explanations of the biology and behavior of the fish that they all rear so lovingly.

Jack, one of the most senior fish culturists in Shallow River, was the main informant in my doctoral research.[1] Trained as a plumber but with proficiency in carpentry and electrical work, he thoroughly enjoyed plumbing as his first job in a big city where he grew up. There he remained for fourteen years until he said that "the trade wasn't a trade anymore. You got plastic pipe and everything getting quicker and quicker and doesn't matter what it looks like—just get it in!" After a brief part-time stint as a park naturalist, he landed a job at the hatchery where he has remained as a fish culturist ever since. A retired DFO biologist once commented that Jack deserved a medal for his rearing work concerning chinook salmon; the number of adult returns has increased so dramatically in Shallow River that scientific research on this species could be resumed. Always the perfectionist to the point of being a nitpicker, Jack resented the new mechanical feeders that were increasingly used by large hatcheries to feed fish for "machines feed ponds but people feed fish" as he often complained. Somehow, these machines according to Jack did not achieve what the expert hand of a fish culturist could do in spreading the feed evenly across the surface of fishponds no matter if machines were able to work tirelessly. Besides, he did not look kindly upon any person who fed fish with headphones plugged in their ears. He tells anyone who cares to listen that it is a fish feeder's duty to be alert to what the salmon are communicating during feeding, or at any other time ("I can put an automatic feeder on there and get the same result. I want somebody to tell me something, what they learned today about those fish so maybe I might learn something"). His reputation as an expert—nearly an icon of Shallow River—was never in doubt, because we often heard his name being mentioned with deep admiration by scientists and other SEP staff.

My ethnographic research revealed considerable expertise among fish culturists like Jack who, upon cursory observation, appear to merely feed fish and clean ponds. Any new staff would in fact face considerable difficulties in competently carrying out many of the activities that old-timers take for granted. Examples abound of such embodied workplace knowledge that includes (a) knowing when the fish are satiated during feeding by observing their swimming behavior in ponds, (b) successfully monitoring water flow and temperature without the use of any instruments, (c) accurately estimating the weight of inch-long fry without instruments, and (d) knowing when fish are ready to be released to the ocean by scrutinizing their behavior and external appearance. From what I gather, there were also few taught courses nor textbooks on fish culture that were provided by DFO in the past, although now there exists a college-level course (with internship opportunities) in this vocation for aspiring entrants. Coupled with the apparent monotony of the daily routine in the hatchery that is dominated by the task of feeding fish about five to six times a day, it seemed difficult to explain how learning and expertise had developed. Certainly, some learning theorists believe that the repetitive and mundane nature of work practices stifle innovation or experimentation by being conformist in both the process and outcome of the action. Suffice to say, it was this glaring incongruity between the high levels of practical knowledge and relatively low levels of formal training among fish culturists that prompted my research in the first place.

LEARNING AND IDENTITY AT WORK

In this section, I briefly explain how learning and identity are rhetorically and publicly displayed accounts of members' (researcher and participant) local, ongoing concerns. Informed by ethnomethodology (Garfinkel, 1967) and conversation analysis (Sacks, 1992), this constructivist perspective stands in sharp relief with ordinarily taking what Jack said/did as the avenue to what he is as a person (i.e., identity) and what he knows and understands in his mind about science (i.e., learning). To elaborate, ethnomethodologists reveal how social order is produced and maintained through members' unconscious though elegant ways. With much of social life being so mundane, we actually forget that greeting a friend, standing in a line, answering a teacher's questions and even talking about learning and identity require a tremendous amount of shared and local knowledge in order to pull off successfully; and because talk is the most accessible means to reveal these local understandings, conversation analysis becomes the method of choice here.

Said differently, in this chapter I elevate interview talk as a topic to be deliberately investigated and not a resource available to generate putative statements about personality attributes and knowledge processes. The alternative readings of the section heading here are apt; the handling of learning and identity through talk is indeed a process of hard work in trying to convince audiences that the speaker is believable, accountable, truthful, morally ethical or being normal. Whether these aspects of Jack's identity and learning at the workplace were self-evident or enduring across all other occasions is nevertheless an empirical manner rather than something assumed. Nor can the presupposition be made that Jack intentionally chose and sequenced his choice of words to demonstrate these aspects of his learning and identity. All I am allowed to show here is how these interrelated phenomena were co-constructed, reified, and made accountable through language during the specific interactional contexts of interviewing. The upshot of this study is that learning and identity in science take the form of situated and contingent accomplishments here, outcomes of the serious business of talk instead of its precedent.

One useful way to see how learning and identity are discursively played out is through what is known as *repertoires*. Because these repertoires comprise related sets of terms, which revolve around one or more core metaphors, they rhetorically serve to organize talk and make it intelligible or believable for participants. For example, it was found that discussions on the topic of marriage among some young people produced both a "realist" and a "romantic" repertoire. Those belonging to the former group tended to give accounts of marriage that described it pessimistically as a hard journey fraught with dangers, whereas the latter group offered theories of ideal marriages as a distinct possibility provided certain conditions were fulfilled. Earlier studies also showed that high-flying biochemists mobilized an "empiricist" repertoire when speaking among knowledgeable audiences but reverted to "contingent" repertoires during informal talk or when putting down rival colleagues or laboratories. Hence, these recurring devices in speech allow analysts to gain purchase into what exactly are people orienting to at that moment in time and how they construct their versions of social reality, which definitely includes issues of learning and identity.

Another conversation analytic tool that will be employed here are *membership categories*. This refers to the idea that people normally carve out their social world into collections of similar things such as family, classroom, punks, and politicians. Thus father, mother, uncle, baby and so forth are all members that can captured within the recognizable category of family, for example. And when a student calls another a teacher's pet, a multitude of commonly accepted inferences can be generated from this act of labeling; hardworking, obedient, well-groomed, punctual, and neat are but a few. Categories thus imply a package of conventionally associated features while the reverse is also true. For our purposes, membership categories demonstrate how identities are *used* during talk whenever people produce categories to align, value, resist or discredit this or that identity for themselves or for others. Prefacing an address with "my fellow Americans" performs a vastly different work of identification compared to "you Americans," "citizens of America," or "comrades." Given the negative connotations associated with the last membership category, it is very unlikely that it will ever find favor as a device to achieve solidarity within American political circles. By examining how identity work is played up using different membership categories, it therefore foregrounds how categories discursively make and remake people if you may.

The Repertoires and Categories in Action

In what follows, we see how Jack used various repertoires and membership categories to work up his accounts of learning and identity in Shallow River. When using a *workplace* repertoire, Jack spoke about the necessity of interest and practical experience to succeed within the occupation of fish culture. Here, the identities of fish culturists like himself were construed as people of modest education but with much hands-on experience who could solve problems on the ground. By contrast, the *school* repertoire treated knowledge and learning as abstract and theoretical with minimal concrete relevance to everyday life in the hatchery. The various people characterized under this repertoire enjoyed high levels of formal schooling—even "too much education" according to Jack—although they usually failed to get the job done. Additionally, I show how Jack attributed the sources of his own widely recognized knowledge of the natural world by means of *nature* and *nurture* repertoires. Talk using the former repertoire credited certain skills and abilities to genetic endowments whereas the latter stated that sufficient effort in learning could increase competence in any domain.

We're just Fish Farmers . . .

The following interview excerpt presents the school and workplace repertoires in action. Here we encounter Jack and another colleague named Eric discussing how they learned at their workplace that was in direct response to the researcher's initial question. Not only do these two employees describe the primacy of practical experience over theoretical knowledge in fish husbandry in their workplace repertoires, they co-construct their identities akin to skilled farmers bent on growing the best product possible, in this case salmon. In final analysis, all that mattered from the interview exchange here was whether one could deliver the goods, not what academic attainment one enjoyed in the past.[2]

Episode 1: 31 January 2003
01 Researcher 1:   So, most of the things you guys just learned on the job, the hard way?
02 Jack: Yeah.
03 Researcher 1:   Like you said, taking the fish with no anaesthetic.
04 Eric: Well, I can, I can only I can tell you what I've seen. What I've seen is that there are guys like us who have wanted to do it, have come in and try to grab what we can. And I've even seen students come in, with the, the theory and, and not the experience and kinda get disillusioned because it's not sitting in an office.
05 Jack: Yeah, I.

06 Eric: It's no different than going to a farm, there's no difference. I mean you're not, it's not just (?), you can put anything on paper, but I mean if you have heart in it, I mean, and it, what I, what I have on paper is immaterial. It's the product that I'm putting out. If I feel that is good then I'm happy. That's what I want.
07 Jack: That's what we're at. And we can go from there, if we wanna go more advanced in a different area then we can eh? But like Eric said we're fish farmers, that's all that we are, basically. And try to get the best product to go out and the healthiest state to come back. And like I was saying, some of the best farmers on the prairies have grade three education. We're just farmers.

Researcher 1 started by asking the fish culturists about the source of their training. Thus, both fish culturists now oriented to this question and not some other, which directed the subsequent trajectory of the discussion. Neglecting the contingent interactional effects of talk and taking these interview responses are objective reflections of people's knowledge and beliefs would accordingly be erroneous. The opening was already marked by wondering if this process was done "the hard way" rather than in some other presumably easier manner. Having affirmed that this was indeed the case in line 2 by Jack, line 3 elaborated on a previous example given by Eric (not shown here) of how incredibly laborious the job was in the past, especially when it came to handling the fish without any anesthetics. Eric then followed up using workplace and school repertoires in line 4 and 6 that was rhetorically strengthened by claiming that it was first-hand experience that he was speaking from. Performing identification work ("guys like us"), he aligned Jack and him as workers who joined the hatchery willingly and tried to grab whatever [knowledge] they could while being on the job.

Eric then deployed the membership category of students that stood in direct opposition to the expert fish culturists. Students carry undertones of youth and inexperience, training under compulsion perhaps even possessing irrelevant knowledge to working life. It now dawns on listeners in that conversation that the two repertoires were inhabited by different kinds of people. One was populated by newcomers possessing highly abstract school-based knowledge and the other consisted of old-timers like Jack and Eric who gained down-to-earth, practical experience while on the job. Whether these kinds of statements find any correspondence with reality is immaterial for these were precisely the discursive resources at work and how audiences would ordinarily hear and interpret them. In turn 6, Eric now realigned the fish culturists with yet another membership category—the farming community. As long as one had the "heart" or interest, paper qualifications did not really matter in these two occupations for the final result mattered more than anything else as portrayed in this workplace repertoire (c.f. Episode 7). Interest in one's work was ultimately the key to a "good" product and personal satisfaction.

The words spoken in line 7 performed interesting rhetorical work in that analysts can recognize what is known as a "showing concessions" tool here. To begin, stating "that's what we're at" rendered this an extreme case, which affirmed the correlation of interest in one's job, getting the best product out and ensuing satisfaction that was possible however minimal one's educational attainment was. In the following sentence, however, Jack conceded that further learning was perhaps possible after all rather than being tightly constrained (by past educational achievement). Without qualifying or mitigating the first sentence—a discursively vulnerable position—would have allowed this claim by Jack to be easily undermined or dismissed by his audience. However, the next sentence embedded with the contrast markers "but" served to overturn the preceding concession. What discursive work the third part did was therefore to actually reprise and buttress the first claim about academic qualifications being unrelated to workplace competence because interest in one's work carried far more weight. Fish culture and farming as membership categories thus shared a fundamental overlap—skilled workers who possessed modest amounts of school-based knowing. To reinforce this claim in his workplace repertoire, Jack reminded everyone that some of the best farmers on the prairies had but an incomplete elementary education—school and workplace knowing were indeed weakly correlated.

The Truly Smart and not so Smart . . .

Besides farmers, Jack used a number of other membership categories including fellow fish culturists, maintenance men, scientists and engineers in conjunction with his workplace and school repertoires. By unpacking six of these episodes whereby issues of identity were at work, we gain a deeper appreciation of how much Jack valued what he called practical experience obtained through hands-on learning versus abstract forms of learning prevalent during formal schooling. I was fortunate to have three parallel accounts whereby Jack spoke of a failed scientific experiment using carbon dioxide, which his colleague Paul had conducted previously. Episodes 2 and 3 below were the two more substantive versions in the transcripts and will be used now to show how despite being employed as a fish culturist, Paul was basically "too smart to be a fish culturist" according to Jack.

Episode 2: 27 February 2002
08 Jack: Same with carbon dioxide years ago, one of my fellow workers screwed up on, on carbon dioxide for an anesthetic and ah, and it went, it did, it, he said it didn't work. Okay so and I got back into it, it worked for me and what happened he made a mistake somewhere but didn't catch his mistake, I wasn't helpin' him so I didn't know, not that I'm tryin' to sabotage his, his experiment and stuff I just wasn't involved in that situation. So when I got it, it worked, I thought well there's nothing wrong here, how simple can we get. It's just put it in and it works. And then I tried it in different water temperatures, all sizes of fish, saltwater, freshwater and it worked fi-, it worked fine.

This particular episode was preceded by a discussion about whether Jack took heed of what the other hatcheries were doing in terms of scientific projects to investigate salmon biology. It was in this context that Jack first revealed the circumstances surrounding an experiment done in the 1980s involving carbon dioxide as a possible fish sedative. This fellow worker in the hatchery was supposed to be somebody equally adept at fish culture but it was not the case. Instead, Jack succeeded in getting results rather easily ("when I got it, it worked") whereas his colleague was continuously frustrated. For some unknown reason, Paul had made a mistake but was unable to pinpoint the source of his experimental error. By stating that there was "nothing wrong here [in the experiment], how simple can we get. It's just, 'put it in and it works,'" it unambiguously painted the failure of Paul to see what a competent fish culturist should have observed.

Episode 3: 11 March 2003
09 Jack: There is a guy here at (?) and he had knowledge that you wouldn't be- but no practic-, nothing practical eh? Smart as a whip, could write papers inside out and backwards, probably could do it in his sleep, he tried carbon dioxide and it was a big disaster. So, something wasn't working right for him but he didn't have that insight to try to figure out this isn't going to work so. Yeah, he's [a] fish culturist and uh that was on carbon dioxide. Well, he, now he's one of the head honchos at, one of the colleges anyway at River College they do adult programs whatever. So he left here and went up there. So that's more suited for him, that's great but he wasn't really a fish culturist because he was too smart to be a fish culturist. He didn't have enough practical, and he couldn't see anything, you know? Like he couldn't see or observe? The lights didn't turn on. So, and then he tried it and I tried it and it worked great!

Using both workplace and school repertoires here, Jack told the interviewer that Paul was a brilliant fish culturist who was able to write scientific papers without any problems. In fact, by giving three continuous examples in his list of Paul's capabilities (i.e., whip, inside out/backwards, sleep), this was a typical discursive listing tool to emphasize certainty. Nonetheless, Paul was said to be "too smart," which disqualified him from being a genuine hatchery worker for he lacked the requisite practical knowledge just as he was perhaps atypical of the average fish culturist who normally did not contribute to formal scientific research as well. If we recall the importance attached to practical experience by Jack, we can immediately understand Jack's dismissal of Paul as a legitimate fish culturist—Paul was only a fish culturist in name. Specifically, Paul could not see or observe in sufficient detail, which in Episode 10 is claimed as one of the biological gifts belonging to Jack. This deficit according to Jack obstructed Paul from making any critical insights into the carbon dioxide experiment ("the lights didn't turn on"). Now that Paul was an instructional leader and teacher at River College, this was felt to be appropriate for he was finally in his element so to speak. Hence, college teaching was the correct category activity and location for Paul for granted that he was knowledgeable in science, it bore no relation to the kind of hands-on skills that fish culturists exhibited and needed on a daily basis.

The next two episodes illustrate how Jack utilizes the membership category of two types of scientific professionals—scientists and engineers—to enable his listeners to better comprehend aspects of his learning and identity as a fish culturist. In a different manner though with the same predictable outcome, some scientists and engineers were portrayed in these repertoires as highly intelligent people though unable to get the job done at the workplace.

Episode 4: 11 March 2003
10 Researcher 1: Um Jack, what jobs do newcomers begin with when they come to the hatchery?
11 Jack: Basically they're doing stuff that we don't want to do you know? (laughter) I was there, I've done it all and uh my philosophy is you start at the bottom and work your way up and I think that just makes you a better fish culturist and a better person instead of coming out of school like some of them do, figure they're going to start at the top, oh man, forget it man, you know? Like I was with engineers at (?) that didn't know one end of the pipe from another and was trying to tell me what to do and I, it doesn't work! You get out and work with the tools for a while and then go back to school, then come out hey it's just going to make you a better engineer, way better than if you're just going to sit on your ass in the university.

This exchange began with the researcher asking about the nature of work that newcomers to the hatchery were allocated, which Jack jokingly replied that it was mainly "stuff that we don't want to do." He then made his explanation discursively convincing by reporting in the first person ("I was there, I've done it all, my philosophy") that one should start at the bottom and work upwards. This means of learning fish husbandry through practical work experience was again consistent with his workplace repertoire that prized these very attributes. In contrast, the school metaphor Jack employed next had connotations of graduates thinking that their educational level permitted them to "start at the top." Giving an example of engineers he once met, this membership category served to reinforce the connotations surrounding the school repertoire: head knowledge and paper qualifications alone had little parallel with solving real world problems. Possibly coming fresh out of university with a wealth of erudition, the engineers unsuccessfully tried to teach this ex-plumber with fourteen years of work history what to do. As such, the contrasts in the membership categories were exploited to the maximum: tertiary level graduates versus high school graduates, decontextualized theories versus abundant functional knowledge. Interestingly, Jack also suggested a means of reconciling the two apparently opposing repertoires in the last part of this episode. He felt that ultimately one could truly be an expert whenever school-based knowing was complemented with practical work in a dialectical manner (e.g., Lee & Roth, 2005). I elaborate more on this rapprochement below.

By now, it should be clear that membership categories perform important identity work in talk by marking and parceling out commonly associated features. The next two episodes elaborate further how Jack deployed a mixture of categories to highlight how practical knowledge is often superior to abstract knowing in the hatchery workplace although it is usually not acknowledged as being so by non-fish culturists.

Episode 5: 11 March 2003
12 Researcher 2: Because I can see you're sometimes you're critical of, you know the big cheeses, the scientists but this is one of those.
13 Jack: If they got any worth, like I'm not negative on everybody, if I see good, then hey I'll pat them on the back, and I see no good I'm going to tell them as much. And

Terry is what I see is a good biologist, and (?) was a animal behaviorist, helluva neat guy to listen to, you know?
14 Researcher 1: So, you can't take a scientist and ask him to raise steelhead? He just cant do it, no way?
15 Jack: No, he wont do it, no. And if it isn't written down, he's going to have to do what I did, trial and error over twenty five years, yeah. And if he's a good farmer, it'll happen quicker for him, and if he's not he's going to have a hard time. But, I look at it really blasé because it's not hard, nothings really hard in fish because you, it's simple, just don't make it difficult? Like keep it simple? And things seem to work pretty good that way...One of my neatest things happened in an aquarium, they put these new pumps in and they were getting mortality. And I thought, I was a guy, he was a maintenance man at Greenfields, and he (?) and he did some maintenance stuff here. Smart, fucking smart, he could take that [camera] apart and fix it and put it back together again! And anyways, I had an idea that these pumps were putting out electric current in the water. So I said to, (?), yeah so I tell you how to do it, so I take a voltmeter, he told me what to do. I went into the classroom I had eight volts twenty-four hours a day on the eggs, coming off those pumps. And that's what it was, we got rid of them, no problem after that. But that was one that I clued into, but stuff, some of it, you never.
16 Researcher 2: So you went to ground one side and just in the water?
17 Jack: Yup. But I didn't know how to set it up. What's his name? His name is gone! He's a great guy and I loved working with him. He said Jack do this and I did, and went in there and shit, eight volts in one, and I think I had ten in another one and all these pumps were giving off a current. . . . But some of the stuff, a lot of the stuff there too is just fish culture practice. The teachers are lazy and they don't do and they try to pull wool over your eyes, you have been at it. I just don't argue with them anymore I say I give you some more. Yeah, yeah that's where they let me down. Even Brad let me down, that was the only time Brad ever let me down and I showed you that paper and they hired a guy. And that's where it's sad eh? And I go back to my boss Collins when I worked for parks and I haven't seen him for a while. And I think I told you this, I bumped into him on the ferry and he never knew what I was doing and I told him what I was doing and he says too bad you didn't have a degree.

In Episode 5 Jack reiterates a belief that workplace-based learning is superior to school-based learning. When Researcher 2 asked if Jack was somewhat critical of certain professionals like scientists, the next line subtly moved to reassert Jack's neutrality in judging who did good work for the proof of the pudding was in the eating, for scientist and fish culturist alike. In line 14 however, Researcher 1 loaded the interview trajectory in a certain manner, which Jack now oriented to and developed his repertoires to good effect. After strongly aligning with researcher 1 that no scientist could rear steelhead salmon successfully, Jack then quickly softened this extreme claim by stating that this was perhaps possible given an extended period of applied work ("trial and error over 25 years") like himself or perchance when the scientist was a "good farmer," again like Jack himself (c.f. Episode 1). The school repertoire here likewise made salient how school-based knowing was often too complex like some Rube Goldberg type of solution ("just don't make it difficult? Like keep it simple"). What followed next was a case in point using the membership category of maintenance persons that extolled the benefits of keeping things simple like how fish culturists and farmers operated as a rule. Jack related the time when nobody could figure out what was causing mortality to fish eggs at an aquarium and how with the assistance of the local maintenance man, he finally managed to discover the cause. Given that the solution was elementary but blind to the experts, it contrasted how simple lateral thinking rather than something abstract or airy-fairy eventually saved the day. That this eureka moment came from people who were not associated with proper scientific competence thus served to align these two membership categories together in Jack's repertoires.

When Jack then stated that "a lot of the stuff there too is just fish culture practice. The teachers are lazy and they don't do and they try to pull wool over your eyes," he was contrasting the aforementioned categories and repertoires once more. Which was more crucial—head knowledge from classrooms or first-hand experience gained on the job? Jack's assessment was that the fish culture teachers had let him down in this respect. When he reminded his audience about an incident with Brad—one of the few biologists whom Jack has expressed genuine respect—it yet again highlighted those people who had let him down and by implication the shortcomings of society placing a premium on educational achievement over practical know-how. To put this particular story into context, there was a mysterious disease that was killing large numbers of hatchery fish in the province about two decades ago. After planning the investigations, doing the fieldwork and collecting data for months, the results were handed over to a hired biologist to write for publication. The rationale was because Jack had no confidence in his own writing ability for a scientific audience just as Brad could not afford the time to do it himself. Unexpectedly, the person resigned before the writing was completed and Jack's hard work never saw the light of day. Above all, it was the lack of getting the proper paper qualifications that prevented him from participating within scientific circles even though the quality of his experiments rivaled that of professional scientists as he related the story about Collins, his former boss. Because it was SEP policy that only graduates could become hatchery managers, Collins was portrayed as a sympathetic person from the "other side" of the organization appreciating that Jack could never be recognized despite all his talents. Although the system of learning and rewards in schools and workplaces is lopsided, Jack did suggest also that these two ways of knowing could perhaps be united as we observe in the next two episodes below.

Episode 6: 31 January 2003
18 Jack: It's like I was fortunate one time I taught for a week at Carnation College at (?) for fish farmers. So they asked me to do it. David the manager was supposed to do it and he put it off on me, and I thought holy shit. You know, I got no experience to teach and I was making big bucks, three hundred bucks a day for five days, wow! So I went through all my stuff, and I had, and I already had about ten or more years of experience and I thought, well that's what I'm going to do. And I went into the library and I saw books were written (?). I put them on the table and I said to the students the first thing you do you get rid of this shit! Put it onto the floor, I says whatever I teach you in five days you take with a grain of salt, okay, if you

don't believe me, you prove me wrong, and if I'm wrong we both learn something, that was my approach. Uh I said the best thing to do is to look at scientific papers, make your own books, this was at this time eight years ago. I think I had one university student in there and he gave me the worst critique of the, of the whole thing. And basically what I, what I did was I taught with my experience, and that's what I did. And showed them things that we did and what to look for, and, and be cautious of books be cautious of even papers because the biologists. So that's what I said to them, "Just take everything with a grain of salt and if you wanna know then you, try to prove that to yourself the best way you can with what, any information that was there." That guy that had the most education, he said, "I didn't want to come here to shoot the breeze." But then you see, he didn't realize that shooting the breeze, there was a lot of things hehheh, which would have saved him a lot of trouble!

It is always analytically interesting whenever a person is described as taking on features that are not normally associated with one's chief membership category. So unnatural is this state of affairs that it requires explanation, which is no exception with this account of Jack and the aspiring fish culture students from Carnation College. As Episode 6 unfolds, we are told that once Jack was asked to stand-in for an absent hatchery manager to teach college students about fish culture. That Jack was invited and that he lacked the necessary qualifications in the first place was rhetorically asserted in various ways such as saying that he was extremely surprised when asked to do the job and that he had to revisit the literature. With more than ten years of work experience under his belt, he then figured that he could perhaps have something to share after all despite the lack of suitable resources from the local library.

In what must have been a truly eye-opening first lesson for the students, Jack reported in direct speech to strongly convey his evaluation of the established literature on fish culture (e.g., "get rid of this shit," "put it onto the floor"). Indeed, the copious use of direct speech here is reflective of investment (e.g., commitment, certainty, determination) in his discourse. Jack then challenged the class to test all knowledge claims about the fish culture, his included, so that if proven wrong, both teacher and student would learn something. Yet, he also did not claim any final authority on the topic ("take with a grain of salt") for he felt it best to first look at the scientific texts out there (Jack consulted his "stuff" and library material), critically assessing these knowledge claims against reality and then to make a final decision. After having gone through this testing process, one could then "make your own books"—an ideal situation ("the best thing"), which synergizes the best from school- and workplace-based learning. Advising the listeners here to make their own books was a typical *discursive device*, which enables the speaker to defend his or her account in the interview especially on those occasions where there were potential conflicts in mediating seemingly opposing repertoires. Without ascribing conscious intentionality to the speaker, totally downplaying book learning would have been unwise for Jack was teaching students in a classroom setting whereas relying on practical experience completely would have required blind faith on the part of the students. Thus, saying that one should make your own books was meant to convey the notion that one should take whatever is available, from schools or from self-experience, and evaluate these claims in the world.

At the same time, the account of the university graduate who wanted to be a fish culturist is a membership category crying for an explanation. To be expected, this student ("The guy that had the most education") found the course most wanting and felt that it amounted to mere chitchatting. This situation whereby the role of the smart and the not so smart are completely reversed (and known only to insiders) forms the cultural basis of much of our humor, which one can observe from this retelling of the incident. Again, whether this assessment was borne out by reality is not as important as the work that the student's identity undertook in the business of talk. What is crucial is that in Jack's version of the event, the student was a prime example of school/book learning, which is abstract, authority-bound and resistant to the practical, problem-solving nature of workplace knowing. I describe another example below of a discursive device that co-opts a mixture of membership categories while delicately juggling the two seemingly opposing repertoires.

Episode 7: 27 February 2002
19 Researcher 2: See this is what we were interested in because often people assume that knowledge just goes this way like scientist to the people or management to the people and yet people like you, they know a lot of stuff that other people ought to know.
20 Jack: Yeah exactly. See this is nice where you guys come in because you gotta have some bookwork but you gotta have some practical work you and I discussed this before like I, as far as schooling to me goes is a waste of time, a lot of it and um common sense. What ever happened to common sense? Some of the best farmers in the world, if they got three, grade three education, they're lucky you know? Now it's different we got computer mod, everything, we got cows hooked up to a, computer. But fifty years ago the cows weren't hooked up to nothin' and they're doin' just quite nicely, you know? And maybe even better because they're not takin' the antibiotics they have to have and stuff like that.

Setting the stage for Jack's reply, Researcher 2 made an opening comment about the unidirectional nature of knowledge flow in the hatchery. By incorporating the categories of people (unspecified user-groups and the fish culturists), scientists (principal generators of knowledge) and management (who often dictate knowledge use), the researcher was asserting that people like Jack really knew "a lot of stuff that other people ought to know." After saying that he was pleased with what the researchers were doing, Jack then invoked another discursive device ("you gotta have some bookwork but you gotta have some practical work"). What did he accomplish by saying this? For one, it seemed that Jack was cognizant of or was referring to our research aims concerning how scientific knowledge moves between different communities. By affirming that one needed different kinds of knowing—the workplace and school repertoires—he was affirming and orienting to our stated research purposes. Alternatively, it can be heard to imply that whether one was a generator or consumer of knowledge, having both forms of knowing was important and could transform or erase the existing one-way exchanges of information. Not having privileged information into Jack's inner thoughts, we have to speculate on the choice for this discursive device.

Whatever the reason, there arose a new problem; Jack, the self-confessed school failure, now had to account for his personal learning situation. Did he heed his own counsel especially since he was widely acknowledged for his expertise in fish culture? He then counterintuitively stated that school was largely a waste of time since education was basically common sense. This assertion that courted rhetorical trouble needed defending, which he did by saying that it was a personal opinion and that it was basically a matter of common sense. Developing the now familiar workplace repertoire, common sense was apparently displaced by the effects of too much schooling and bookwork. He supported the argument by (a) invoking the membership category of farmers with minimal schooling but with loads of common sense (c.f. Episode 1) and by (b) criticizing how modern technology (derived from the school-based learning) was undermining traditional reliance on common sense (and practical work) in farming. As an example of too much education getting in the way, the practice of feeding antibiotics to cows was described as a menace to public heath. Taking all these strands together, it is reasonable to conclude that Jack was still preaching the same message here—practical experience/common sense trumps over school-based learning—by exploiting the two repertoires in his accounts.

Either You Have It or You Don't—Nature and Nurture

Uplifting practical over abstract forms of learning through the use of contrasting workplace and school repertoires was the spotlight in the previous section. These organizing metaphors in talk were further sustained by various membership categories that typified the kinds of learning and knowledge expected for the respective identities invoked. We now turn our attention to survey Jack's explanations for the basis for his vast expertise: was it learned or innate? No longer are we concerned with which forms of knowing are more valued but rather we locate the source(s) of his many abilities. His justifications are interesting for Jack had utilized, as discursive resources, one of the canonical polarizations used by psychologists to account for the origins of human intelligence or skill. By no means do I claim to have solved the longstanding problem of nature or nurture in cognitive development. I am merely content in showing how these psychological constructs were unconsciously appropriated by Jack to reach his own unique purposes during talk-in-interaction.

Episode 8: 11 March 2003
21 Researcher 1: So you saw that, then how did you learn about fish?
22 Jack: How did I learn? Just through interest. Everything I learned is self-taught, pretty well. And it's just, the things that interests me, like mathematics that they don't interest me and computers pfoof, it just doesn't go in there! But the stuff that I like to do like aquatic insects or fish or whatever nature, it's just boom, it's there, and I remember all this stuff. But computers and maths I don't remember. . . . My dad wasn't a fisherman so I had to do basically what, myself, so. All the way through

life basically. And neither was my mum. And so I got into aquatic insects and I got into photography then I got into macro-photography then (?) for insects and I just kept going and going and going. So where I am now at the hatchery like it's never been a chore coz I always knew my species of fish. I got, seen people here that are managers and assistant managers and they are standing here with a fish in their hand didn't know what the hell it was! But I just never, just I knew, you know?

When asked how Jack acquired his knowledge about fish, he opened with a modified showing concessions tool; the underlying strong claim ("just through interest. Everything I learned is self-taught, pretty well"), the elaborated softener ("it's just, the things that interests me," mathematics and computers were the exceptions) and the elaborated reprise of the initial claim ("But the stuff that I like to do . . . it's just boom, it's there, and I remember all this stuff"). Having established this commonsensical relationship between interest and the ease of learning, he launched into what I call a nurture repertoire when he gave examples of how he learnt about fish, insects, and macro-photography. Here, mastery in a domain depended on the degree of past exposure to learning and instruction, which Jack brought in the membership category of his parents to strengthen this assertion. In general, the home environment is widely understood to be the prime venue of a child's lessons in life-skills although this particular excerpt highlighted the absence of parental skills being transmitted. For instance, Jack stated that his father did not pass down any know-how in fishing (and other things) because he was not a fisherman neither did his mother teach Jack anything related to his current talents (see also line 33 in Episode 10). If this was true, then Jack the expert fish culturist and bricoleur was indeed a self-made man in many respects. Because of his interest in the natural world Jack "kept going and going and going" until he arrived at where he was today. By declaring that this learning process was never a chore, it served to underscore how some managers and assistant managers perhaps gained their position other than reasons of expertise in fish culture. Whether Jack was revisiting the school repertoire and its connotations of head knowledge and academic grades or simply implying how some people rose through the ranks for other reasons I am uncertain. All I can say is that Jack is performing important identity work—for others and himself—by telling an anecdote about some so-called experts who could not even identify a fish when it was literally thrust into their hands.

Although reusing the nurture repertoire in the immediate episode, Episodes 9 and 10 now introduce what I call the nature repertoire. The source of one's abilities in the latter is portrayed as a genetic endowment, something that is a given from birth. Of course, the nature and nurture repertoires stand on the extreme ends of the spectrum in establishing the basis of one's abilities and thus Jack occasionally resorted to a discursive device to mediate the two. This device shown in Episode 10 allowed Jack to simultaneously assert, on the one hand that some skills were innate, and, on the other hand, state that given enough time and persistence, anybody could perform compensatory actions for the absence of inherited skills.

Episode 9: 1 December 2003
23 Jack: Yeah. When I was working and I didn't have teevee coz I didn't want teevee when my kids were growing up and uh my daughter was hell of a reader she would read, unbelievable the number of books that she would read. And I would read to them, you know what's it like reading to kids eh? And my son, he didn't read, he's kind of like me you know? And uh I don't think he's dyslexic though he didn't seem to have any problems that way in school. His math was uhm, neither one of them was scholastic but they're doing great you know? And so, some of the stuff, my daughter was hell of a reader you know? I thought she was going to be a writer and she might yet, you never know boy it's all up there, I'm sure she could do it.

The nurture repertoire is apparent when we hear Jack saying "And I would read to them, you know what's it like reading to kids eh?" I take this sentence to imply that learning through reading activities is amenable to change and improvement. That is, reading to kids can result in something positive, which Jack assumed was common knowledge when he used the Canadian expression "eh?" meaning "right?" In the very next sentence nonetheless, the membership categories of his children made it evident his kids were chips off the old block with respect to scholastic aptitude—the *nature* repertoire. That Jack's son in his youth was similarly challenged as the father in the domains of math and reading was referenced and mention likewise made to dyslexia, which Jack told us he probably suffered before while in grade school. Thus, this nature repertoire gives the impression that abilities (and deficiencies) are inheritable. Yet by quickly qualifying this admission with "but they're doing great you know?" (and also in line 25 later) served to lessen the force of this negative claim. In fact, it can be heard as part of the workplace repertoire in action for the researcher knew that his son was a highly skilled technician and his daughter a feted hair stylist.

Jack's daughter was exceptional in this account for she was both a voracious reader and potential writer unlike Jack who found reading and writing very painful, as he frequently made known to us. The logic supplied for this anomaly was that "it's all up there"; genetic potentials can be fulfilled if given the opportunities and inclination. Can these two repertoires co-exist so close to each other? Are they used consistently? The answer is that talk is oftentimes contradictory for depending on the interaction contexts and goals that speakers are orienting to moment-by-moment, different statements and claims are not to be seen as contradictory. In the final episode below, we can witness how a form of rapprochement is achieved between these two repertoires in Jack's interview talk.

Episode 10: 19 November 2003
24 Researcher 1: For a person who is not formally trained in science, you're actually behaving like a person who is.
25 Jack: Well, that's, that's cool. Like if people see that in me. And you know you go back, you can, I can take you back to my grade school, what my principal used to say about me, you know? But he's never saw what the result was even though I was a failure in school up to a certain point and then I smartened up and then I started to do pretty good yet I wasn't academic. And both of my kids aren't academic but they're both doing great you know? I don't have a problem with that. And, to think

like that, I don't know, I think it's just the mind you know? And like I tell you, my asset is my power of observation, if I didn't have that, then there wouldn't be a whole lot there, you know, with the amount of education that I have.

26 Researcher 1: Well, observing is one thing but making connections is another too.
27 Jack: Yeah, that's, you know I don't know that just happens for me.
28 Researcher 1: Ahuh
29 Jack: It's like when I tell you I build that cabin I don't have a drawing, I just, I see it all back in here, it's just there, I just boom boom boom it's done! With the science thing it's not quite, it's cut and dry but there's little windows that you look into? And then you got to get that, you got to get in there, and that's when you start to get things going you know?
30 Researcher 1: Meaning that, what do you mean by get into the windows?
31 Jack: Well, you see you get an opening, so you want to find out what's behind that, or wall if you want to look at it, you know there's a little window and there's a question. So you got to get in there and you got to do something to answer those questions.
32 Researcher 1: Okay, and you've said it a few times today, and even in the past that your predisposition, or your asset which you have was observing. I'm just thinking that it also has to do with your experiences and background.
33 Jack: Yeah, and I think it's just there too, like it's there? You know like I come out of the womb it's just probably there? But it was never really, it was never really brought out? Like my parents eh they had but grade three education. So I'm not getting, I'm not blaming my parents, that's got nothing to do with it but if you took a person that had, I don't know if this would be a good thing or not but let's say parents they were academic, and you grew up in that and they're feeding yer, on a different level you know how that makes you? Like to me I didn't have that but still if you say I kinda think like a scientist then I say well, maybe there's a gene in there or something, you know? Who knows? Like you don't train, I never went to train, I never read a book, and nobody ever told me this, it's just there.
34 Researcher 1: Oh, wow.
35 Jack: And it's like when I do blacksmithing, it's there. I came out of the womb, it was already there.
36 Researcher 1: Did somebody teach you?
37 Jack: No, no.
38 Researcher 1: You just, then how did you?
39 Jack: When I get near it, it's like I've done it before.
40 Researcher 1: Hmm?
41 Jack: When there's that fire going and a piece of steel in there red-hot and an anvil and hammer, I pick it up and I go!
42 Researcher 1: Hmm, how do you explain that?
43 Jack: I don't know! See, when that happens to me then I'm good at that.
44 Researcher 1: But with fish it's a different.
45 Jack: And that's a different thing. And if it does more with the hands, it's different, like I don't think there is a lot of brainwork in the iron work, there is a little bit you know? It's mostly taking it and doing yer thing....But the thing is that I don't let that stop me, and the same with my kids, uh if you wana do something, don't let that stop you! I feel everybody is capable of doing it, and I might be capable of doing more on the computer but I've got to keep doing it and doing it and doing it. And it's the same I think with a person that's good on that but can't do nothing

with his hands but if he kept doing it and doing it, he'd get fairly good at it even though he might not like it.

As I repeatedly show in this chapter, the analyst who forgets the interactional nature of the interview event falls into the trap of accepting participants' talk as nuggets of information to be mined so as to reveal something about the world. Jack in line 25 of Episode 10 only responded in this manner because of the previous turn whereby researcher 1 complimented him on his likeness in behavior to practicing scientists. This was apparently all the more commendable for a person "not formally trained in science," which then became a discursive resource for Jack to coconstruct his account of learning and identity. The exchange unfolded in this manner possibly because it was jointly relevant and of concern to both participants. Developing his nature repertoire in this turn ("I wasn't academic . . . both of my kids aren't academic"), Jack repeated that his premier natural asset was his power of observation (c.f. Episode 3), which mitigated his lack of formal learning. As Jack asserted in line 33, his scientific behaviors were something congenital ("out of the womb . . . it's just there," "maybe there's a gene in there") for he never experienced any instruction in this area ("Like you don't train, I never went to train, I never read a book").

Other exemplars of tacit expertise within Jack's nature repertoire in Episode 10 included: making connections, blacksmithing, cabin construction and myriad everyday tools and objects that he crafted all by hand. As many accomplished musicians have said, playing music is not so much the brain telling the hands what to do but more of the hands assuming control over the body (see line 45). A moment's reflection would make it clear that there certainly are parallels here with the tension between the workplace and school repertoires explained earlier.

However, claiming that either a person has it or does not have it is an extreme rhetorical stance that needs defending in order to be heard as rational and trustworthy. Thus in line 45 we notice Jack accommodating another discursive device to balance these rival nature and nurture repertoires: With time and persistence as obligatory ingredients ("doing it and doing it"), anything can be mastered although one might not develop much genuine affection for it. Supporting this device with the membership category of his children again, he said that everyone was indeed capable of learning anything including himself ("I don't let that stop me") if he wanted to learn more about computing for instance.

CONCLUSION

To summarize the findings, we have observed how the properties of discursive repertoires creatively organized talk to account for what kinds of learning are valued in Shallow River Hatchery as well as the source of Jack's incredible expertise. These repertoires were used in conjunction with numerous membership categories that performed decisive identity work by rhetorically defining persons, creating alliances, and drawing boundaries. Suffice to say that through all these examples of recorded interview talk, Jack came across as a committed worker who cherished practical experience gained on the job over bookish forms of learning that was often highly theoretical and failed to get the work completed successfully. It was not that he was utterly against the latter for he occasionally suggested that the path towards true proficiency lay in combining hands-on and conceptual forms of learning in the science of fish culture. In the same way, although Jack attributed the source of his knowledgeability (e.g., power of observation, thinking like a scientist, craft-making skills) to internal factors, competency in any domain was possibly within the grasp of everyone whatever their given abilities.

It seems rather odd that in setting out to better understand learning and identity in science, I have sidetracked the reader into digesting long stretches of interview transcripts. Science educators might still be scratching their heads and asking, "Where is identity?", "Was *that* identity?", "How does this impact my teaching" or "What did I learn about the learning of science?" I am reminded here of the anecdote of a man holding up a sock separately in each hand and asking his wife to search for his missing pair of socks. With talk being so mundane, educators often do not fully grasp the significance that the work of identity and learning was there all along *in* talk, just like the man who failed to recognize that two socks already comprised a pair. Because this chapter defines issues of identity and learning in science as situated accomplishments enacted in talk, it overturns some fundamental notions held by many science educators today. One incorrect premise is that the object of interest (e.g., learning and identity) lies *behind* talk rather than being analyzed as a real-time process of talk-in-interaction, which I believe is the foremost contribution of this chapter.

A tradeoff of this discursive framework is that we have to abandon any notion of fixed truths or stability concerning the social phenomena under scrutiny. That is, these constructs are like shifting sand dunes and thus one cannot say that Jack is definitely like such and such a learner or with that identity or not. Nothing about Jack's description of his learning and identity is permanent for all of his recorded talk was oriented towards the prevailing contexts and demands of the speech event. Change any one of those and we encounter different aspects of Jack. All I have established is how learning and identity were made accountable through language, how these were "done" discursively by participants. More optimistically, it highlights the intrinsic value of the ethnomethodological stance for it proves that the discursive use of these problematic concepts is not the exclusive privilege of analysts. Furthermore, any accusation that this kind of analysis suffers from researcher biases loses its force because the examination of actual transcripts brings us empirically closer than any other method into the actual turn-by-turn moves of interaction. The reader is thus allowed an opportunity to verify for him- or herself the stated assertions and claims and the quality of the investigation now stands or falls with the degree to which the reader is persuaded by the researcher's analysis—this is the lively spirit of conversation analytic research.

I imagine once we conduct research in these distinctive ways to examine learning and identity in talk, it correctly repositions our participants as the true practitioners that they are. This kind of method is emphatically not a panacea though it adds something fresh in terms of the analysis and conduct of research that I feel science educators ought to be aware of. At no time does this chapter pretend to be exhaustive in its coverage of these important social phenomena and it should be regarded as the beginning of a long (heated) conversation. Further work along this framework that is of great theoretical interest could delve into how culture (e.g., societal expectations/definitions of an educated person or good student) "speaks" the individual into existence just as agents can resist or transform these social practices and institutions. Glimpses of how individuals match, evaluate or characterize their worth via these cultural templates were demonstrated in the skilful use of various repertoires, categories and discursive devices in this chapter. Jack's accounts of his learning and multiple achievements without doubt constitute a beautiful life in science, which I count myself very honored to have participated in for some time. This was a very successful story in science, and one that I will definitely add to my collection.

NOTES

1 This was part of a large federally funded project on knowledge exchanges among coastal communities and scientific institutions in which Wolff-Michael Roth was the principal investigator. The period of my research in SEP lasted from 2002 to 2005.
2 The following transcription conventions have been used: Numbers within the interview excerpts (e.g., 06, 07) represent turns during the conversation and (?) indicate inaudible words. Ellipsis is shown by (...) while words within square brackets (e.g., [camera]) are missing words or objects marked out by the speaker during the interview.

REFERENCES

Antaki, C., & Widdicombe, S. (1998). *Identities in talk*. London: Sage.
Garfinkel, H. (1967). *Studies in ethnomethodology*. Englewood Cliffs, NJ: Prentice Hall.
Harper, D. (1987). *Working knowledge: Skill and community in a small shop*. Chicago: University of Chicago Press.
Lave, J. (1988). *Cognition in practice*. Cambridge: Cambridge University Press.
Lee, Y-J., & Roth, W.-M. (2005). The (unlikely) trajectory of learning in a salmon hatchery. *Journal of Workplace Learning, 17*, 243–254.
Lottero-Perdue, P. S., & Brickhouse, N. W. (2002). Learning on the job: The acquisition of scientific competence. *Science Education, 86*, 756–782.
Mintz, S. W. (1974). *Worker in the cane: A Puerto Rican life history*. New York: W. W. Norton.
Roth, W.-Jack., Hwang, S.-W., Lee, Y. J., & Goulart, M. I. M. (2005). *Participation, learning, and identity: Dialectical perspectives*. Berlin: Lehmanns Media.
Sacks, H. (1992). *Lectures on conversation, Vols. I & II*. Oxford: Blackwell.

BRYAN A. BROWN & GREGORY J. KELLY

13. WHEN CLARITY AND STYLE MEET SUBSTANCE

Language, Identity, and the Appropriation of Science Discourse

In this chapter we identify how when talking about the trajectories of curveballs, minority students demonstrate rich understandings of physics through the discourse of baseball. We describe how they are able to enter into detailed explanations about complex physical phenomena, while maintaining their identity as linguistic and racial minorities. We argue that out of school literacy practices offer identity constructions that differ from those available in schools. We contend that learning the physics of baseball can provide educators with opportunities to consider how to engage students with specialized discourse practices, such as those of baseball or more traditional physics talk. We show through this example how appropriation of specialized discourse practices and the construction of identity reciprocally develop and support each other.

DISCOURSE AND IDENTITY

As people affiliate over time they develop common ways of being. An important aspect of group identity is participation in discourse and activities that mark a group as unique. Acculturation into a discourse community entails a transformation of identity through appropriation of social practices. This identity modification suggests that learning to engage in the discourse of science requires developing new repertoires for interaction with people, texts, technologies, knowledge, and assumptions about the world. The consideration of identity then becomes crucial for understanding the socialization processes of education and how affiliation or alienation might occur. This view suggests that identity is situational, contextualized, and becomes evident through discourse and interaction—members of groups make decisions about how to position themselves with discourse that draws from a repertoire of ways of interacting. Thus, the interaction of the uses of scientific discourse and group affiliation is an important research area for science education.

The stylized use of language serves to mark speakers as members or potential members in discourse communities employing such discourses. As described by Brown (2005), the linguistic choices of the scientific register can be interpreted to provide members clarity regarding technical knowledge. Achieving communicative success presupposes sharing a common understanding. Through shared experiences in common activities, speakers and listeners come to mutually define what counts as a successful explanation. New meanings are constructed from common assumptions and understandings. Achieving clarity in communicating is a joint construction of speaker and audience. The speaker selects a style of discourse by assuming it represent the idea. This choice from a repertoire of discursive styles not only contains the content of the utterance, but also signals the social and expressive functions of language. Thus, the listener interprets the interchange for meaning, while concurrently using the cues embedded in talk as symbols of group affiliation and social status. When both the speaker and audience come to agree on a shared conceptual understanding, clarity is achieved. Success at such deemed clarity may serve to build affiliation and modify the identity of the participants.

Although clarity may be sought and interactionally accomplished at the propositional level, the means to achieve such understanding will be accomplished through choices in linguistic style. A speaker can choose from several ways to communicate the same idea. Baugh's (2001) work on variation explored this phenomenon by noting how the selections of some genres are more privileged than others. Although the message communicated between individuals is based on a commonly understood phrase, linguists have identified how the choice of those words can privilege individuals in certain contexts (e.g., Eckert & Rickford, 2001). For example, in choosing to greet someone, a speaker can select from phrases like "Hello," "Hi there," "What' up," "Howdy," or "How do you do." Although each of these phrases accomplishes the act of greeting, they can be received differently, and some are more privileged than others. There is great variation in how a speaker may seek intersubjective understanding by making stylistic choices among discourses. Nevertheless, such a choice entails bids to position themselves in or outside of the cultural norms of a group of speakers.

The Style of Discourse in Learning Science

The notions of clarity and style are particularly important for considering ways that discourse features of science limit access to relevant knowledge and practices of groups. Beyond the choices of propositional (cognitive) information, science is communicated through choices in the style of communication. Science educators have identified ways that "talking science" provides access to scientific knowledge. In the broader education community, the conflicts minority students experience while using academic talk have been explored, yet research on the role of discourse conflict for minority students have been limited in science education with a few exceptions (e.g., Rosebery, Warren, & Conant, 1992).

One such example of ways that scientific discourse may be at odds with minority students' identity concerns notions of objectivity in science discourse. Science has developed ways to convert complex processes, which are generally described through clauses, into simple nouns that represent complex ideas in a concise manner. Through this process of nominalization, scientists' discourse converts processes like photosynthesis into tangible entities (Halliday & Martin, 1994). Rhetorical processes like these provide an efficient means from which to convert intricate biological processes (e.g., standing water, evaporating water vapor, and falling rain) into a singular abstract phenomenon (e.g., the water cycle). In general, the collection of stylistic discursive practices in science, including the removal of the narrative voice, use of taxonomical terminology, and specific practices like nominalization, have created lexically dense, impersonal text as the norm for communication. Whereas this style of scientific discourse can provide an efficient means to promote clarity amongst group members, it creates barriers to people with different discourse practices.

Given this perspective, one must consider how using the style of science discourse affects those who are new to the community of science. Unfortunately the schooling process implicitly assumes the processes of learning these norms to be uniform across the student population. The failure to consider the relationship of students' identity and their learning poses a problem for science education. If science is interpreted as apolitical and objective then those who experience identity conflict when employing such practices will find their ways of speaking and being at odds with science. The cues associated with a style of discourse are inherently symbolic of intelligence, sophistication, political positioning, and identity. Therefore, the ways that science is talked into schooling processes impacts students.

Acquiring the discourse of science may be especially difficult because students must learn to use specialized language practices to describe phenomena they do not understand. In courses like Spanish or French, students learn a new language and vocabulary, but they employ new language practices for familiar concepts. In science, students are expected to learn new language and vocabulary about ideas with which they may have had little or no experience. To further complicate the issue, research on minority students in science education suggests that using science discourse creates an identity conflict for minority students (Varelas, Becker, Luster, & Wenzel, 2002). In light of this position, science educators must begin to explore how issues of language, identity, and classroom learning are connected to the relationship between students' identity as expressed through language.

This chapter engages in two primary tasks. First, we offer a theoretical explanation of the relationship between language, identity, and science learning. Second, we analyze students' learning in the sport of baseball as a means to demonstrate the continuity between how concepts, language, and symbol systems in sports offer a model for science learning. We offer this analysis of how students demonstrate an ability to manage complex conceptual, linguistic, and symbolic learning in an out-of-school context as a means to explore how science teachers can learn from these practices.

Language, Identity, and Science Learning

Identity provides a lens through which individuals reason about the world and their role in it. Individuals use information about themselves and people they encounter to make decisions about what type of reasoning and behavior are appropriate in given social circumstances. Although individuals have agency in determining "who" they understand themselves to be, external forces also construct their identities. The interaction between the individuals' perceptions of themselves and the external worlds that define them provides the framework through which identities are constructed. The intersection of these influences makes identity construction a sociocultural interaction. Scholars have begun examining how these sociocultural identity interactions influence learning, including those created through language. As a result, contemporary research on identities and learning suggests that individual identities are influenced by sociohistorical patterns at the level of curriculum, pedagogy, and language. The challenge for analysts is to identify how identities are shaped and changed, by whom, and under what conditions. Our analysis proposes examining the sociocultural impact of identity by examining identity across three theoretical domains, source, scale, and trajectory.

Dimensions of Identity

We draw on three dimensions of identity: sources, timescale, and trajectory to explore how language is an artifact of identity. We use the perspectives to explore how taking on an identity through language provides a framework for accessing the concepts, language, and symbol systems of a community of practice.

Gee (2001) offered a comprehensive assessment of the notion of identity by providing four primary sources for determining identity. First, Gee defined the role of nature in identity by describing the *Nature-Identity* (N-Identity). In Gee's description, a Nature identity is determined by the characteristics that define someone out of the person's control. Thus, to an extent, physical characteristics of a person afford a certain identity by nature (e.g., an African-American in contemporary U.S. culture). In Gee's *Institution-Identity* (I-Identity) the source of understanding identity is provided by an institution. If one is to become a high school baseball player, the institutional gatekeepers (i.e., a coach) determine who is able to participate on the team.

In Gee's *Discourse-Identity* (D-Identity) the source for determining identity is the discourse used to define people. Thus, for example, one's identity as a "clever" person is ascribed by the use of the term *clever* as a means to define that individual. Accordingly, the source of a D-Identity is a socially constructed descriptor. In the Affinity identity, or A-Identity, the source of one's identity is shared cultural practices. For example, the source of a "surfer's" identity involves the shared practices and cultural affiliation of the sport of surfing. An alternative source for ascribing and taking on an identity involves using discursive symbols and cues to interpret and communicate "who" an individual is understood to be. The term *Discursive Identity* is another source of identity that examines how people use the symbols of an individual's style of discourse to interpret who they are. This source of identity is highly symbolic and involves interpretations based on social interaction and history. For example, a telemarketer may call someone named Pat Smith. Upon conversing with Pat, they may use the tone and style of Pat's discourse to infer that Pat is a man. If calling an individual named Patrick O'Connell, the telemarketer may interpret the tone and style of the discourse to indicate that Patrick is an African-American male. As an analytical lens, Discursive Identity suggests the source of identity construction and interpretation is the subtle cues embedded in daily talk.

In conjunction with the source of one's identity, we contend that each source is impacted by a timescale that provides meaning to what it means to be a particular person at a particular time. In support of the notion of source, one's identity is also defined in terms of the scale from which identity assessments are made. Lemke (2000) suggested that the formation of identity and subsequent changes in them do not take place on short timescales. According to his perspective, identity is a long-term process, which must be explored across multiple timescales. Nasir and Saxe (2003) offer a three-strand approach for including the subtext of timescale that included (a) viewing identity in local interactions, (b) examining those same identities over developmental time, and (c) remaining aware of that identity and its meaning in a historical context. If one applies these frames, one can conceive of identity as something that has different symbolic and cultural meanings over time. Therefore, researchers must examine identity development across all three of these contexts. For example, if the source of one's affinity identity is that of an African-American professional baseball player, consideration of timescale becomes essential for analysis. What it means to be a professional baseball player in the Negro League (when African-Americans were excluded from Major League Baseball) may be significantly different than what it means to be on a professional team in 2006. Therefore, the notion of identity bears the influence of both the source of one's identity and the scale that gives it a particular meaning.

The third domain, trajectory, incorporates the notion of individual agency into assessing one's identity. As individuals engage in multiple communities of practice, they select pathways to achieve the identities they pursue. A student with aspirations of medical school may take on the cultural practices of a ninth-grade science classroom in pursuit of a loftier goal. The term "trajectory" captures how the direction of one's identity development plays a role in how individuals reason about practices associated with particular identities. Wenger (1998) provided a framework that can be applied to understanding the role of trajectory by explaining how learners exist as a part of larger cultural systems. As an artifact of cultural membership, individuals learn sets of shared practice. However, existing in these cultural worlds is often problematic and individuals may choose to take on or avoid particular practices to signal membership in specific cultures. In this way, the decisions one makes about one's own identity shapes the direction of how one will take or avoid particular identities.

The intersection between the source, scale, and trajectory of one's identity provides a framework from which to understand how identity is constructed and how it affects student learning. This perspective provides an insightful lens from which to examine how students appropriate science discourse. The notion of trajectory suggests that those who fail to learn the discourse practices of a science classroom may fail to do so for different reasons. Some students may simply be unable to grasp the complex discourse practices of science due to a variety of reasons, including their previous experiences. Other students may simply choose not to appropriate the discourse practices of science due to an identity trajectory that purposefully rejects the use of science discourse. From this perspective the question of how students are using science discourse may be a matter of what it means to use a particular discourse.

Discursive Identity: Vernacular vs. Non Vernacular Language

When the issues of trajectory are incorporated to an identity analysis the political subtexts of language use and identity affiliation become a foundational issue. Given that all people develop everyday or vernacular ways of communicating, learning specialized discourses may be symbolic of identity frameworks that are potentially oppositional to an individual's identity trajectory. Thus, scholars of the relationship between language identity and classroom learning must ask how do the discourse practices of classrooms frame opportunities for teaching and learning. More specifically, applying this lens to an analysis of science education has the potential to highlight how the dynamic role of language use and its association with identity play a significant role in science learning.

The classic work of Labov (1972) establishes a framework to understand the complex language of everyday practice. As people interact and form communities, they develop common or *vernacular* ways of communicating. Over time, these patterns of communicating becoming increasingly complex as individuals use specialized talk, writing, and symbol systems to efficiently achieve clarity. As individuals travel from community to community, they learn to use a variety of communicative patterns ranging from general (e.g., talking to strangers) to more specific ways of using language for special purposes (e.g., talking with teammates or club members). Labov defines this general approach to communicating as vernacular or "everyday," ways of communicating. The natural opposite to this vernacular mode of communicating is Labov's more specific non-vernacular language. This type of communication uses talk, writing, and symbol systems to efficiently communicate to a smaller community of individuals with shared expertise. In order to transition from either vernacular or non-vernacular modes of communication, the speaker must assume the audience shares an understanding of the specific communicative practices being used.

Vernacular and non-vernacular modes of communication are essential components of students' lives. For example, a baseball coach may use a detailed system of symbols to explain a play to his players. He might use hand signals to touch his hat, hip, and nose to select the appropriate pitch. The catcher may then interpret the symbols to make meaning of them. If the catcher understands that the last sign the coach provided is the one he must interpret, he can make the quick assessment that the coach has requested that he inform the pitcher to throw a curveball. He may then carefully place his hands so his competition cannot see his signs and signal two fingers to the pitcher. The pitcher may then identify this symbol as the sign for the curveball and nod his approval. Concurrently, the shortstop my read these same signs and yell out to his fellow defenders "Blue 13." Such an action tells the defenders what pitch is coming next and lets them know where they should position themselves. All of this complex signaling and communication happens in seconds and requires nine baseball players and one coach to share a rich set of non-vernacular discourse practices. The coach must assume that his ball players share the same non-vernacular understanding that will enable them to correctly interpret the complex discourse involved in this type of communication. If our coach was to attempt to provide a similar explanation to another audience, he must assume what levels of shared understanding are common between the speaker and the listener. In sum, the lines between vernacular and non-vernacular are always contingent upon individuals' common understanding of the language and culture guiding community participation.

In line with Labov's work, Gee (1999) offered an explanation of the relationship between language dimensions of discourses and the ways that specialized discourses are tied to culture. Gee refers to the specialized discourses as "D" discourse. The similarity of their work is apparent: All people share genres of communication that are specialized for a small group of people, but they also use a generic version of discourse effective for communicating with a larger community. In light of these perspectives, we make the assertion that what counts as a non-vernacular, or specialized discourse, is highly group dependent. Discourses are non-vernacular discourses only for cultural outsiders.

If students are able to master complex (non-vernacular) discourses in certain contexts, why do they experience difficulty in mastering the discourse of science? The Discursive identity dimension becomes important in answering this question because the individual's role as a cultural insider or outsider is imperative in their acquiring a non-vernacular discourse. This issue is particularly significant in considering some contexts, such as sport, that allow students to master complex discourses that are rich in scientific thought.

How does a non-vernacular genre of language become vernacular for new users? In the remainder of this chapter, we propose that gaining insight into answering this question stands at the heart of improving science learning for minority students. If language is an essential component of science learning, how are students learning to master science concepts and complex discourses in alternative contexts? How can the science learning (including the learning of complex discourses) accomplished in non-school contexts be useful for improving classroom science learning? These theoretical questions serve as the foundation for the analysis that follows.

The Science of Baseball—The Non-Vernacular/Vernacular Transition

The transition from vernacular understandings of phenomena to use of specific, non-vernacular language practices is common in students' experiences. In fact, students are constantly involved in practices that are analogous to learning science. Like science, these practices include conceptual, linguistic, and symbolic components as a part of their discourse. For example, the sport of baseball includes conceptual, linguistic, and symbolic features that require participants to take on a discursive identity that is consistent with those practices. Examining these practices provides a valuable framework for understanding how students make the transition from vernacular understanding to more complicated non-vernacular understandings.

The language of sports often includes conceptual components. In order to understand how to perform, athletes often develop rich conceptual understandings. For example, baseball players may come to learn the science behind why a curveball curves in an effort to improve their performance. This conceptual understanding may include rich understandings of velocity, drag, Magnus forces, and air pressure. Although the language practices of the science they are learning may not include the basic terms scientists use to describe these phenomena, these concepts often carry analogous terms that are framed in the genre of baseball discourse.

Along these same lines, the language practices of sports provide an example of how discourses provide users with a set of specific terms to describe phenomena in an efficient manner. Students participating in baseball learn to apply specific terminology as a means to communicate complex ideas in an efficient manner. Although the linguistic components of baseball are largely spoken, they also include a set of symbolic practices that are both written and communicated through hand gestures, code words, and a numerical system for defining each position. This Discourse is largely incomprehensible to cultural outsiders—and when coded, incomprehensible to opponents.

The complex signal system of baseball has long been considered one of the most distinctive features of the sport. Coaches, umpires, and players communicate a diversity of messages through the use of a dynamic set of symbol systems. The umpire communicates the number of outs as well as the numbers of balls and strikes through hand gestures that support their verbal explanation. These gestures and explanations include numeric codes for each of defensive and offensive positions on the baseball field. Ultimately, the sport of baseball provides fertile ground to explore the potential continuity between how students learn to appropriate conceptual, linguistic, and symbolic knowledge that maintains continuity to those types of knowledge appropriate for science learning.

CASE ANALYSIS

This case focuses on Stokley Carmichael High School, one of six high schools in a school district in California, USA. Carver averages slightly over 1,500 students per year. It is located in an upper middle class neighborhood in a large city with a population of over 400,000 residents. Most residents in the school's local vicinity have children well beyond school age. As a result, the majority of students at Carmichael are bused to school. Academically, the school has improved its score on statewide assessments in each of the past three years.

The baseball team (varsity and junior varsity) that participated in this study is as ethnically diverse as the school itself. Of the 27 members of the baseball team, eleven were African-American, seven were Hispanic American, seven were Caucasian-American, and two were Asian American. English was a second language for all of the seven Hispanic-American players. Of the 15 varsity baseball players interviewed five were African-American, five were Hispanic American, three were Caucasian-American, and two were Asian American.

The excerpts that follow provide an exploration of how students participating in baseball develop discursive identities while learning to play the sport of baseball. Their mastery of the discourse of baseball demonstrates how the discourse of baseball, much like that of science, includes complex conceptual, symbolic, and linguistic features. As we examine the complex non-vernacular language practices of baseball, we engage in a comparative analysis of how students demonstrate an ability to master the concepts, language and symbol systems of baseball. We claim that the mastery of these practices provides evidence of how these discourse practices in students' everyday lives are similar to those of physics.

Twelve high school baseball players were interviewed about their understanding of the physics of a curveball. A curveball occurs when a pitcher throws the baseball with a particular spin to speed ratio that causes the ball to drop downwards and away from right hand batters. Learning to hit a curveball is a key step in maintaining active participation in competitive baseball. Therefore, all the players had an inherent interest in understanding how to hit a curveball. Furthermore, the type of pitch thrown requires fielders and base runners to adjust their roles and positions almost instantaneously. Pitchers need to learn to throw a variety of pitches, often with different types of spin to confuse batters. Therefore, across multiple roles (hitter, pitchers, and fielders), players need to learn about what a curveball does in its flight.

The players were first asked to explain how a curveball moves in its flight to the batter. Later in the interview, they were given a diagram (Figure 13.1), which identified the general path of the ball, the direction of spin for a curveball, air pressure differential, and the direction of spin. The interviews were coded according to discourse features by identifying the discursive style—science or baseball. Both the science and baseball styles of discourse, or registers, had unique terminology and ways of accounting for curveball phenomena.

Talking about Curveball Science: Concepts in Vernacular Form

In this section we consider the ways that the students talked about the science of a curveball. Through the coding process, we identified a range of explanations using physical phenomena to account for curveball trajectories. Two predominant accounts concerned the spin of the baseball and the relationship between the seams of the baseball and the air surrounding the ball.

Explanation One: The direction and rate of spin The most common reasoning for why a curveball curves was the role of the spin of the ball. Damon suggested, "it's gonna spin around, so when you get to that point, like half way—it's gonna start dropping." Adam explained, "If it had enough spin, like a front spin on it, it would drop." Damon and Adam identified the importance of the ball's spin—from the perspective of an observer in the plane and orthogonal to the direction of the pitch. Each thought the ball spun with topspin moving in the same direction as the ball. They used the baseball term "drop" to connote the change in vertical height (from the already otherwise curved path from the pitcher's hand to the location of the batter). Damon noted further that the break occurs as the ball reaches the second half of the ball's path. A similar, but more extended explanation, is offered by Nick, a senior outfielder:

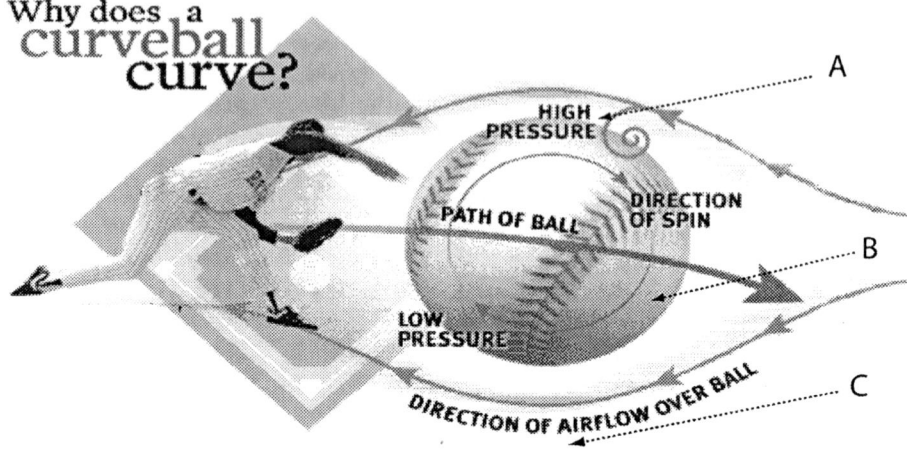

Figure 13.1. Why does a curveball curve?

> Based on whatever I learned and experience seeing a curve ball and you're just throwing a curve ball—when a pitcher throws a certain way, at a certain angle and with like a certain amount of force so that it – when he throws the ball, the ball spins a certain way and breaks air differently than a fast ball.

Nick's explanation provides an example of how the students' experiences playing baseball led them to attribute the spin as the cause of the ball's deviation from its (otherwise non-curving) trajectory to the plate.

A second interview question asked students to describe the difference between an effective and an ineffective curveball. Sam attributed the curve to the ball's "front spin." This provided him with an argument about the differences between effective and ineffective curveballs:

> An effective curve ball probably has more spin on it so that the wind will push it more. And an ineffective curve ball would have less spin so it wouldn't curve as much and wouldn't move as much. An effective curve ball would curve more and so it'd move more, which would make it harder to hit.

Players also explained their understanding of the role of spin in effecting movement of the ball. Although players used everyday talk, their descriptions revealed an understanding of how the spin affects the physics of throwing a curveball. This understanding is connected to how the players understood the role of air pressure in causing a curveball to curve. Manuel suggested that "the ball is gonna, first go up while it's still spinning, therefore the low pressure would go under it, the high pressure would go on top." Manuel used his knowledge of the differential pressures to explain the downward movement of the curveball. Jimmy reversed the causality, stating, "A curve ball is actually dropping. So what it shows is the pressure on how the ball rotates." In these two examples, the students are beginning to include physics explanations into their everyday baseball discourse.

Explanation Two: The relationship between the seams and the air Another conceptual understanding involved the players' conception of the role of the seams of the baseball on its movement. The players' explanations of the relationship between the seams and the ball's movement were of three kinds. While some players merely mentioned a role for the seams in creating the movement of a curveball, others considered the relationship of the seams to the passing air. Others emphasized how manipulating the ball helped produce the appropriate spin to speed relationship. For example, Javier thought the seams interact with the air resistance by stating, "the air takes the seams and makes it curve." Another way the seams are implicated in the movement of a curveball concerns the way that the pitcher manipulates the seams. Daniel explained, "so the air gets more turn on the ball. And then on a curve ball you throw it with two seams so it gets less air on the strings – on the laces. And so it gets more movement." Daniel's understanding reflects an understanding of how the air affects the seams. Adam, a sophomore pitcher and outfielder, provides another example of this position by stating: "If it didn't have any seams, then everything would just be a fastball 'cause it couldn't spin as much. The seams allow you to like grip it, and spin it with more force."

Adam's comments provide an example of how the players developed rich understandings of how the seams effect the ball's movement and how the seams enable the pitcher to spin the ball. Jimmy, a senior pitcher and catcher explained how the seams and their relationship to the air cause the ball to curve. Jimmy explains,

> The ball is not completely flat. You know, it's not a smooth surface. So I mean, the lace, the seams on the ball, are gonna cut through the air and the air around the ball is obviously that's gonna slow the pitch down.

He continues his explanation by offering the following description of the role of the seams:

> That's what the seam is for—the seams catch the wind, and so when the seams catch the rim, I mean the wind, then it affects the rotation of the ball.

Taken together, the set of explanations make clear that the players have a fundamental understanding of the curveball's movement based on the ways that the seams interact with the air.

Talking about the science of a curveball Although players held a rich understanding of science as expressed in their everyday language of baseball (demonstrating initial understandings of relevant phenomena), they also began to integrate science concepts as expressed in detailed non-vernacular science discourse. These explanations can be seen as building on the everyday terms and beginning to enter into the register of physics. The players were able to appropriate science discourse in their explanations that included explanations of the relationship between the spin and the seams, while providing an explanation of the role of pressure on the ball.

Relationship between spin and seams Similar to the players' answers in everyday terms, scientific terms were employed to emphasize the role of the spin of the ball on its trajectory. The players used the idea of pressure in a variety of ways to explore how the spin of the ball affected its relationship to the wind. Sam provides a rich example as he explained, "the top pressure is pushing it down in kind of an arc." He continued by explaining, "the top one's high pressure and the bottom one's low pressure and it's pushing it [the baseball] down so that it looks like it's curving—so it is curving." This notion of pressure was applied richly throughout the players' interviews. James attempted to explain what baseball players refer to as a "hanging curve ball" (lacking proper movement in plane orthogonal to the direction of the pitch) by stating, "If you don't have the top [air] pressure, the ball's not gonna go anywhere, it's gonna stay flat." He continued by saying, "it's gonna float because the air pressure going away from the ball is not gonna catch the seams." As a group, the baseball players demonstrated an ability to integrate conceptual understandings as expressed through the language of baseball with the science discourse. For example, when asked to provide an explanation after viewing Figure 13.1, Sam offered the following:

> A is the high pressure. And that's I think probably what gives it the spin or makes it curve. B is the rotation of the ball. C is the low pressure. And so when you throw the curve, the spin of the ball directs where the—what the air pressure is gonna do to the ball, which pushes it down in a kind of a curve shape. And that's why it's called the curve. And the low pressure can't—it's just holding the ball in the air, but it can't like—it can't outdo the high air pressure, so the high air pressure makes the ball curve.

This example demonstrated the players' understanding and appropriation of non-vernacular science discourse. More specifically, this image is reflective of the players' ability to incorporate the notion of air pressure to their understanding of why a curveball curves.

Baseball and Scientific Language

In their responses players also offered scientific explanations for why a curve ball curves. Some even noted a distinction between their comments and an official scientific explanation. The lines between what was considered baseball talk and what was considered science talk were often blurred. John began to explain his understanding of a curveball by stating, "instead of the forward spin, like this, of a fast ball, [a curveball] is spinning more like that *(spinning a ball at an angle)* So, I think—I don't know about physics or anything—I think that motion naturally brings it this way if you're– a Righty throwing [the ball]." His explanation is representative of how the players saw their explanations as non-scientific despite their use of science content and language.

In their response to a picture of a curveball students used science vernacular language to explain the image of a curveball (Figure 13.1). Daniel, explained how the seams of the ball were "pulling away from the wind resistance." In all, students employed a variety of science terms in their explanations including the terms "resistance," "air pressure," "wakes," "deflections," and "velocity."

The application of these terms demonstrated students' ability to appropriate scientific terminology when learning the basics of baseball. An intriguing component of this appropriation is the fact that some of the terms that are common to science vernacular talk hold different meanings in the context of baseball. For example, the term "velocity" is often used as a substitute for the term "speed." In addition, the term "movement" is often used to describe the velocity or changing trajectory of the baseball. In many ways, the terms used to represent scientific ideas in baseball are similar to, but not exactly the same as the terms as they would be used in science vernacular talk.

FROM AFFINITY IDENTITY TO SCIENCE IDENTITY

Students who are traditionally marginalized in science are quite capable of mastering complex discourse. This analysis indicates how minority students are gifted in mastering complex discourses. The irony is the linguistic, conceptual, and symbolic features of baseball share continuity with the linguistic, conceptual, and symbolic features of science. Table 13.1 provides an analysis of the continuity that exists between the science students are learning in baseball and their scientific equivalents. For example, although the terms movement and velocity are used in baseball, they do not represent the same concepts. In baseball, *movement* is the equivalent of the scientific term *velocity*. The continuity lies in the fact that the students have a fundamental understanding of the concepts, despite the use of alternative terms. Educators must question why players are able master complex discourses in out of school contexts, while they encounter difficulty appropriating the discourse of science. The contrast between students' use of science learning in baseball and their classroom science learning must become a site of further research. Despite the player's socioeconomic status and race, they readily engaged in the discourse of science as a component of their baseball identity. The differences between the technical Discourses of baseball and the technical Discourse of school science may best be understood as a matter of identity.

Identity Domains

Examining the learning that takes place in the context of baseball across the identity domains of source, scale, and trajectory provides insights into how science learning occurs. The source aspect of identity is perhaps the most salient influence in their use of science discourse, as players' affinity identity provides an overarching framework from which to reason about their behavior and actions. As players, they deem conversations about the science of a curveball a reasonable component of their interaction. Through participation with others who need to know about the physics of baseball, players come to make such understandings commonplace, resulting from shared practices and interests. Their identity as baseball players supercedes other cultural identities that may constrain how they use science discourse.

Table 13.1. Continuity of science terms and concepts

Baseball Register	Description	Science Equivalent	Analysis
Linguistic			
Movement	Players defined movement as the sudden movement or change in direction of the baseball as it travels towards the plate.	The term " is a more appropriate term in that it addresses the change in direction of a object.	Although the players use the term 'movement' where science would use the term 'velocity' players are able to make a distinction between the speed of an object and its movement
Velocity	Players used the term velocity as a way to describe the speed the ball traveled from the pitcher to the place	The term speed would be appropriately applied more in science to explain the ratio of distance per time interval.	Similar to the above, the context of baseball provides access to two ideas regarding movement: per time and change in direction of a moving object.
Baseball Register	Description	Science Equivalent	Analysis
Conceptual			
Spin to speed relationship	Players suggested the curveball's movement was due to the rate of spin and the pressure produced by the spin to speed ratio.	A scientific account of the curveball would include the forces in relation to movement (Magnus Forces/Drag) and the spin of the ball.	The players uniformly understood a basic concept of the spin of the ball altering the ball's trajectory.
Seams to air and resistance	Players reiterated that the seams met with air to produce a force that pushed the ball down.	The aerodynamics of the ball are greatly altered by the seams. If the seams are spun at an appropriate angle, the seams change air pressure.	The emphasis on the seams implicates the shape of the object by producing desired aerodynamic

If the identity perspective is applied with respect to timescale, the baseball context provides additional insight into how the identities that support learning the science discourse of baseball are significant. In the 50s, 60s, and 70s baseball was among the most popular sports for African-Americans. Since the 80s the sport has seen a dramatic decrease in popularity among urban youth. In recent years, the lack of inner-city youth in baseball has spurned programs designed to promote greater interest in the sport. This issue becomes especially important given these players' willingness to engage in the discourse of baseball despite its marginalized position among contemporary sports. Thirty years ago, taking on the identity of a baseball player would be seen as a mainstream framework for minority students. Today, taking on this type of identity requires players to take on an identity that is no longer a mainstream framework for them.

Applying the notion of trajectory to this analysis may help to understand this identity appropriation further. The players desire to play baseball creates a trajectory from which to appropriate the discourse of baseball. During their early instruction, players are taught baseball tasks (hitting, throwing curveballs) that provide a pragmatic reason to learn why a curveball curves. After their instruction, they are provided activities to build localized expertise (drills). The baseball games provide a performance assessment that enables them to enact their knowledge. As a performance assessment, the games provide an incentive for improving performance and knowledge. If the players perform well they will continue to play, if they underachieve, they may lose those privileges. The trajectory of their baseball identity provides a framework that is supportive of their learning science discourse.

Learning and Identity

Theories that emphasize a situated approach to understanding learning provide support for the way baseball players come to master the discourse of baseball. A situated approach to understanding learning would suggest that learning occurs where knowledge is situated in contexts where it is valued (Lave, 1993). Learners acquire new knowledge in the context of a broader cultural system where that knowledge has pragmatic functions. The baseball examples, demonstrate how the act of hitting, defending, and pitching provide pragmatic contexts for understanding why a curveball curves. In these contexts the identities that develop support the acquisition of such knowledge.

Cognitive apprenticeship (Collins, Brown, & Newman, 1989) also provides a valuable lens from which to view the identities players take on in playing baseball. The instructional process that coaches and players experience establishes expert and apprentice roles that promote player learning. The coaches often delegate instruction to older players whose primary task is to ensure others share a common expertise. The instruction that results can be seen through the lenses of modeling and scaffolding. Fellow players often explain by demonstrating the appropriate technique and then offer their peers opportunities to execute the activity on their own. In the end, players co-construct expertise, conceptual knowledge, while learning to take on baseball identities that share common conceptual, linguistic, and symbolic expertise.

Education and Identity

What can educators learn from exploring this context? First, research on minority students in science often fails to access the knowledge and language skills students bring with them to the classroom as resources for science instruction. Research in literacy and mathematics has called for greater examination of the knowledge resources available for learning (e.g., Silvia, Moses, Rivers, & Johnson, 1990). Exploring these issues for instructional reasons is important for science because they have the potential to provide evidence for how identity construction occurs in these contexts.

Second, in the setting of baseball, the communal nature of the discourse and instruction is a natural component of the culture. The coaches provide instruction, other expert peers provided modeling and scaffolding, all of which promote player learning in the context of baseball. The learning that takes place has multiple dimensions including instruction (coaching), application small-scale (drills), and pragmatic performance assessments (game). There is a strong collective responsibility for individual achievement. Players are supported and perform in "low-stakes" exercises, before being placed in competition, where they nonetheless must perform high stakes tests of ability. The instruction and learning that takes place also reflects a distributed expertise and a hierarchy of instruction. The head coach, pitching coach, and more expert players all play prominent roles in promoting the type of identity construction that promotes students' learning and identity development.

CONCLUSION

Science instruction in inner-city communities must learn to pay particular attention to the process of discursive identity construction as a precursor for, and continuous contributor to, classroom instruction (Moje et al., 2004). If language use is conceived as a component of identity, successful science instruction must include a pragmatic framework for constructing identities that will support student learning. We would be wise to learn from the success of athletics. The detailed discourse of sport operates much like the discourse of science by applying intricate terms that promote clarity. The result is a style of discourse that has the potential to mark its users as particular types of people. The confounding difference is the way sport supports those identity constructions and the manner science makes no effort to promote identity construction for minority students.

REFERENCES

Baugh, J. (2001). Variation. In A. Duranti (Ed.), *Key terms in language and culture* (pp. 260–263). Malden, MA: Blackwell.
Brown, B. (2005). The politics of public discourse: Discourse, identity and African-Americans in science education. *Negro Educational Review, 56,* 205–220.
Collins, A., Brown, J. S., Newman, S. E. (1989). Cognitive apprenticeship: Teaching the craft of reading, writing, and mathematics. In L. B. Resnick (Ed.), *Knowing, learning, and instruction: Essays in honor of Robert Glaser* (pp. 453–494). Hillsdale, NJ: Lawrence Erlbaum Associates.
Eckert, P., & Rickford, J. (Eds.). (2001). *Style and sociolinguistic variation.* New York: Cambridge University Press.
Gee, J. (1999) What is literacy? In C. Mitchell & K. Weiler (Eds.), *Rewriting literacy: Culture and the discourse of the other* (pp. 77–101). Westport, CT: Bergin & Garvin.
Gee, J. (2001). Identity as an analytic lens for research in education. In W. Secada (Ed.), *Review of Research in Education Vol. 25* (pp. 99–125). Washington, DC: American Educational Research Association.
Labov, W. (1972). *Sociolinguistic patterns.* Philadelphia: University of Pennsylvania Press.
Lave, J. (1993). Situating learning in communities of practice. In L. Resnick, J. Levine & T. Teasley (Eds.), *Perspectives on socially shared cognition* (pp. 63–85). Washington, DC: American Psychological Association.
Lemke, J. (2000). Across the scales of time: Artifacts, activities, and meanings in ecosocial systems. *Mind, Culture, and Activity, 7,* 273–290.

Moje, E., Ciechanowski, K., Kramer, K., Ellis, L., Carrillo, R., & Collazo, T. (2004). Working toward third space in content area literacy: An examination of everyday funds of knowledge and Discourse. *Reading Research Quarterly, 39*, 38–70.

Nasir, N. S., & Saxe, G. B. (2003). Ethnic and academic identities: A cultural practice perspective on emerging tensions and their management in the lives of minority students. *Educational Researcher, 32* (5), 14–18.

Rosebery, A., Warren, B., & Conant F. (1992). Appropriating scientific discourse: Findings from language minority classrooms. *Journal of Learning Sciences, 2*, 61–94.

Silva, C., Moses, R., Rivers, J., & Johnson, P. (1990). The algebra project: Making middle school mathematics count. *Journal of Negro Education, 59*, 375–391.

Varelas, M., Becker, J., Luster, B., & Wenzel, S. (2002). When genres meet: Inquiry into a sixth-grade urban science class. *Journal of Research in Science Teaching, 39*, 579–605.

Wenger, E. (1998). *Communities of practice: learning, meaning, and identity.* New York: Cambridge University Press.

NANCY W. BRICKHOUSE & PAMELA S. LOTTERO-PERDUE

14. CONSTRUCTING CRITICAL SCIENCE AND SOCIAL IDENTITIES IN A GIRLS' AND A BOYS' SUMMER SCIENCE BOOK CLUB

How do children engage in science critically? How might they question science in ways that lead to robust understandings of science and an empowered relationship with science and scientists—enacting what we call a critical science identity? In what ways might children use (and be used by) science texts as they go about forming a critical science identity and continuously constructing their broader social identities? We believe that "identity" is a particularly useful way of examining critical engagement in science because it is a construct that includes much more than competence. A person's identity is built over a long period of time and includes not only what a person is able to do, but also what a person is likely to do in a given context. Thus, dispositions, habits, and tendencies to act in certain ways are critical features of identities. We suspect that one could directly teach learners how to competently analyze science texts, but such instruction would be of little value beyond school if the disposition to approach science texts critically was not taken up and made a part of a person's identity.

We also know from prior research that high school students are not typically critical of science texts encountered in the media (Zimmerman, Bisanz, Bisanz, Klein, & Klein, 2001). Similarly, students rarely ask questions that might facilitate critical analysis, such as: Are the researchers qualified to do this research? Who paid for this research? Was this research ever published in a peer-reviewed journal? (Korpan, Bisanz, Bisanz, & Henderson, 1997) Nor did students frequently ask questions that might aid their consideration of an alternative explanation for the data.

Advocates of conventional approaches to learning might also argue that one should not teach children to be critical of science until they know enough to do it well. For example, students with strong understandings of the nature of disease were better able to detect erroneous news reports regarding HIV infections (Keselman, Kaufman, & Patel, 2004). While we find Keselman et al.'s report convincing regarding the relationship between substantive content knowledge and the ability to competently critique science texts, we also think that there may still be value in encouraging critical practices at a young age. If such practices are to become part of who these children are and are becoming, then it may well make sense to provide opportunities to develop such practices and dispositions early and often.

In this chapter we share our experiences with a summer science book club where we provided young children with opportunities to engage critically with science texts. Here, we take science texts to include both scientifically themed written texts (e.g., that may be found in tradebooks or through other forms of media like the Internet) and spoken texts (e.g., which, in the case of book club, may come from a peer or a discussion leader). Fundamentally, we understand critical engagement to entail the children questioning, challenging, or supporting the validity of science texts. As we observed in our book club, critical engagement also includes *positioning*.

Here, we take positioning to be a discursive production of self in the course of children's conversations about and interactions with science texts (Davies & Harré, 1990). Positioning may be of two forms: (a) interactive, in which children position one another, children position other science texts, or science texts position children (see also Freebody & Luke, 1990); or (b) reflexive, in which the children position themselves. These positionings are not necessarily known to the positioner or the positioned. The process, then, of examining positioning is one of seeking out articulation or awareness of positioning by participants, and in the absence of these things, inferring positioning using discursive clues in the autobiographies of participants' speech:

> People will . . . be taken to organize conversations so that they display two modes of organization: the 'logic' of the ostensible topic and the story lines which are embedded in fragments of the participants' autobiographies. Positions are identified in part by extracting the autobiographical aspects of a conversation in which it becomes possible to find out how each conversant conceives of themselves and of the other participants by seeing what position they take up and in what story, and how they are then positioned. (Davies & Harré, 1990, p. 5)

By examining patterns of positioning across multiple meetings we intend to show how children's critical engagement with science texts—i.e., the discrete trying on of performances of critical science identity within book club—is related to the longer-time-scale construct of social identity as evidenced by additional data regarding these children at school and at home. We understand identities as continuously under construction. We agree with Davies and Harré that humans are "characterized both by continuous personal identity and discontinuous personal diversity." Our analysis foregrounds the former over the latter. Towards this effort and in what follows, we (a) highlight key features of the book club design and participation selection process; (b) describe data collection and analytical methods; (c) share results for the girls' book club, followed by results for the boys' book club; and (d) make connections across the girls' and boys' book clubs in the discussion and conclusion.

BOOK CLUB DESIGN & PARTICIPATION

The book club was part of a larger project. In this section we discuss the relationship between the book club study and the larger project, the process of selecting participants for the book club, designed features of the book club environment, and text selection.

Relationship to Larger Project

During the 2002–2003 school year, the authors of this study and others were engaged in the first year of a three-year project investigating third-grade girls' reading preferences (see Ford, Brickhouse, Lottero-Perdue, & Kittleson, 2006), girls' interactions with science texts, and the way in which science texts can enhance inquiry-based science instruction. This project included interviews and classroom observations of four classrooms across two schools, with roughly 22 children per classroom. Although these data collection methods gave us great insight into the girls' reading preferences and interactions with science texts, we wanted to know more about how they would respond to a broader range of science texts than they had seemingly been exposed to thus far. Towards the end of this first year of the project, we sought to create a summer science book club for the girls in the study so that they could read and talk about science texts with which they were likely to be unfamiliar—especially science texts that dealt with physical science and engineering topics. Furthermore, we wanted to see the girls interact with these books in an environment with fewer constraints than is typical in school.

Selection of Book Club Participants

Our selection of book club participants was influenced by our interest in (a) being equitable to the boys and girls in the classrooms of the larger project; (b) keeping the number of book club groups we would facilitate reasonable, and the number of participants per book club optimal; and (c) creating spaces in which children would be most apt to participate. First, as we prepared to invite girls from the project to participate in our book club we became concerned about excluding boys from participating who might also be interested in and benefit from reading science texts. Thus, we decided to invite entire classrooms, including both girls and boys, to participate. Second, we decided to invite two classrooms based upon our (a) ability to facilitate at most two simultaneously-running book clubs; (b) desire to limit each book club to ten children so that all participants would be able to contribute to book discussions; and (c) a prediction that fewer than half of those invited would be interested and able to participate. Third, we decided to establish one book club for girls and one for boys. This allowed us to have a girls-only group in which girls' voices would not be overpowered by the voices of boys. Also, since most of these children formed their closest friendships with someone of the same sex, we wondered if children might be more likely to participate in friendship groups that were sex segregated. Ultimately, we invited all 44 children from the two project classrooms in one of the schools to participate in either a girls' or a boys' book club to run parallel throughout the summer. Of those invited, thirteen returned permission slips indicating interest in participating. Due to family vacations and other conflicts, the children's participation fluctuated week by week. On average, six girls and four boys attended each meeting. The girls met for eight sessions, while the boys met for seven.

Book Club Environment

Another significant design decision had to do with the environment that we sought to create in which children would want to participate and to which they would want to return. Our book club sessions included pizza, drinks, and snacks, allowed children to socialize freely before our book club discussions, and although we asked the children to read before coming, did not penalize those who accidentally read a different text or read only part of a text because they disliked it. In addition, we designed our roles to be facilitators of discussions, not teachers with standards to address or content to convey. Making book clubs fun and free of many of the constraints of school was recommended by literacy researchers focusing on out-of-school book clubs (e.g., Chandler, 1997). That being said, book club sessions were structured with respect to time; each session lasted approximately one hour, including at least fifteen minutes for snack and socialization. We also imposed some structure with regard to the texts that girls and boys read for book club.

Texts and Text Selection

Prior to an introductory meeting for all book club participants and their parents, the authors and a librarian selected approximately 35 science texts (science trade books, in this case). In addition, most, but not all of the texts we selected had won awards (e.g., receiving Horn Book and Orbis Pictus awards). Participants voted for their favorite three books from our collection during an introductory meeting with participants and their parents. This was, at times, a social event with some children encouraging others to vote on his or her favorite book. We generated a list of the most frequently voted-upon books, and compiled children's preferences. These preferences, coupled with issues of availability, resulted in a schedule of readings for the girls and boys. During some meetings, the children read and talked about the same book. During other meetings, children read different books around the same theme (e.g., water, experimentation, or volcanoes). A list of the books featured in the discussion is presented in Table 14.1.

RESEARCH BACKGROUND

We used multiple data collection methods and two analytical approaches to locate children's critical contributions (and positionings within those contributions) during book club meetings, and to get a sense of children's social identities. Additional analytical methods were used to compare the patterns of the former with enactments of the latter. In what follows, we discuss these methods and approaches as they pertain to these goals.

Data Collection

Book club meetings were videotaped and audiotaped to capture critical contributions, and field notes were taken to support these data collection methods. In addi-

Table 14.1. *List of books children read in their book clubs and featured in the discussions*

Busenberg, B. (1994). *Vanilla, chocolate, and strawberry: The story of your favorite flavors.* Minneapolis: Lerner.
Cole, J., & Degen, B. (1997). *The magic school bus: Inside the Earth.* New York: Scholastic.
Kramer, S., & Kunkel, D. (2001). *Hidden worlds: Looking through a scientist's microscope.* Boston: Houghton Mifflin.
Markle, S. (1999). *Outside and inside kangaroos.* New York.: Atheneum Books for Young Readers.
Markle, S. (2003). *Outside and inside big cats.* New York: Atheneum Books for Young Readers.
Nicholson, C. P. (2001). *Volcano!* Tonawanda: Kids Can Press.
Robinson, R. (1999). *Science magic in the bathroom: Amazing tricks with ordinary stuff.* New York: Aladdin Paperbacks.
Simon, S. (1988). *Volcanoes.* New York: Morrow Junior Books.
Wells, R. E. (1995). *What's smaller than a pygmy shrew?* Morton Grove, IL: A. Whitman.
Wick, W. (1997). *A drop of water.* New York: Scholastic.
Wulffson, D. L. (2001). *The kid who invented the trampoline: More surprising stories about inventions.* New York: Dutton Children's Books.

tion, as book club facilitators and discussion leaders who attended each session, we were participant-observers in the thick of book club—interacting with the children around books and pizza, scientific ideas and humor, questions and, at times, answers. Our conceptions of children's identities were constructed through our engagement in the aforementioned larger project, and via Nancy's personal knowledge of the book club participants. As we gathered field note, video, and audio data from the project classrooms, we observed the girls and boys doing school science and, more broadly, doing school. We watched as they interacted with one another in small and large groups, manipulated science equipment, read books, conversed with the teacher, misbehaved and were praised, were on- and off-task, and so on. End-of-year teacher interviews for the project created opportunities for us to hear teacher perspectives about each child's participation in school and school science. Triangulating these more traditional data collection methods was Nancy's role as a mother of a child in one of the classrooms invited to participate in book club. She was a parent-member of the school and classroom community in which the children participated years prior to both the larger study and the book club. Thus, she informally observed the social and academic dynamics of many of the book club children and their peers by participating in guest reader programs, field trips, and other school-related events. Additionally, she interacted with some of the children outside of school (e.g., sports events, birthday parties).

Analytical Methods

The two approaches that we used to analyze the data attended to children's social identities and critical contributions (and positionings) in book club meetings, respectively. In either case, we began by utilizing a *grounded theory* approach (Glaser & Strauss, 1967) to search and re-search the data for themes that informed these constructs. Teacher interview transcripts, classroom observations, and the first author's experiences helped us to generate descriptions of the children's identities. These descriptions were emergent during the analytical process, being written and rewritten to accommodate the sometimes consistent and sometimes complex identities of the children. During this process, we sought to balance our interest in characterizing children's identities apparent in the data with our disinterest in stereotyping the children or making a caricature of their identities.

More stepwise was our approach to analyzing book club data for critical contributions. With the aid of QSR NVivo we coded the data for "critical episodes" or topic-consistent conversations in which at least one child made a critical contribution (i.e., asked a question, made a comment, or responded to another child by questioning, challenging, or supporting a science text). Both authors scrutinized each episode to ensure that it included substantive talk about science, and demonstrated at least one critical contribution by a child. Altogether, we identified 37 episodes, including 23 for the girls' group and 14 for the boys' group.

After identifying and then naming critical episodes, we used the tabular software, Microsoft Excel, to break each episode into a sequential series of critical contributions. We did so to see more clearly patterns of critical engagement and positioning across critical episodes. For each critical contribution that was located in a row of our table, we recorded in respective columns: (a) the child making the critical contribution; (b) a quotation or paraphrase of this critical contribution; (c) whether this critical contribution involved questioning/challenging or supporting; (d) the text or texts that were being questioned/challenged or supported; (e) the warrants the child used, if any, to support their question/challenge or support (i.e., a book club tradebook, and outside book, authority of an author, experience or experimentation, other warrants); (f) the context in which the critical contribution was made or what initiated the contribution; and (g) identifiers that connected the critical contribution to the episode name and book club discussion number. We documented a total of 104 critical contributions across both book clubs: 77 of these were made in the girls' group, whereas the boys made 27.

The tables we created to document critical contributions were highly informative, yet were so massive as to necessitate one more level of analysis to synthesize the data further and to help us navigate these critical contribution tables. The Excel software enabled us to quickly generate a smaller table whereby each child was represented in a row, and the following was calculated for each child in columns: (a) total number of instances (i.e., across all book club sessions) of questioning/challenging a tradebook, and the particular tradebook(s) questioned/challenged; (b) number of instances of questioning/challenging a child/children, and name(s) of the child/children being questioned/challenged; (c) number of instances of supporting a tradebook, and the particular tradebook(s) supported; (d) number of instances of supporting a child/children, and name(s) of the child/children being supported; and (e) kinds of warrants used to support the question/challenge or support. Additionally, we transposed this table to generate another one, whereby we documented the number of times that each child was either questioned/challenged or supported as well as who was doing the questioning, challenging, or supporting.

Highly aware that the numbers we calculated for these synthesizing tables were not quantitatively meaningful, these tables were suggestive of patterns of critical contributions, positionings, and interactions among children that we verified by going back to the larger critical contributions tables and the original critical episode transcript data. It was these patterns—consistent in each product of analysis that we generated—that helped elucidate the intersections of critical contributions and children's identities.

THE GIRLS' BOOK CLUB

The three girls who emerged as the most consistent in their interactional and reflexive positioning are Maude, Hui Ying, and Joli. Maude questioned and challenged the written texts more than any other book club member across both boys' and girls' groups. Hui Ying was not as likely as Maude to challenge or question the book, but she was the most consistent questioner/challenger of the spoken texts of other girls in the book club. Finally, Joli was interesting because she was far more likely to support her peers or a book than she was to challenge them.

Maude

Maude was the most outspoken member of the girls' book club. Missing only one meeting, she was always eager to contribute. She was the only girl from Dr. Debbie Tasker's class in the girls' book club. Unlike most of the other girls who cliqued in dyadic or triadic friendship groups, Maude was relatively independent. She was also a year older than most of the girls because she repeated third grade. Dr. Tasker characterized Maude as the student she would pick from her class who most enjoys science, and shared that Maude was a slow but competent reader, reading on about the fifth-grade level. Dr. Tasker described Maude as "conscientious, perfectionist, . . . a bit . . . compulsive . . . and . . . careful," "creative," "opinionated," and "a cool little kid." Her teacher also discussed Maude and her mother's decision to have Maude repeat third grade:

> I think she and her mom decided that another year in third grade would make her stronger and give her confidence and so she's been a leader in the classroom, which was really, I think it worked for her, you know she, kids look up to her, she's, she's just very accurate, careful worker.

Repeating third grade was a choice made with Maude rather than a decision made by someone else and imposed on her. It is unusual at this school for a student to repeat a grade voluntarily and suggests that Maude and her mom have a strong desire to take charge of Maude's education in ways that others would likely view as risky. The fact that Maude repeated third grade is well known amongst Maude's peers, perhaps due to the fact that Maude talks about it with no hint of embarrassment. Dr. Tasker explained:

She wasn't shy about, you know, I was pretty sensitive, like I don't want to say anything that would make the other kids realize that she had been retained, and from the first week of school when we were doing some test, she goes, "I remember doing this last year." You know, it was the same story or the same test, that . . . you'd kind of think she'd wanted to hide that, but she never, never did.

In addition, Dr. Tasker shared that she was not shy about "going to speech."

The first book read by many of the girls was *A Drop of Water*; the girls often referred to this text as, simply, *Drop of Water*. In this award-winning science book, high-resolution photography is used to show some of the characteristics of water, such as surface tension. Included in this book is a photograph of a needle floating in a glass of water. Two weeks after reading and discussing *A Drop of Water*, the girls talked about experiment books they had read. Maude read *Science Magic in the Bathroom: Amazing Tricks with Ordinary Stuff* (henceforth referred to as *Science Magic*). In this book the author tells his readers that they can amaze and baffle their friends with magic tricks, including one where you can make a needle float on water by making a paper raft for it. Maude initiates the discussion:[1]

1: Maude: . . . these magic tricks aren't really magic / . . . so when they say what happened, like, I sort of already know what happened. So it's like telling me something that I already know sort of. / Here's something that I found out / it's not exactly a mistake . . . a needle really can float.
2: Joli: I read that in the *Drop of Water* book.
3: Maude: Yeah it, it really can float.
4: Discussion Leader: Did you try it out Maude?
5: Maude: Yeah I did, I did with the / *Drop of Water* book / it [the *Science Magic* book] says that you make a raft and stuff and sneak it in there to make the needle float because it, but, it really can happen.
6: Discussion Leader: You don't have to do that?
7: Maude: Yeah I know it really can happen / you don't have to do that!

Maude acknowledges that although *Science Magic* did not necessarily make a mistake, the book incorrectly assumes that it is necessary to use a raft to make a needle float on water. As the discussion continues, Maude questions the author of *Science Magic* and even suggests that perhaps they do not know that the raft is not needed, while Joli and Hui Ying try to rationalize what appears to be an inconsistency between the books.

8: Joli: And, like, maybe they are trying to make the book, like, fiction but nonfiction.
9: Maude: I don't know if they know that it really can happen. I think that they don't know because they're acting like, it's because they wrote in here, here let me find, here // it says actually in here "a needle cannot float on water."
10: Joli: It could be how big the needle is.
11: Hui Ying: Sewing needles.
12: Discussion Leader: You think that some needles couldn't float?
13: Hui Ying: Yeah you need a big one.

14: Maude: . . . it says // about it being impossible because metal, because metal is heavier than water, but it can / it can happen / but it says in here that it can't. It's wrong.
15: Discussion Leader: It's wrong? Can I see that for a second?
(Maude hands her the book. Discussion Leader reads silently.)
16: Hui Ying: It might be by the size.
17: Maude: They think they are experts and a needle cannot float on water but guess what, [the needles] can.
18: Discussion Leader: So why is it that you believe what you read in the other book rather than what you read in this one?
19: Maude: I believe in that, I believe in the *Drop of Water* book because I tried it.

In this passage, Maude challenges the *Science Magic* text quite strongly, and in addition, reflexively positions herself as an authority over the *Science Magic* text (turn 14). *A Drop of Water* is accepted as an authority since its pictorial results were confirmed by Maude's own investigation (turn 19).

Joli and Hui Ying have not actually read *Science Magic*. They seem to be looking for a way to reconcile the apparent conflict by explaining why the author of the book may have thought that creating a raft was necessary. These girls suggest at different points in the discussion that needle size may impact whether a needle floats. Maude largely ignores Joli and Hui Ying's attempts to maintain the authority of the text. The conversation continues:

20: Discussion Leader: (Reading aloud from *Science Magic*) . . . "challenge your audience to float a needle in the bowl of water when they can't manage it / they will mutter something about it being impossible because the metal is heavier than water / that is when you make your announcement."
21: Maude: Yeah, yeah, so if you actually, if you actually try that / if you actually try that, like, and you do the experiment [the audience members] will say 'look I made the needle float on water' . . . I don't, I really don't think the magic trick part will work because they will be able to do it ...
  . . .
22: Maude: . . . They [the author of *Science Magic*] should have tried it / they should have tried it / I don't think they tried it / I think they just wrote it down in the books that it was impossible.
23: Hui Ying: Maybe it's because they couldn't do it.

Hui Ying implicitly seems to accept Maude's point that the trick would not work if the audience actually attempted the foil. However, she is not ready to concede that the authors would be willfully ignorant. The conversation then turns to a judgment on the part of the girls that *A Drop of Water* is more "professional" and thus more credible than the author of *Science Magic*. The girls considered the photographic evidence in *A Drop of Water* to be better and more direct than drawings in a book with comic features.

24: Joli: If everyone had the same book of that bathroom book and it told you / how they / that it's impossible for / the needle to float on top of the water / well /

they're possibly lying because first of all Maude did it and um / like // that book looks, looks more cartoon than a *Drop of Water*.

25: Maude: But I mean I think that see, I think that if they can't do it, like, they tried, like once or maybe two times and if they can't do, then they think it's impossible.

26: Joli: That's, that's not how you should do it / 'cuz I bet you the people in the Drop of Water did it a million times.

27: Maude: They should get a professional! They're not even a professional. If a kid can do it then they should be able to do it. . .

28: Discussion Leader: So you think that *Drop of Water* looks more professional or does it look older, or why is it?

29: Maude: Because it looks like it's more real.

30: Joli: Yeah because the pictures, and like, you would either have to use the computer for that / but you really wouldn't, it's, it's, it's real pictures . . .

31: Maude: Yeah they even have a picture of the needle floating on the water in the *Drop of Water* book.

32: Joli: And it had to be a camera / because in the one part of the *Drop of Water* book / there is a faucet in it and it showed the water dripping from it / and it had to be a camera.

The intertextual conflict is thus resolved with Maude asserting the authority of the *A Drop of Water* text over the *Science Magic* text (turns 29, 31). Maude's initial questioning of *Science Magic* may never have taken place if she had not read *A Drop of Water* and adopted for herself the text of the needle floating on water.

To Maude, *Science Magic* positions her as someone who is not knowledgeable about floating needles in water. She adopts an authoritative text and leads the other girls in a discussion questioning the authority of *Science Magic*, elevating both her authority and that of the *Drop of Water*. This intertextual conflict eventually led to agreement that there were ways of deciding what to believe, including Maude's direct observation, the second-hand evidence depicted in the photographs in *A Drop of Water*, and a judgment that since *A Drop of Water* gave the appearance of a book that was more carefully put together than *Science Magic* it was more credible. At the same time that this cognitively demanding assessment of science texts is happening, Maude, a girl who has not been particularly successful academically, seems to reflexively position herself as better informed than the authors of the *Science Magic* text and, less directly, than the other girls in the book club.

Although the episode described above was one of the longest sustained discussions in which there is disagreement with a text and perhaps shows Maude performing a critical science identity better than any other episode, it is certainly not the only time Maude questions an author or a peer. For example, in the following episode, Maude responds to Louisa's question about the photographs in a book about big cats. Louisa thinks that taking the pictures must have required the photographers to get very close to tigers. Maude responds by saying that maybe the tigers are not wild—they are trained and tame.

33: Louisa: Well just looking at your book there are so many close ups how can they do that without being scared?

34: Discussion leader: How can the photographers take the pictures?

35: Louisa: Yeah.
36: Maude: They probably trained/ trained once/ the trainers tell them like what to do / not tell them but they probably train them.
37: Louisa: But they are in the wild.
38: Maude: And then they might not actually be / I mean they are in the wild in these pictures but they may not actually be wild in these pictures.

Here, Maude questions whether the animals that are depicted as wild in the book really are wild. Looking across episodes, we see that Maude is more likely to question or challenge a book than the other girls, and that once she does, she is unwavering to the attempts of her peers—including Hui Ying (see below)—to compromise her stance. Her peers typically respond to her respectfully, with as many cases of peers supporting as challenging her.

Interestingly we found no cases of Maude supporting a peer. She does not seem to be terribly influenced by the ideas of her peers. She occasionally questioned or challenged a peer. For example, in examining a photograph in a book, Hui Ying suggested that one would need a "fast video camera" to take such a picture. Maude responded by suggesting that you would not need a special camera—that you can view any of them "in slow motion." These exchanges, however, are not as sustained as Maude's challenges to book authors. This suggests to us that Maude is simply more invested in her engagement with the ideas in the books that she is with the ideas of her peers.

Hui Ying

Mrs. Michaels described Hui Ying as a "blessing—in so many ways!" Mrs. Michaels identified her as the top girl student in science and as one who "has a very strong science background" since her parents are scientists. She further described her as a very avid, strong reader, who is "always very focused," self-motivated, and "a phenomenal student." Hui Ying enjoys both chapter books and fact books. She moved to the US from China when she was three. Although she is fluent in English, her parents are not—a matter about which Hui Ying routinely expresses embarrassment and frustration.

Hui Ying was the only member of the girls' book club who was never absent. She always contributed during book club sessions. Although she socialized with children like Joli and Krista, she consistently communicated—verbally and through body language—her desire to distance herself from Maude and other lower status participants.

In the prior episode with Maude challenging the *Science Magic* book, Hui Ying at first attempts to reconcile the controversy between the two books by adding qualifications to the issue of what kind of needle can float (turn 11). Although she concedes that Maude's challenge has merit, she struggles to maintain the authority of the text over Maude's challenge to that authority (turns 11–23). This happens during other episodes as well, including an episode from a meeting that Maude missed. In this episode, for example, when the girls are trying to understand a diagram in a book, one girl suggests that the illustrator had been sloppy in the way that s/he drew the diagram. Hui Ying defends the illustrator by stating that the shadings that they find confusing appear to be intentional by the illustrator.

At times Hui Ying appears to be especially exasperated by Maude's seemingly constant questioning of the books. In two episodes, Maude's critique of a book begins when the book is first introduced—before she has even read it. For example, Hui Ying defends the title of a book, *The Kid who Invented the Trampoline*, criticized by Maude:

> 39: Maude: I have a question / I don't think a kid could invent a trampoline.
> 40: Hui Ying: (. . .) Walt Disney was a kid when he got the idea and when he grew up he decided to become a cartoonist.
> . . .
> 41: Maude: Well I think he may have thought of the idea but he didn't make the first trampoline.

Hui Ying was the girl most likely to challenge or question a peer. At times she is helpful, such as when she helps Zora, the youngest member of the group, interpret a world map that shows where chocolate grows. Zora interprets the brown shading in the ocean to represent chocolate growing in the ocean. Hui Ying helpfully points out that the brown shading is on islands where chocolate is grown. During another meeting she helped another girl who was struggling to read orally by quietly taking her book and reading for her. Hui Ying also engaged in three instances of supporting her book club peers, in one case supporting Joli (turn 11) and in two other cases, supporting Maude (turn 13 and another turn we did not share in the girls' discussion of the floating needle).

Although she supported Maude on two occasions, Hui Ying more frequently demonstrated that she had little patience for her. For example, during one session the girls read *What's Smaller than a Pygmy Shrew?* and *Hidden Worlds: Looking Through a Scientist's Microscope*. (These texts are hereafter referred to as *Pygmy Shrew*, and *Hidden Worlds*, respectively.) *Pygmy Shrew* has drawn illustrations and is intended to give the reader a sense of scale to imagine how small things like bacteria, atoms, and quarks are. *Hidden Worlds* is a larger book with smaller font words, but large computer colorized photographs of microscopic creatures like mites. In spite of the challenging content, *Pygmy Shrew* appears to be intended for a younger reader than *Hidden Worlds*. When we asked the girls which book they liked best, Hui Ying chose *Hidden Worlds*, in part because she liked the colors, whereas Maude chose *Pygmy Shrew*.

Although Maude liked *Pygmy Shrew*, she reveals how challenging it was for her to understand the scale of the smallest objects in the book, eliciting a sustained response from Hui Ying:

> 42: Maude: Because if you think about it for a second like you might think / wow that's pretty big / and then if you think about it for a second and then you look at those things and they you're like / wow / look how small those things are.
> 43: Discussion Leader: What are those things?
> 44: Hui Ying: It's bacteria it's bacteria.

45: Krista: Did you read this?
46: Maude: Yeah I read this / it's the other half of my brain that is talking right now / the part that doesn't know.
47: Discussion Leader: So you found surprising because it's hard to think of things that small?
48: Maude: They show it.
49: Hui Ying: I have something to say I have something to say.
50: Discussion Leader: What do you have to say?
51: Hui Ying: Everyone knows how big an elephant is and that's a pygmy shrew standing right next to it / so you can imagine how small that is.
52: Zora: It looks like a little dot.
53: Hui Ying: and then he goes over to an acorn and there is a tiny little lady bug that is standing on it and everyone practically / everyone knows how small the lady bug is / it's about that big / and then they magnify it / until it is that big / it's like that huge // and there is this drop and if you look at it its eyes are bigger than golf balls and then there is this huge mound and there is little fuzzy things in there / and you can imagine how small these were / these could get in like a lady bug's eye like five times / and so you can hardly see a lady bug / and you can't see its eyes / and its eye would be like a tiny speck compared to you / and here when it's like magnified there is a little bacteria there and you can imagine how small the bacteria is.

Maude is not embarrassed to admit that imagining these scales was both challenging and fascinating (turns 42, 46). Hui Ying, however, responds by providing a page-by-page explanation to Maude in hopes that Maude will understand it like she and "everyone" else understands it (turns 51, 53). She is reflexively positioning herself more as a knower than as a wonderer, and in so doing, interactively positions Maude as someone who does not know what is obvious to most folks. Hui Ying's is an expression of confidence, not doubt, which seems consistent with her high academic value in a conventional school setting.

Joli

Joli's teacher, Mrs. Marisa Michaels, described her as "a very creative child" who loves to dance. She often needed support in comprehending what she had read; her teacher recalled that she and Joli's mother were very concerned about Joli passing the state reading test. Mrs. Michaels shared that her "reading ability and her comprehension have improved dramatically," and that she "did quite well on the test." Also, her teacher said that Joli enjoyed science, like the other students in her class. She mentioned, however, that Joli often needed the support of others to work through science activities.

Joli participated in six of the eight book club meetings. We were often impressed that she would contribute to book club discussions by asking questions of other girls about the books that they read. Although she was cooperative and often eager to participate in book club discussions, she was comfortable enough in the book club environment to share her distaste for a schoolish-looking experiment book about rocks and minerals. She exclaimed quite proudly at the book club and later in her interview "I didn't even read it!"

In our analysis of the book talk data Joli emerges as giving far more support of her peers than any other book club girl. While she occasionally challenges or questions a peer, these were not sustained challenges. She never presents her own ideas forcefully. Joli is often in the position of someone who supports the ideas of others. She is a strong listener who seems primarily interested in understanding the points of view of other girls in the group. For example, the following discussion occurred when Zora shares what she learned about kangaroos in Sandra Markle's (1999) *Outside and Inside Kangaroos*:

> 54: Zora: I liked it when they showed the pictures of the kangaroos hopping because it shows you exactly how the kangaroos hop.
> 55: Discussion Leader: Oh ok / so the hopping kangaroo.
> 56: Zora: and hopping helps them save their energy.
> 57: Discussion Leader: Oh does it / is that why they hop? Did you know that before you read that book?
> 58: Zora: No.
> 59: Louisa: I thought it would take up more energy.
> 60: Joli: That's what I thought too.
> 61: Zora: It actually saves their energy because when they run it takes up their energy / to walk you take the back legs and put them in front of your front legs.
> 62: Discussion Leader: Did you hear that? They take their back legs and put them in front of their front legs.
> 63: Joli: My dog does that when she runs so fast she just goes crazy and her legs will be in the front and she'll be running.
> 64: Louisa: You mean her back legs.
> 65: Joli: The back legs in the front.

Here Joli first supports Louisa's questioning (turn 60) and then supports Zora (turn 63) by stating that her description of the kangaroo hop is consistent with her observations of her dog. Unlike Maude, Joli is a consensus builder and can be convinced of a point of view not her own. She was frequently a supporter of Maude. This is particularly evident in the conversation of the floating needle where Joli attempts to reconcile the books with Maude's criticism of them. She suggests that maybe both Maude and both books are correct—that whether a needle will float or not depends on the size of the needle (turn 10). As the discussion progresses, she becomes more convinced by Maude's argument and drops her support of the Science Magic book (see turns 24, 26, 30 & 32).

In these data, we see in-the-moment positionings of the girls with the text and with one another, which can be examined across multiple episodes to illuminate repeated positionings. It is these repeated positionings that help us to understand the longer-scale issues of identity that structure and are structured by the book talk episodes. For example, the academic status of the girls in school certainly plays a role in the book club meetings. However, one can also see how the book club environment allows the girls to shape their participation in somewhat unanticipated ways. Maude, for example, repeatedly positions herself reflexively as an authority over the science texts. In this way, we believe we see how individual agency can shape identity, as Maude develops the competence and disposition to question the authors of science books.

In the girls' book club meetings, the girls position themselves with and against other science texts. The science books were typically the focus of the discussions. While there were times when the girls were using science books in order to compete for social status, this was much more subtle in the girls' group than it was in the boys' group. The point of many of the exchanges of the boys was the boys' social standing. The content of the books often seemed relatively peripheral. Thus in discussing the data from the boys' group we will focus on how the science books were used by the boys—and how the science books used one of them—in the construction of social identities.

THE BOYS' BOOK CLUB

The boys' book club was interesting in that there were far fewer episodes in which there was a science-related challenge or question. The boys were more likely to use the texts to interactively position themselves against one another. Here we pay most attention to the positioning of Adam, Darren, and Timothy. Of the boys, Adam was the most likely to question or challenge a text. Adam and Darren were the boys most likely to challenge or question a peer. Finally, Timothy most frequently positioned himself as a support person for his peers.

At the first meeting of the boys' book club, the boys brought books we provided on volcanoes. Adam and Darren had both read *Volcanoes*; Timothy had read another book on volcanoes. The scene opens with Adam recounting some of the details of his volcano book:

66: Discussion Leader: So what do you remember from reading your book?
67: Adam: . . . It like tells us about some of the sea gods and fire gods / gods and goddesses / and it tells . . . about how volcanoes blow up / and it tells about the four different kinds of volcanoes . . .
68: Timothy: There are three different types of volcanoes.
69: Adam: No, there's four.
70: Timothy: Nuh uh / there's three.
71: Adam: There is four in here.
72: Discussion Leader: What are the four in yours? Let's take a look.
(Time elapses while Adam searches for the four names)
73: Darren: They might take place in different places on the earth.
74: Adam: // Okay, there are shield volcanoes // cindercone volcanoes // strato volcanoes // and // I'm looking, I'm looking // Where the heck is it? // And dome volcanoes.
75: Discussion Leader: Now what does your book say?
76: Timothy: It said I have a shield volcano, a cylinder volcano, and a composite volcano.
77: Discussion Leader: So it didn't mention dome and it didn't mention one other [strato] right? . . . / Do you think there is different information in those books?

Timothy notices a contradiction between his text and Adam's (turn 68). However, the boys do not progress on resolving the apparent contradiction until a discussion leader steps in and prompts them to look more closely at specific information. The conversation then jumps to questions regarding judging the authority of texts and how it is possible for conflicting texts to be written by scientists.

> 78: Darren: I always believe the Seymour Simon books.
> 79: Discussion Leader: And why is that, Darren?
> 80: Darren: Because they got some really good facts in them.
> 81: Discussion Leader: They do.
> 82: Timothy: This is by Cynthia, Cynthia Pratt [Nicholson].
> 83: Darren: But she probably didn't do really good research. (laughter)
> 84: Timothy: Yeah she did.
> 85: Adam: Well Seymour Simon was wrong, too. There are more than four, there's five (counts silently on his fingers).

Darren begins by asserting the authority of the book's author, Seymour Simon (turn 78). He speaks with considerable confidence, as though he expects that his views will be taken seriously. Darren has had considerable exposure to science books. Although Darren justifies his faith in this author based on "good facts," we suspect that Darren recognizes this author as one whose books are credible at both school and home. Cynthia Pratt Nicholson, the author of the other book, is not nearly as well known for writing science books, which may be the reason why Darren questions her expertise. For Darren, resolving the conflict requires that Cynthia Pratt Nicholson be wrong and Seymour Simon be right about volcano types. Adam—perhaps merely looking for a way to counter Darren's argument—suggests that both authors may have been incomplete, and assumes that the volcanic categorizations represented in the two books are most correct and complete when added together.[2] The discussion continues:

> 86: Discussion Leader. There might be some different information in them [the books]?
> 87: Timothy: I think this was a long time ago book. His name was David Johnson and he already died so . . .
> 88: Discussion Leader: So do you think sometimes books about science might have different information in them?
> 89: Adam: Mmm hmm. (expressing agreement)
> 90: Discussion Leader: Why do you think that happens?
> 91: Timothy: Because you have different scientists.
> 92: Adam: 'Cause the author finds out different things, the scientists.
> 93: Darren: Yeah probably some authors don't do the same . . .
> 94: Adam: They go to different places to find out different things with the volcanoes.
> 95: Discussion Leader: That might be how they learn different stuff.

The boys suggest that credibility can be judged by who the author is and when the book was published (although Timothy's book actually has a more recent copyright date than the other book). With prompting from a discussion leader, they discuss why it is possible to have books with different information. They do not, however, reflexively position their own personal texts over the authors of the written texts. It simply seems sensible to them that different authors would know different things and thus write about different things and that some authors may be better researchers than others (turns 91–94). This conversation is relatively shallow in terms of substantive science talk. They never attempt the more cognitively demanding work of figuring out what the real differences are between volcanic classification systems. The conversation continues only briefly, with Darren suggesting that finding out what another book says might help.

> 96: Darren: I read one with the *Magic School Bus* books and there only three volcanoes that they mentioned.
> 97: Adam: You still read *Magic School Bus*?
> 98: Darren: I read it a long time ago (very annoyed).

Although Darren was not consistent in arguing that three types of volcanoes existed, he used his prior experience reading *The Magic School Bus: Inside the Earth* book to suggest that there were three. Adam disregarded this book as babyish, diminishing its authority and hence, Darren's. The conversation ends; however, the themes return several minutes later, but then deteriorate into a power struggle between Adam and Darren.

> 99: Adam: I want to know something / how do they erupt? // I mean like.
> 100: Darren: It's a simple simple.
> 101: Adam: Like how are they even created?
> 102: Darren: Well okay / my book says here right away.
> 103: Adam: Your book says nothing / it's the same as mine and I didn't find anything on that.
> 104: Darren: Oh well/ you need to read other books about volcanoes.
> 105: Adam: I watch the science channels all the time.
> 106: Darren: Well the science channels aren't very good.
> 107: Adam: Are you kidding / it talks about volcanoes all the time.
> 108: Darren: Ok well / if you don't know/ I don't know anyone else who doesn't know how volcanoes are created / it's when two plates on the earth collide together.
> 109: Adam: You are boring me.
> 110: Darren: Well good.

Although this conversation is nominally about the authority of texts (books vs. television), texts seem to be used primarily as instruments for social positioning. Darren asserts that the real problem with Adam is that he does not draw upon the same resources (books) as he does (turn 104). Instead he relies on what Darren regards as a lower status popular outlet like television. (Darren's family does not have network television at home.) Darren concludes with an exhibition of his own knowledge and an "everybody but you knows" kind of claim that is reminiscent of Hui Ying's interactions with Michelle (turns 51–53). Darren is using his knowledge of science in general, and science books in particular, to gain social advantage over Adam. Adam responds by insulting Darren—this time with "you're boring me." The boys' primary concern was with their power struggle and how they were positioned interactively—in this context, in terms of being an authority on volcanoes. Perhaps this is not surprising given the antagonistic relationship between the boys; however, what was interesting was how each boy aligned himself with various science texts that were positioned against the other. Unlike Maude in the discussion about the floating needle, none of the boys addressed the way in which the written texts positioned them (turn 1).

The identity building we see in this episode is in many ways consistent with what we see throughout book club meetings. Darren consistently uses science texts to construct a high status social identity. Adam attempts to position himself as someone who is interested in and knowledgeable about science—but Darren consistently challenged such positioning. Adam is not able to fend off Darren's challenges and therefore resorts to a personal insult. Timothy, as we shall see later, plays a supporting role for Darren and appears dependent on Darren's authority. While this episode portrays the positionings of the boys that we see repeated throughout book club, we now turn to some additional exchanges to see how they can inform our understandings of these boys' identities.

Adam

Three of the parents of the book club boys expressed some concern about Adam's participation in book club. These parents regarded him as a bully at school. In our own observations at school, we saw Adam be physically aggressive with other boys. (He hit other children when the teacher could not see him.) This history of animosity transferred to book club. Twice at book club he had to be restrained from hitting another child. Despite this physical aggression, Adam contributed some thoughtful insights at the three book club meetings that he attended. He ceased coming to book club after three times because of a family-related problem. Mrs. Michaels alluded to Adam's challenges with behavior coupled with his ability to participate academically. She mentioned that he had trouble focusing during class discussions, and benefited from the hands-on approach during science investigations. She shared that "it worked well for Adam when [he was] paired with very serious, focused partners." In addition, she suggested that Adam "generally did quality work even though he often would not volunteer to share his findings."

Although Adam often questioned or challenged both books and other boys in book club, only Darren challenged Adam. Marcus, a very high achieving boy who had unpleasant experiences with Adam at school, mostly ignored any input by Adam including Adam's direct questions. Although Adam was not considered a poor student at school, he distanced himself socially from the boys with high academic status. For example, when Timothy describes using bread as an eraser, Adam responds with "a good thing about that is they spend less time writing and more time erasing and you got out of some work." Later the boys questioned how far they were supposed to read in one of their books.

111: Darren: I didn't know whether to stop here or read on.

112: Adam: So you read on.
113: Marcus: I did read on.
114: Adam: I knew it.
115: Marcus: I was just telling you.
. . .
116: Adam: Most of this book was kinda boring / I skipped to the ones I wanted to read.

Here Adam is making a distinction between himself and the strong academic identity of Marcus. He is selective regarding what he reads, but also frequently talks about watching a lot of TV, especially the Discovery Channel (see also turn 105).

117: Discussion leader: Can I ask you something different? Have you ever seen a volcano?
118: All: No.
119: Discussion Leader: Have you seen one on TV?
120: Timothy: Oh yeah, I have.
121: Adam: I watch a lot of the Discovery Channel.
122: Darren : Yeah, yeah, yeah (condescendingly).
123: Timothy: I've seen one erupt on TV.
124: Adam: I've seen plenty / I've seen inside.
125: Discussion leader: Oh inside the volcano.
126: Adam: I've seen a guy die in there.
127: Discussion Leader: Oh you saw him die in there.
128 Adam: Magma went over his feet and started burning and he got stuck there because it's sticky.
129: Discussion Leader: Do you think it would be scary to study volcanoes?
130: Darren: No.
131: Discussion Leader: Why not?
132: Darren: I think it's cool how the magma and lava flow.
133: Adam: You said it's not scary? I saw a guy die in a volcano / no duh/ it's scary.
134: Discussion Leader: You think it might be?
135: Adam: Would you really go inside a volcano and tape the magma flowing / with it on your feet and up to your legs melting you?
136: Darren: Cool.
137: Adam: Bone through bone.
138: Discussion Leader: Do you think that happens to all scientists who study volcanoes?
139: Adam: Yep / it could happen.
140: Darren: Well / probably when the volcano erupts they probably had a helicopter.
141: Adam: No they don't / when the volcano erupts there is . . . the magma goes right over and becomes sticky and then they die.

Once again, Adam relies on his knowledge gained through television to tell a story that also positions him against the other boys as one who is relatively "worldly." He knows about death and violence and the other boys would do well to listen to him. Once again Adam spars with Darren, who has little regard for Adam's gruesome story, perhaps in part because its source is TV and perhaps in part because it is Adam telling the story.

Darren

Darren's teacher, Mrs. Michaels, shared that if she had to pick one student from her class who was *the most* interested in science, it would be Darren. This interest, she said, was coupled by a strong background knowledge that she attributed to his scientist parents. A strong reader, Darren was apt to choose science texts during self-selected reading time. He participated in class discussions readily, and was a well-behaved, conscientious student. With respect to his peers, Darren was largely independent, at times bullied, and regarded as a sensitive child; for example, recess was *not* his favorite time of day.

During the four of seven book club meetings that he attended, Darren demonstrated his interest in science and science books by sharing the details and highlights from his reading. His thoughtful insights were intermingled with silly episodes, usually also involving his very good friend, Timothy, in which he was caught up in the social atmosphere of the boys' book club—an atmosphere which we were often challenged by with regard to balancing fun with talking about books, with the former often superceding the latter. Although Darren seemed to enjoy reading the science books, he did not typically challenge or question them. Nor did he challenge or question the oral science texts of his peers (with the notable exception of Adam). He often seems to position himself as an authority about science, by challenging Adam (e.g., turns 99–110) and accepting the support of Timothy (e.g., turns 142–148, below).

When Adam was present, a tension regularly was evident between Darren and Adam through both body language and words—with Adam being more physically antagonistic and Darren responding by asserting his own scientific authority. Although Adam and Darren were in the same class at school, we noticed few interactions between them at school. Their teacher did not comment on their relationship. Although Darren has certainly seen Adam's physical aggressiveness at school and at book club, he may be more fearful of him at school than he is at book club, where there are many trusted adults present. Here, Darren engages Adam in a way he apparently does not at school, by using science texts to gain an intellectual advantage over him. Whereas Darren often uses his background knowledge to justify his claims, Adam has fewer resources and often merely restates his claim.

Timothy

His teacher Dr. Tasker described Timothy as a really nice kid who was an average to low average student. She did not believe he read very much outside of school. He liked science in Dr. Tasker's class, as did all of her kids according to her, but tended to avoid any writing in science, as in other subjects. He is a highly social boy who was frequently distracted from his academic tasks by his social interests. He is also a competitive athlete. Timothy participated in six of the seven boys' book club meetings, making him the most regular boys' book club attendee. He routinely answered our questions and shared his own insights, although he often responded and contributed in silly ways, or simply repeated what others said on a given topic. For example, in the discussion about volcanoes, the discussion leader points out that two books gave conflicting numbers in terms of the number of dead following a volcanic eruption. She then asked them how they might find out who is right. Timothy's refers to the authority that Darren previously suggested.

142: Timothy: I would agree with the Seymour Simon books.
143: Adam: You go find out yourself / do some research.
144: Timothy: I've looked at . . .
145: Adam: (interrupts Timothy) You go find the answer from all the scientists there are on the Internet or something.
146: Darren: I also read another book about Mt. St. Helens and doing research on / and it also told me that fifty-seven people died.
147: Discussion Leader: Which one do you think would be more right?
148: Timothy: Fifty-seven because Seymour Simon says that.

Here, Adam suggests independent research, this time using the Internet. Darren continues to perform the role of a knowledgeable science kid by providing information gained from independent reading. Timothy defers to Darren. Unfortunately, we never see Timothy gain much control over either the written or oral texts in book club meetings. Unlike Darren and to a lesser degree Adam, he seems unable to use the science texts to gain social position among the boys. In fact, we argue that this lack of authority over science texts leaves him to often be used by the science texts of others. He is dependent on others to tell him what to believe with few resources to bring to bear on any controversy.

TEXT, POSITIONING, IDENTITY

Questioning, challenging, or supporting a science text—wondering if what it says is correct or asserting that it is not—engages readers in positioning themselves not as passive recipients of information, but as active, critical readers. Here we see Maude as a good example of someone who seems to be taking on this identity of a critical science reader. She repeatedly questions the science texts she reads. It is this repeated reflexive positioning of herself as an authority over the science texts that is somewhat suggestive that she is forming a critical science identity. However, can Maude's activity over a couple of months in a single setting tell us enough for us to make claims about her identity? After all, an identity is constructed over a lifetime (Lemke, 2000). How would we know if the actions we observe indicate something more than a blip on the long timescale of identity development?

One way of trying to ascertain whether these positionings in book club meetings are an expression of identity or merely a "blip," is by examining these actions in light of what we know about Maude from school and home. Her teacher described her as "opinionated," and as a careful reader. At school she showed considerable confidence in talking about issues that other children would likely have found embarrassing. We suggest that these characteristics, as well as the observations in book club, at least suggest that Maude is constructing herself as someone who can and will critically analyze science texts. It seems to us as though this is not a change in who she is so much as it is a new context in which she can express already existing dispositions. To say this is not to diminish the significance of Maude having a forum such as the book club to express this identity because it provided her with an opportunity to extend her critical competence to science texts.

We see similar continuities in the kind of girl Hui Ying exhibits in school and at book club. She is a very strong student and is accustomed to being the knowledgeable one. We saw confident expressions of knowledge in book club as well—actions that at times were helpful and at other times condescending. Similarly, the social identities of Darren and Adam observed at school were expressed in the book club setting.

The boys' book club brought to the foreground the fact that the primary agenda of eight- to nine-year-old boys is learning to be an eight- to nine-year-old boy (Lemke, 2000). The fact that their talk about science was primarily intended to compete with one another and to present themselves as particular kinds of boys (e.g., scientific boy vs. worldly boy) frustrated us. We wanted a different kind of participation in book club. We wanted the boys to be nice to each other and to talk about the books. However, we were not successful at subverting the boys' agenda. I think this should remind us, however, what *is* important to eight- to nine-year-old boys. They have far more at stake in being particular kinds of eight- to nine-year-old boys than they do in pleasing the book club leaders.

Finally Joli and Timothy were primarily interactively positioned as supporters of the ideas of others. There is, however, a very important difference between the two. Joli impressed us with the way in which she very carefully listened to the ideas of other people. She seemed to be sincerely interested in weighing the ideas of others even though she rarely put forth her own ideas with much conviction. Joli seems to us to be taking up a traditional feminine role of listening and mediating, much like a matriarch at a family gathering. Timothy, however, chooses "sides" quite readily, albeit with little or no justification: he is being used by the science texts. He seems to exert little or no authority over either the books or his peers.

We learned from these children that they can and do critically engage with science texts, and that whether and how they critically engage is more than a matter of creating an environment that presents children with opportunities to ask questions of, challenge, or support texts. Creating such an environment—e.g., by welcoming a wide range of responses to science texts, or by having children read multiple and sometimes conflicting texts—appears to be important in fostering the construction of critical science identities. However, we suggest that who children are and who they want to become—their social identities—frame this construction as much as book club facilitators' intents and strategies to provoke critical engagement.

NOTES

1   Excerpts of transcripts from book club sessions include numbering that is indicative of turns of speech (e.g., this turn is numbered "1"); these turns have been re-numbered from the original transcripts for ease of reading throughout this chapter. The following conventions were used for tran-

scripts of book club sessions: / indicates an "um" "ah" or brief pause; // indicates a pause of 2 seconds or longer; **bolded** words indicate emphasized speech by the children; [words in bracketed text] have been added to transcripts for ease of reading; (words in parentheses) are authors' descriptions of the nature of speech; and . . . indicates speech that has not been included and is not necessary to understand the transcript.

2 Adding the categories of both books would actually result in a total of six, not five, categories: (a) shield (in the Nicholson & Simon books); (b) cylinder (Nicholson); (c) composite (Nicholson); (d) cindercone (Simon); (e) strato (Simon); and (f) dome (Simon).

REFERENCES

Chandler, K. (1997). The beach book club: Literacy in the "azy days of summer." *Journal of Adolescent and Adult Literacy, 41*, 104–115.

Davies, B., & Harré, R. (1990). Positioning: The discursive production of selves. *Journal for the Theory of Social Behavior, 20*, 43–63.

Freebody, P., & Luke, A. (1990). "Literacies" programs: Debates and demands in cultural context. *Prospect: The Australian Journal of TESOL, 5*(5), 7–16.

Glaser, B., & Strauss, A. L. (1967). *The discovery of grounded theory: Strategies for qualitative research*. Chicago: Aldine De Gruyter.

Keselman, A., Kaufman, D. R., & Patel, V. L. (2004). "You can exercise your way out of HIV" and other stories: The role of biological knowledge in adolescents' evaluation of myths. *Science Education, 88*, 548–573.

Korpan, C. A., Bisanz, G. L., Bisanz, J., & Henderson, J. M. (1997). Assessing scientific literacy: Evaluation of news briefs. *Science Education, 81*, 515–532.

Lemke, J. (2000). Across the scales of time: Artifacts, activities, and meanings in ecosocial systems. *Mind, Culture, & Activity, 7*, 273–290.

Zimmerman, C., Bisanz, G. L., Bisanz, J., Klein, J. S., & Klein, P. (2001). Science at the supermarket: A comparison of what appears in the popular press, experts' advice to readers, and what students want to know. *Public Understanding of Science, 10*, 37–58.

YEW-JIN LEE, BRYAN BROWN, NANCY BRICKHOUSE, PAMELA LOTTERO-PERDUE, KENNETH TOBIN, WOLFF-MICHAEL ROTH

DISCURSIVE CONSTRUCTIONS OF IDENTITY

Michael: It appears to me that despite the common reference to discourse analysis and identity as a discursive achievement, there are some real differences between the three texts in what they presuppose and how they theorize identity. For example, Bryan and Greg suggest at one point that speakers have the option to select among alternative ways of greeting someone. Yet in everyday life, we hardly ever think about choosing to greet someone but simply greet, just as we do not plan to place our feet to walk but simply walk. This aspect of language, to be used without reflection and as providing patterned ways of talking about certain issues comes out much stronger in Yew-Jin's chapter.

Yew-Jin: Certainly, my chapter privileges the immediacy without timeouts of the business of interaction, which contrasts with notions of choice in language use. At places, Bryan and Greg's study seems affiliated with a view of language that it is primarily informational in function, a medium of exchange with identity being an outcome. Likewise, Nancy and Pam's analysis looked at how conscious and unconscious talk about science texts did positioning work. I'd like to think that these are one half of a good story for I'm presently placing my bets on language as a useful tool/prosthesis for getting around the world, like pulling off a successful interview with a researcher or discussing a science text in class with somebody. By trading words in everyday life, identity is *being* done and is intentionality located within the flow of talk. My deliberate choice of theoretical frameworks grounded in ethnomethodology and conversation analysis reflects this keen desire to get to the bedrock of identities-in-talk. To me, it's the basic starting point for analyzing learning and identity in science.

Bryan & Greg: There seem to be two salient issues here. First what is the function of language relative to identity? Second, what is the role of consciousness in the language-identity medium? In addressing the first, we certainly agree with Yew-Jin's notion that language is more complicated than a simple medium for the functional exchange of ideas. In fact, we hold the belief that although the content is offered at the same time as identity cues are being sent, language is primarily functional. The social constructions stand as a secondary component of its natural interaction. Michael Agar's work on *languaculture* provides a vivid analogy that is pertinent to the second question (Agar, 1996). He suggests that our languaculture, and by default discursive identity, are invisible until you encounter a culture that is not your own. For a cultural insider, languaculture is often invisible because the rules, rituals, and connotations of language use are familiar to you. So familiar, in fact, that the rules of language use become invisible. What about cultural outsiders? For these individuals, they are continuously made aware of their cultural differences by the rules of languaculture that are in conflict with their own. The foundation of our chapter is using this understanding as a way to assist students in their transitions from cultural outsiders to unconscious cultural insiders.

Nancy: We are talking here about what is conscious and what is unconscious discourse, but is this not really about the structure|agency dialectic that poststructuralist scholars recognize within discourse? What is unconscious and structuring are those aspects of discourse over which we have no control, whereas in those moments when we exercise choice over discourse is when we enact agency.

Ken: My perspective is a bit different. I think it is a question of focus. It is impossible to notice all aspects of social life and at the micro level almost everything that happens is beyond active control and awareness. I think of social life in terms of many dialectical relationships including conscious|unconscious.

Nancy: While it may be true that we could not possibly seriously contemplate every discursive move, we do think consciously about those that are important to us. While I agree with Bryan and Greg that we want youth to know scientific discourse sufficiently well so that they need not contemplate each discursive move, I worry about what happens if the rules of the game become so invisible to them that they cannot stand outside the rules and critique them. I think my point is a refusal to take sides here. Whereas denying the structuring force of discursive practices is a dangerous fantasy, without the possibility of choice and control, the possibilities for change are severely curtailed. There may be times when our research foregrounds one over the other, but these kinds of analyses need to be tempered by the dialectic of structure|agency.

Michael: I hear you caution against an either/or approach, and I agree with such a caution. Knowing has both implicit and tacit aspects and explicit articulated aspects. I agree with hermeneutic philosophers who suggest that practical understanding (of language, culture) is developed through explanation; but explanation presupposes practical understanding, which precedes, accompanies, and concludes the former.

Yew-Jin: Interestingly, scholars in cultural studies have also located identity at the unstable interstices where structure and agency collide! Toying around with dialectics has clouded my recent foray into Foucault where I'm interpreting him through similar lenses: Power and discourse mold us into speaking positions while they also enable us to enact power and identities. My feeling is that we are all trying very hard to make sense of these conundrums and it's plainly visible in our own writings here (c.f., positionings, cultural repertoires, sources, scales, and trajectories of identities etc.). I'm not sure if we will ever get to the bottom of this but the effort of working through it will be important in itself.

Pam: Whether we construe of identities as somewhat continuous, yet situated and (re)constructed in contexts like science book clubs or interviews, or as moment-by-moment constitutions in language—differences that are significant and ripe for theoretical argument—the greater point for me here is that the process of making sense of identity-in/as/and-language within science-rich contexts informs the way in which people identify with, and thus learn and do, science. Interestingly, in the course of studying identity-in/as/and-language, we have all chosen to: (a) explore "informal" science settings; and (b) examine the talk and learning/doing science of those who are somehow traditionally outside of the scientific community (c.f., a fish farmer, minority baseball players, and third-grade girls and boys). Why? What does this reveal about our use of identity-in/as/and-language to understand science learning/doing (or our agendas)? For Nancy and I, the girls and boys are learning about science, learning to be citizens, and—perhaps—learning to be scientists, all of which can benefit from enacting critical science identities.

Ken: In what sense do you use critical here?

Pam: Nancy and I use "critical" engagement with written, visual, and oral science-related texts to mean taking a questioning stance towards those texts, inquiring about the validity of those texts and not assuming that those texts have objective authority. This way of being critical has been called for by others in the science education field (e.g., Norris & Phillips, 2003). Importantly, we also understand critical approaches to include instances in which readers and listeners: interpret the ways in which they (and others/reality) are positioned or potentially used by science-related texts (c.f., Freebody & Luke, 1990); and use science-related texts to position other people or texts (Davies & Harré, 1990). The idea that texts may be agents or subjects of positioning, and that we as readers and listeners can question this agency or subjection, is not far from us (i.e., the authors of this metalogue). While we fundamentally aim to unpack and critically examine ideas common to and different among our chapters—ideas that may be deemed more or less valid by readers and writers alike—the metalogue is also a place of positioning. Are there not instances here of aligning ourselves with particular texts, people, or scholarly traditions (e.g., citing particular scholars), or questioning the ways in which others position us (e.g., as stage theorists or not) or subjects (e.g., the agency of the boys versus the girls in our book club, a matter that Nancy takes up later in the metalogue)? In our writing about ideas about discourse and identity, do we not also attend to a collective desire to position one another in respectful ways and as having something thoughtful and interesting to contribute, while at the same time being aware of the range of expertise and authority in the group? My point here is to use the metalogue space as a way to expose the way in which critical approaches to texts cannot simply entail examination of the validity of the message or ideas within, but must also include unpacking the inevitable positionings of people and texts.

* * *

Ken: For me any story about identity has to address emotions. Let me personalize this. How do I feel here? Of course that is tied to the issue you have all raised—to what extent can I do what I want to do and succeed? So how is it useful to theorize these questions? First, I like to look at the emotions of participants, their valence and strength—exploring the range expressed during the myriad transactions that unfold during the episodes under analysis. Then, as I examine agency|structure relationships, very many other dialectical relationships present themselves—the most central ones that have been alluded to so far being individual|collective, conscious|unconscious, and the triple dialectic micro|meso|macro. Of course, there is more, and even a small portion of social life commands a framework that examines many sets of relationships. What becomes most salient for me, at the bottom line, are questions about affiliation—Do I belong here? Do I like it here? Can I be successful here? At the same time my answers to such questions (which are rarely posed explicitly) are dialectically interconnected with similar questions raised by the collective other—Does he belong here? Does he like it here? Can he succeed here? Responses to questions such as these can mediate the symbolic capital I can accrue in a field (e.g., status, respect) and whether or not I can create social bonds easily—to use in the production, transformation and reproduction of culture and thereby to succeed in the field.

Michael: And belonging probably is strongly related to and mediated by emotions. I have come to think that emotions play a central role in consciousness and being, both the immediate emotions deriving from our bodily states and the long-term ones that are related to the pay-offs linked to the outcomes of the personal goals we pursue and the collectively motivated activities in which we participate.

Yew-Jin: Well, the notions of "choice" and "selection" used by Bryan and Greg in their chapter can be understood as "emotions and affect" from a third-person perspective. Those repertoires that I refer to in my chapter do not explicitly address the role of affect although I believe that they are coextensive phenomena. How could it be otherwise? These discursive tools that facilitate understanding will develop certain affects for speakers and hearers alike. For one, after interviewing Jack at the hatchery we always left with a feeling of respect and awe at this man's accomplishments despite how school had failed him. We felt outrage when he told us how "the system" was causing committed employees to leave the organization because of the intense politics surrounding the hatcheries. Likewise, we believe that our interviews afforded Jack many opportunities to recount his beautiful life story in science that would have otherwise circulated among his colleagues and friends. These emotions are at work in identity formation with the children in Nancy and Pam's chapter too.

Nancy: Of course they are. How one feels in any given social situation will influence how one participates in the activity and what risks s/he is willing to take. One of the reasons why Maude was fascinating to us was that she did not appear to need a strong sense of "belonging" in order to participate in a critical analysis of science texts in this context. Although we have every reason to believe that she enjoyed book club meetings, she did not have a strong affiliation with the other girls. We find this kind of resilience to be quite interesting. Adam, however, probably did not feel a sense of affiliation with the other boys and likely was also the only child who was forced by his parents to attend our book club meetings.

Michael: However, it is not all about my intentions and me. This agency|structure dialectic Ken invokes here lacks an essential element to describe and explain human experiences—passivity. And this passivity, too, is associated with emotions and emotional valences, and therefore, an integral aspect of identity. Passivity is an issue science educators will have to confront head on in their research. Life does not merely consist in realizing intentions, for example, wanting to construct a certain identity, but also is characterized by a passivity that each person is subjected to. This begins with the intention to speak, which, all the while expressing agency, is itself given to us, as we do not intend our intentions. More so, when we speak without reflecting while drawing on repertoires, we are subject to the way these repertoires constitute who we are as much as we actively deploy repertoires and therefore contribute to the nature of our identities.

Pam: It is this issue of the construction of critical science identities by those outside of elite science that makes Jack—and folks like him—so endearing. A sort of grown-up Maude, he is critical of science that presumes authority in the world "just because." Both Jack and Maude demonstrate agency by rejecting some scientific texts and accepting others, often using scientific reasoning (e.g., about the voltage given off by a pump, or about experimental evidence documented in a photographic account of floating needles) in their work as fish farmer or science book reader, reflexively positioning themselves or momentarily belonging to membership categories aligned with science. In this way, Jack and Maude seem to be smart consumers of science, employing critical science identities and thus agency to accept scientific knowledge/practices (or not).

Michael: I would be more hesitant in employing the term science as if it denoted something unitary and stable rather than something characterized by multiplicity and heterogeneity.

Ken: I agree. In the different chapters it might be illuminating to think about the creolized forms of science being enacted by participants. From such a perspective more of what is done might count as science—in which case identities affiliated with science might be constructed more readily.

Bryan & Greg: The power dynamic in science is clear. Science discourse provides its users with analytic, epistemic, and descriptive tools for knowledge construction. The authors of this power dynamic lean heavily upon an assumption that all students must learn it; and it is therefore universal. By default, it can be learned and is assumed to become passive for the science community. However, the "passive component of being" is not always passive. To achieve passivity is a major accomplishment. When children begin to walk, they stumble. At some point, they become increasingly stable, but continue to fall from time to time. There is a time that they become so comfortable with walking that it becomes unconscious.

Michael: Let me interrupt you for a second to say that I view passivity more radically. You don't have a choice to be born, but you are born; you don't have a choice about the culture you are born into, but one day you find yourself being conscious in a way that you have not chosen. Even your intentions are not intended, but whatever you intend, you take at face value. You like chocolate, and you eat it. You do not intend your preference for chocolate but you accept it passively. Sorry for the interruption, please go on.

Bryan & Greg: The passivity of walking is an accomplished activity. In this same way, minority students often use their "stumbling" through using science discourse as a symbol for their belief that science is not a culture appropriate for their participation. If students are to transition from their astute awareness of culture conflict as made manifest by the language of science, science educators must become aware of how minority students are making these transitions. Our chapter models how students regularly make this transition in other contexts that have cultural components similar to science. If students are able to acquire complex discourses and take on the identities associated with them outside of science, then why do they encounter such difficulty making the same transition in science?

Ken: I thought one of the really interesting issues from your chapter is the potential for the baseballers to use their knowledge of baseball inside physics to create a diasporic identity—a home-away-from-home—whereby they use their knowledge of baseball culture to create creolized forms of science culture—a bricolage of baseball speak and science speak. The advantages I see of the creolized culture is that what students learn of physics is likely to be useful to them in other parts of their lifeworlds, especially when they play or watch baseball. Their diasporic identity can segue into learning more physics and building interests in science that transcend applications to baseball.

Yew-Jin: Oh oh, my alter ego and the job that pays me hard cash to teach science prods me to say that for many science teachers, myself included, we are not very optimistic that learners can perform those border-crossings across contexts (i.e., school talk-sports talk) easily. Instead of these discursive styles and concepts (e.g., velocity, speed, movement) acting like passports, they sometimes behave like fake ID that gets you into serious trouble the minute you use them. One the one hand, your teacher can get mad at you for using sloppy language in your homework and, on the other hand, nobody at the baseball diamond can understand what exactly you are trying to say. I'm not denying the value of motivating kids by means of identity appropriation in science, I'm just not so hopeful that this can be readily undertaken by out-of-school literary or creolized science practices. Some things don't get translated easily into classroom practices and I'm reminded of Jack's frequent use of oppositional school and workplace repertoires here and what was just said about power dynamics. In fact, these contrasting repertoires about the value of what people really learn in school versus what they learn on the job seems to be a well established way of speaking across many industrialized cultures as I've come to realize.

Ken: In some ways we have different positions here. I am encouraged by the idea of using cultural resources to build hybrid forms of culture that are science-like—creolized forms of science. In some respects these are stops on the way, not the terminus. We owe it to students to provide them with full access to power discourses—including canonical science. However, the creolized versions of science will likely serve them well in their day to day lives. Without going into it here, some of the research in urban classrooms is very promising in this regard. Perhaps we can agree that the possibilities can be explored in further research.

Nancy: I think these are some of the key issues with which Pam and I were struggling. For too long women and minorities—scientific outsiders—have had very little agency in their dealings with scientific institutions. That is why we were looking for ways to provide children with opportunities to critically engage with science and why in our analysis we examined issues of positioning—we were really attempting to understand what agency might look like in this particular context. We believe that the way in which Maude positioned herself as an authority over a science book was perhaps an interesting case of "agency." She not only engaged in the science in the books, she also questioned what the authors of these books were doing. Similarly, the case of Timothy is interesting (and distressing) because of his passivity in dealing with science texts. He seemed to have no voice—no agency. While I do not disagree with Michael and others regarding how much of who we are appears as passively accepted, I'm rather hesitant to celebrate it.

Michael: I wasn't intimating that we ought to celebrate passivity but that we have to accept it as a fact of life. Also, there are nuances in the way the term is used.

Ken: I wish to comment on your theoretical point about preferring an agency|passivity dialectic rather than an agency|structure dialectic. I do not see the usefulness of having one or the other when I can have both. I do not see it as a useful idea for each construct to have one and only one dialectical other. I accept your ontological point about passivity and further address some theoretical points later in the text.

Michael: The one I am trying to articulate is ontological. I therefore have suggested repeatedly to theorize not agency|structure, where structure itself has a dialectical form (schema|resources), but to build a theory in which agency and passivity also are dialectically related, which leads us to a threefold dialectic: agency|passivity||schema|resources. The concept I am proposing, agency|passivity||schema|resources includes passivity as a core moment of activity, and therefore redresses the imbalance this important notion has received in the scholarly literature.

Ken: I accept your point and note that I am reminded of fundamentlal and derived units in physics. I am sure there will be times when we depict relationships among constructs just as you have articulated. Also, there are other times when agency|structure will be a useful starting point. Depending on the points we want to make, unpacking of structure and agency inevitably involves most of the relationships you have laid out. Having said that, I do think that the agency|passivity dialectic is a brilliant way to think about macro schema as structures that are closely related to identity, unconsciously "writing" identities as individuals enact social life without being aware of salient ideologies, some of which are hegemonic.

***

Ken: A question to consider is why Timothy was not involved in the ways you wanted him to be. Is it because he did not have the culture at hand when the opportunities to get involved presented themselves? What is it about the culture that afforded what Timothy did? What did he do? How were his transactions consistent with his identity? What were his emotions like during the episodes you analyzed? If the purpose is to better understand Timothy's identity the way forward might be to better understand what he did and try to figure out why he did what he did.

Michael: And what he was undergoing while engaged in the face-to-face transactions, of which his identity and his Self are the result. If conversations and activities are collective processes, then we are not entirely in control of our identities in the same way that we are not in control of the conversation, for example, as it eventually is available to us in transcribed form. I think that my position is the direct consequence of speech act theory, where the speech act is distributed across turns; and if this is so, then who we are is the product of a relation of which we inherently cannot be in control.

Nancy: I think part of the reason why Timothy did not engage in the ways we would have liked is because the culture was simply too unfamiliar. I also think, however, that he had very little at stake with regard to his participation in science or in book club. Unlike Darren, he never attempts to position himself as a science-knowledgeable kid nor to position himself as having authority over the other boys in this context. He was happy at book club and voluntarily attended almost every meeting, but the substance of the books and the talks had little to do with his identity. If this had been a baseball game, however, he would likely have positioned himself as an athlete—and as the best athlete present. Timothy might have been far more engaged with Bryan and Greg's baseball physics. Baseball plays a much bigger role in Timothy's overall identity than science.

***

Michael: Nancy and Pam write about rather young children, which raises identity formation as an important issue for science educators to pursue given that children do not come to the world with identities. Philosophers such as Paul Ricœur have shown that our understanding of Self is mediated by our understanding of the Other, which has led him to the formula of *oneself* as *another*.

The notion of Self, and even understanding emotions and attitudes, cannot exist prior to our competency to view ourselves through the eyes of the other, which, following Jean Piaget, does not happen until the second half of a person's first decade in life. Perhaps the frequently small influence of childhood memories on identity construction points us to the greater roles played by discursive developments and interactions that we have later in life? Are, as Nancy and Pam ask, experiences of positioning and being positioned in book club meetings at an early age mere blips in the life of a person? Do they constitute mediating influences that we forget but which nevertheless tremendously influence the direction identity development takes? Or do they play no role at all?

Yew-Jin: In this chapter, positioning was understood as a rhetorical construction of the Self involving science texts. I wonder if identity construction among the kids occurred more frequently outside of these topic-consistent conversations? The verbal jousting between Adam and Darren (turn 99 to 110) for example was fascinating as it was sad from a conversation analytic perspective; their personal competencies in science were being challenged by the other at every single turn. But change the topic away from science texts, change the setting, and we get the same manifestations of social identity or pre-existing dispositions between them I dare say. This was also mentioned in the final section where we hear that the young boys and girls just really wanted to be young boys and girls. Thus, identity issues arising from the book club meetings are like smaller streams that feed into a larger river.

Ken: The example of Adam and Darren is a reminder for me of the porous nature of the boundaries of social fields. Adam and Darren did not come to the book club just to analyze books and talk science—and the relevant history was not just what happened in the last book club meet or the last time they watched a TV program on volcanoes. When Adam and Darren leave the book club they will no doubt meet in many other fields in their respective lifeworlds and when they do there may be an accounting.

Bryan & Greg: The age dynamic of these studies provides a very specific context from which to understand the relationship between discourse and identity. As young people, students learn what identities and associated discourse are appropriate for specific context. We learn to take on "inside" discursive identities that are specifically different from their "outside" discursive identities. In a state of rebellion or due to a lack of concentration, a child can easily take on an identity that is not supported by his or her current environment. Where science learning enters the picture is in how these identities are constructed in science classroom. More importantly, we must become concerned about not providing avenues for all students to construct discursive identities that support their science learning.

Nancy: The developmental question is indeed a very interesting one and gets at the very heart of what we mean by "identity." Does one have to be able to articulate an identity in order to have one? I would argue "no." One can act in certain ways and build particular competencies that are recognizable to others but are not part of any intentional identity building on the part of the child. A six-year-old child learning to play soccer would be unlikely to articulate the relevance of this activity to her identity as a girl. However, if she continues to play around with soccer over and over again, her experiences can become much more substantive than a mere blip in her life. These blips can become a pattern and thus become relevant to her identity whether she intends them to or not. Very recently a group of eight-year-old girls at a local elementary school explained to me how the four of them were the sports girls. They talk about sports and they dress differently from another group of girls that they derisively referred to as "girly girls." Clearly these girls are beginning to articulate a sense of themselves and how others may see them, but this competence will clearly grow in sophistication as the girls mature.

Although the way that we understand and use the construct of "identity" does not require a fully developed sense of "self," working with young children does pose some methodological challenges. Namely, interviews yield very little information. I would never claim that the children in the book clubs could have articulated for themselves the positions we described in our chapter. I would be very surprised if they even remembered much of it. But again, that does not mean this is all irrelevant. As long as there is enough short-term memory for experiences to have a cumulative impact, then they are at least potentially significant. I like the metaphor Yew-Jin uses—"identity issues arising from the book club meetings are like smaller streams that feed into a larger river."

Bryan & Greg: We tend to agree that the members of the book club, Jack at the hatchery, and the baseball players do not articulate an identity in an explicit academic manner. Nonetheless, for each of these chapters the identities are constructed through interaction, discourse, and positioning. Through being in the world identities are accomplished. The boys' book club identified for us a highly relevant issue not yet raised in the metalogue—the agenda and agency of the boys. As with previous research we have conducted, the boys in the book club have much at stake, and what is at stake for them is highly relevant to their lives and, thus, their emerging identities. The baseball players from our chapter were able to develop ways of talking about physics that was highly contextual and perfectly adequate for the purposes of understanding the complex trajectory of a curveball. This was in contrast to previous studies where Bryan found (even despite his efforts as a teacher) that students' perceived risks with talking science in school far outweighed for them any perceived advantages of school success. Whether students alienated from school science make conscious decisions in each instance is unlikely, but frankly, not easily known. Some students do not engage with science, not because of any inherent aversion, but because the perceived cultural costs are prohibitive. As academics we may parse out the nuances of a structure|agency dialectic, but such distinctions are far from bringing understanding to what it means to be a nine-year-old boy, a fish culturalist, or a high school baseball player.

Nancy: What the boys highlighted for us was that we were not in control of their agenda during book club. Bryan and Greg are absolutely correct that there was much at stake in the boys' book club and their agency in enacting their agenda could not be easily thwarted by the women leaders. We wonder if the ways the boys expressed their agency in titillating, physical and provoking ways—sometimes heading towards violence—is more easily recognized or harder not to notice than the more subtle exchanges in the girls' book club.

We were more successful in asserting some control with respect to the girls' agenda, and enjoyed some of the substantive conversations around science that occurred there (also present, yet rarer in the boys' club). That being said, the girls also have agency that may not be as easily recognized as it was for the boys. The agenda and agency is there—particularly pronounced in Hui Ying's interactions with Maude. Hui Ying at times seems to want book club to be very much like school—an environment in which she is comfortable and successful. Maude, on the other hand, is trying to change the agenda of book club in ways that make it far less schoolish.

Michael: I think I agree, for some of the ways our identities are shaped does not have to be conscious at all and this does not prevent you and me from articulating the processes from which identities emerge as products.

Ken: This is an example of what you mean by passivity—identity construction can be articulated through an agency|passivity relationship.

Pam: A relevant point here has to do with agency and the design of our book club. We hoped to create a space for the children that did not look like school, and thus did not have the same structures inherent in school systems (e.g., from hand-raising to snack limitations to acceptable answers when a teacher asks children to read and "respond to" a science book). We wanted the children to have agency to respond to the books in any way possible, opening up the possibility for children to question and, yes, be critical of the science books we read. Indeed, some children seemed relieved to be able to share with us their distaste for or emotive responses to a science book (e.g., Joli said "I didn't really read it—it didn't really interest me"), and, as we shared in our chapter, we witnessed the children critically responding to the texts. Agency! Voice! Yet the structure of school—as we admitted to ourselves—was omnipresent: we met in a university library, Nancy and I look like teachers and asked teacher-like questions, many of the children were third-grade classmates, we are adults and they are children. So the book club was not school, but so many of its structures remained that we wonder if the children's agencies to enact critical science identities were fully realized.

Bryan & Greg: An important aspect of this book club was clearly the desire to develop repertoires of critique. As suggested in Yew-Jin's chapter, repertoires develop over time through participation—Jack spent decades working to develop relevant knowledge and ways of being at the hatchery. Unfortunately, the repertoires of school are often compartmentalized so that students' otherwise relevant knowledge is not invited and evoked, and the continuity over time lost to the urgency of the banal moments of everyday life in school. Our chapter attempted to show how emerging identities, constructed over time and through interaction, were highly context specific, developed as needed to solve particular problems, and were accomplished through participation in a small collaborative community. Yet, we do not know how the game will proceed for the baseball players. While we would like to believe that the development of the baseball register (at least potentially) provides ways of developing other repertoires, this remains to be seen. We thoroughly agree with the stance developed in the book clubs that sought to foster repertoires of critique of science. Indeed, such critiques may form ways of being that develop competency across contexts and allow for border crossings.

Yew-Jin: Yes, as I said earlier, border crossings are much harder than we think! And since we're all talking about the effects of time on identity development, I revisited Jay Lemke's (2000) paper about timescales and he seems to say that identity develops on timescales that allow a sense of identity or habits of sign-interpretation to develop. Its growth may not be such a long term phenomenon although it cuts across different timescales with varying effects. Later on he differentiates between persona (literally "masks") and more stable identities ("selves") in which emotional investments are parked. If this is true, whether the reading clubs are blips that form patterns depends really on what and how long the kids make use of them, which was different for the boys and girls. This identity-over-time notion is similar to what Bryan and Greg meant when they talk about trajectories although their sense of the concept comes across as more agentic (i.e., selecting pathways) than the other two papers here. What I found was that Jack's use of workplace and school repertoires seems to be consistent (and probably an established aspect of his identity) over the years that we've interviewed him, which can be partially explained by the nature of the interview. Now, defining what *was* Jack's identity took second fiddle to when this or that identity became salient or how it was accomplished, although the former question is always lurking beneath somewhere.

Pam: Yew-Jin's insights about the *timescale* as it relates to this intersection of discourse and identity led Nancy and I to discuss the somewhat disheartening notion of the blip. Thinking again of our participants, we suspect that many of the children will approach science-related texts in much the same way as they did prior to our book club. Our knowledge of Timothy in various contexts and over many years suggests that he will largely approach texts uncritically, and is likely to be subject to them. Maude's independence and critical spirit evident in our book club was encouraged at home and operated frequently in school; she is likely to continue to be critical. We cannot know for sure if and how the seemingly stable identities in these children were influenced by our relatively short time together. We wonder if others, like Hui Ying, may be more apt to change, as she seemed to try on a slightly different and more critical identity in book club than we had known her to take on otherwise. Nancy and I wonder, too, about the baseball players about which Bryan and Greg wrote. Are their "tried on" identities as physics-talking baseball players constructs of the nature of interviews and physics classrooms, with future discourses likely to emphasize the more colloquial talk of baseball (of games, players, statistics) once the constraints of research and school are removed? We hesitate to suggest that physics will either be a focus of or absent from baseball talk for all of these children; as with our book club, variation is likely. At the same time that we acknowledge that baseball and book club experiences, discourses, and persona may be blips, we wish not to sound like disgruntled teachers in the faculty lounge. We value scientific understandings of the everyday and the need for students and citizens to approach science-related texts critically. While we simultaneously acknowledge the notion of blip, we also want to remind ourselves and others that the ideal of formal science education is to create many "blips," leading to more repeated trying on of identities and use of discourses that in time may fit.

REFERENCES

Agar, M. (1996). *Language shock: Understanding the culture of conversation.* New York: William Morrow.

Davies, B., & Harré, R. (1990) Positioning: The discursive production of selves. *Journal for the Theory of Social Behaviour, 20,* 43–63.

Berry, K. (2006). Research as bricolage: Embracing relationality, multiplicity, and complexity. In K. Tobin & J. L. Kincheloe (Eds.), *Doing educational research: A handbook* (pp. 87–115). Rotterdam: SensePublishers.

Freebody, P., & Luke, A. (1990). "Literacies" programs: Debates and demands in cultural context. *Prospect: The Australian Journal of TESOL, 5*(5), 7–16.

Lemke, J. L. (2000). Across the scales of time: Artifacts, activities, and meanings in ecosocial systems. *Mind, Culture, & Activity, 7,* 273–290.

Norris, S. P., & Phillips, L. M. (2003). How literacy in its fundamental sense is central to scientific literacy. *Science Education, 87,* 224–240.

KENNETH TOBIN, WOLFF-MICHAEL ROTH

IDENTITY IN SCIENCE

What-for? Where-to? How?

We planned this book as a forum for (a) presenting advanced scholarship on the relationship between science, learning, and identity an exchange between leading scholars and for (b) engaging the authors in a debate about the differences and similarities concerning their mutual understandings of this research. Perhaps sometimes explicitly, perhaps sometimes more implicitly, the different authors and author teams have articulated the ways in which science, learning, and identity are related and articulated. In this chapter, we sketch (a) some of the fundamental trends that arise for us from the chapters in this book and (b) some of the possibilities that the presented work opens up for future research and praxis in science education.

THE SALIENCE OF IDENTITY IN SCIENCE EDUCATION

Why the interest in identity? A perhaps most important reason is that there are many members of society who not only say "I am/was not good at science" and, more disconcertingly, "I hate(d) science." Such utterances are not just statements about some anonymous electron, proton, or other scientific object. These are statements about the speaker himself or herself. These are attributions used to make statements about Self, about identity. In part, such sentiments and attributions have arisen from forms of science education that reproduce hegemonic scientists' (journal-article) science and, in the process, imparts disinterest and even hatred in some parts of the population more so than in others. Among those who have suffered especially are female, Aboriginal, African American, and working-class students. All too often, the very structures by means of which science education comes about in schools—bells that arbitrarily chop the day into periods of equal length, rooms that disconnect this from that assemblage of students produced for institutional rather than learning purposes—also bring about the idea of knowledge as something disconnected from everyday experience. The regimes of knowledge are deeply aligned with the regimes of power (Foucault, 1979). Not only science education but also schooling as such, because they constitute a form of violence for many of those they supposedly serve and thereby mediate the identities they develop, need to be rethought and reworked in praxis. To do so requires us to understand how the dynamic and static aspects of identity emerge from activity and are constituted by means of different cultural-historical and sociocultural means individuals-in-situation have at their hand.

As science educators we seem aligned with the view that those who study science education can learn and build identities that reflect an affiliation with science. It is also possible that, through the study of science, participants might resist affiliation and reject what it stands for. Perhaps then it is about choice. Through science education individuals get to choose whether to affiliate with science or not. That seems all too rational and makes it seem as if one really can choose which identity to wear and which to cast aside. Whether or not we like it, we live in a world in which all of our identities have been inscribed by science just because we have been born to live in the world at this time. Just as we are gendered and inherit culture due to our ethnic trajectories, so it is that science has inscribed the identities of those who populate the planet. Though we can choose about many of the paths we follow, how to be in the world, there are aspects of identity that are inscribed without us being aware of the inscriptions or the ways in which they structure our lives—that is, our agency to write our own identities sits dialectically with social structures that also inscribe, structures beyond our consciousness that contribute to who and how we are. In each part of this volume we have shown that agency and passivity are dialectically related and it is through this dialectic that identities flow.

As science educators we adhere to a motive that science education can make a positive difference to the quality of life on earth. Yet there are many contradictions, which can be identified and become part of science education. For example, for many decades science educators have addressed issues associated with science and society and especially the interconnections between what is done locally and global phenomena. At all levels of schooling science has taught many *inconvenient truths* and students have apparently learned them in ways that may or may not have led to changed lives. Some have bought into recycling programs and many have explored ways to conserve energy and natural resources. However, global consumerism has rapidly accelerated and the destruction of fragile ecosystems has occurred in much the way scientists and organizations like the Club of Rome predicted so many years ago. The agency of individuals does not offset the practices of a collective that flow in accord with global ideologies that seem beyond the control of individuals and the reach of science education. It seems like only yesterday that highly influential politicians, like President Bush, denied the potential problems of global warming and promoted policies of consumerism that aligned with ideology of capitalism, providing contradictions to the exhortations of scientists. Some leading scientists claim that efforts were made to silence them when their research and scholarship provided arguments against political ideologies. Where does scholarly work on identity fit into the grand scheme of things? If scenarios like these are true and widespread we would hope that science education would play a role in resisting their hegemonic potential. Although science offers so much to contemporary life, also there are costs attributable to the advances of science. Can science education play a role in creating identities among citizens that would create new forms of science culture to address the problems of our age—including inequitable distribution and use of resources, poverty, warfare, and destruction of species and ecosystems?

As students participate in science they create identities that reflect the success of their transactions and the emotions associated with the enacted culture. Identities are coproduced with other forms of work and no individual's identity can be produced in isolation of others. Having said that, science-related moments of their identities are not only produced in science classrooms and participation in the lifeworld of an individual is salient—a reminder that insights into identity and participation in science cannot be fully understood by studies of classrooms in isolation of the other fields in which social life is enacted. However, what is done in a science classroom can become resources in the production and reproduction of identities and afford the enactment of science culture in other fields that comprise the lifeworld.

THEORIZING IDENTITY

The authors of this volume are in agreement that identity is not something that is fixed and stable, but that it varies with place and time and that there are some parts that seem to change more easily than others. In our theorizing of identity we have adopted sociocultural theory to illuminate issues in different contexts. One cluster of issues that is central to identity involves Self in relation to Other, individual in relation to collective and singular in relation to plural. What is clear is that for every currently living Self, Others came before; but even then, Self and Other can be understood only when thinking of the two in terms of multiple dialectical relations (see Figure 1 in the introduction). Hence being is always contingent on Others who came before and the associated historical culture into which a person is born. For this person, Self and Other co-emerge along a trajectory that subsequently, and partially, is captured in life narratives. Hence, identities are formed as life is lived in lifeworlds that already are shot through with meaning. It is not through agency that a baby learns to point and ask for what she wants. Embedded within a culture the baby inherits intentions, preferences and ways of being—at the very same time that she develops agency and pursues her individual goals and collective motives. The sense she makes of her day-to-day life, and hence her unfolding identity, flows from an agency|passivity relationship. Passivity occurs at macro- meso- and microlevels and occurs because structures that an individual is not conscious of interconnect with her agency. The dialectical relationship guards against determinism. Hence, the social class of the family into which an individual is born does not determine her identity, though it does play a part. Similarly, identity is not determined by ethnicity or gender, though they, too, may be salient as are processes such as racialization. For example, based on appearance alone, those with whom she interacts might regard a Filipina as "Asian" or even "Chinese." In fact, to better capture the essence of identity constructs, we need to think of "Filipina," "Asian," and "Chinese" as both resources and products of transactions rather than as stable adjectives that go with a person (physical body). Accordingly, she might emerge from a transaction as "Chinese"; hence, unconsciously racializing her and thereby altering her agency.

The problem with much of past and current theorizing of identity arises from the fact that learning and development are viewed in terms of stable cultures to which newcomers are acculturated or enculturated. Scientific cultures, too, thereby come to be theorized as containers into which students enter and of which they become a part by adopting the knowledgeable ways of discoursing and materially acting of those who already are present. With our theorizing we need to head toward ways of conceptualizing culture dynamically, which means, theorizing even the youngest members as constitutive of the collective. We can observe changes in the practices of couples and groups of people when a woman is pregnant; and these changed practices signify changes in culture as well. More so, the identities of individuals and the identities that are possible in a collective cannot be theorized independently because of the ways in which each of us is accountable to others for what we do. But because what we do is a resource for who we are, we are also accountable for our identities. Our actions, however, emerge in and as mediated by collective activity, which means, that we are both actively involved and passively subjected to the individual|collective production of our identities. Future theorizing of science, learning, and identity needs to develop better ways than we have today for capturing the aporetic aspects of identity. It can be expected that such theoretical developments allow us to understand ourselves in new ways, and therefore, allow theory to bear directly on practice. For example, if we not longer adhere to a model of the Self as the intentional monad that pursues and realizes its wants and desires but understand Self, interests, and desires as products of transactions then we no longer can expect to "have it our way," whether this is a science classroom, a science department meeting, or a school-based science teacher enhancement project. Rather, we—e.g., as science teachers, science methods professors—begin to understand the lessons as transactional events from which we (our Selves), as the students, emerge as products. As a teacher, I cannot *determine* the quality of the lesson, because it is the product of a collectivity. Theorizing and recognizing the radical passivity that is located at the very heart of agency provides us with new resources for bringing about science, learning, and identity in new ways.

RESEARCHING IDENTITY

It appears self-evident that the research methods we employ need to be consistent with the theoretical frameworks we select for our work. Thus, throughout this book the authors variously emphasize the irreducible nature of person-in-setting as the unit from which identities emerge. In cultural-historical approaches, the relevant unit theorized is *activity*, which is understood as a cultural-historical formation that serves the production and reproduction—therefore the survival—of society. Because of the existing divisions of labor, we may contribute to society in many different systems of activity (farming, fishing, schooling, manufacturing tools, trading, doing science research) and thereby develop different forms of identity. Most importantly, if activity is the irreducible unit, our research methods need to be such as to capture how forms of identity emerge from and are the results of the activity system as a whole, which nevertheless is realized concretely in the microlevel details of mundane everyday participation. Researchers need to have an eye both on this micro- and mesolevel detail that participants are conscious of and on the macrolevel structures that are not salient in the lifeworlds of participants but the effects of which become structures that reach into the lives of people. Thus, for example, we employ discourses that thereby are constitutive moments of the activities in which we participate. But because these discourses have evolved and have been shaped outside our lifeworlds, they are like Trojan horses, leading us to contribute to the reproduction of the inequitable conditions that are sources of our suffering (e.g., Smith, 1990). Institutional ethnography is one of the ways of investigating these Trojan horses designed for investigating the coordination of the different levels in which society as a whole and identities in particular are constituted. Institutional ethnography is a way of investigating "how the everyday world of our experience is put together by relations that extend vastly beyond the everyday (Smith, 2005, p. 1), that is, it also concerns those aspects that are not visible to us. It shares with Marxist and ethnomethodological approaches the commitment to begin and develop inquiry rooted in the very world that we constitute with our living bodies. It also resists the dominance of theory and therefore constitutes as much a social ethnography for people as it is about people including ourselves. Membership categorization analysis and the allied ethnomethodological and conversation analytic approaches constitute research tools adapted to finding identity-constitutive processes that simultaneously take into account the very micro-, meso-, and macrolevel aspects of situations that members draw on in the continuous production and reproduction of everyday, mundane, and "immortal" society.

Considerable work has been done in a domain referred to as *membership category analysis* (MCA), which reveals the different minute-to-minute ways in which coparticipants in everyday walks of life classify others and themselves (for a review see Hester and Eglin [1997]).[1] As members in society, we develop and have ways to find the very resources that we need to produce the regular and stable objects—including types of people—that we need to be able to anticipate the outcomes of the events in which we are constitutive parts and which are in part the results of our actions. Because we suggest future theorizing to move away from the static to more dynamic ways of theorizing identity, the ways in which we research needs to be consistent with the new theoretical resources that we build in the process. We must not only attend to the discourses people employ in the discursive constructions of identity, but also to the origins of these discourses and how they come to produce and reproduce ideologies and societal class structures.

PRAXIS OF SCIENCE EDUCATION

In traditional societies, children learn not by sitting in specially designed buildings and spaces where they produce and reproduce selection mechanisms, but by participating in the everyday practices that produce and reproduce their society. For example, girls in Mayan societies participate in gatherings around the bed of a pregnant woman or of a woman that recently has given birth. These girls already learn what they can draw on when they eventually become midwives (Jordan,

1989). Inuit children grow up quickly, through their greater responsibilities, through partaking in the hunt at a very young age, and, without doubt, through the production and reproduction of the binding ties of the traditional Inuit family (Waterman, 2001). Society, family, and individual are closely tied together and it is recognized that who someone can be is closely linked to an understanding of the collectivity as a whole. In "developed" nations and societies, however, an activity in its own right has emerged that is made responsible for the education of the young: schooling. The praxis of schooling does play a tremendous role in the production and reproduction of identities. There therefore is an utmost need to rethink and change our school-based practices and resources that contribute to the production and reproduction of the identities of our students.

There are ways in doing school science differently, as shown, for example, in chapter 8, where seventh- and eighth-grade students participate in the environmental activities in their community. Contributing to the community and to its salient needs that others also participate in addressing provides different forms of opportunities and makes available different resources for producing and reproducing identities—those of students and teachers alike. Much like the Inuit children, these students have opportunities to participate in producing and reproducing the ties that link members of this town in particular and society at large more generally. The lens of identity gives us a different perspective on the everyday teaching of science: participation in school science tasks for the sole purpose of producing grades and diplomas to be traded for other forms of symbolic, cultural, and economic capital (Roth & McGinn, 1998) is replaced by meaningful contribution to the working of a local community. The kinds of identity students develop by enacting caring relationship for the environment that constitutes a context for their community provides very different resources for the emergence and development of identities than memorizing concept definitions, facts, and theories from science textbooks.

As science educators consider their roles, especially in contexts characterized by diversity, a key opportunity arises to create social solidarity based on difference, rather than sameness. At issue is finding ways for persons having vastly different cultural repertoires to create legitimate culture that is interstitial and accepting of its insurgent character. In relation to the authorization of cultural hybridities, Homi Bhabha (2006) notes "borderline engagement of cultural difference may as often be consensual as conflictual; they may confound our definitions of tradition and modernity; realign the customary boundaries between the private and the public, high and low; and challenge normative expectations of development and progress" (p. 3). Moral solidarity may be central when differences within a field are strident, as often happens in urban schools where students differ in regard to categories such as ethnicity, social class, and gender—thereby having quite different cultural resources on which to build, for example, knowledge of science. In the process of transactions identities are at stake and the emotional climate seems paramount. How can all learners maintain their differences and hence individuality while creating interstitial forms of culture that are legitimized as creolized sciences? Sameness is not central to these questions. In contrast, success leading to the production of positively valenced emotions and receiving and giving respect seem paramount. At the heart is acceptance of difference as legitimate and normal—acknowledging that all participants have a right, not just to visit, but to stay, succeed, and retain their differences (Derrida, 2006). We see the creation of moral solidarity around differences as a challenge because so much of science education has focused on producing canonical forms of science (i.e., sameness)—as if there is only one viable form. The epistemological and ontological shifts needed to embrace differences will surely shake the foundations of where, when, and how science education is practiced and even what counts as science education. Given the lamentable history of science education, a history grounded in producing coherence and stamping out contradictions, a history reflecting oppression and inequity of cultural others the roads toward equitable science and science education might be hard to construct. Projections extending forward from the present argue for cultural brokering of a sort that acknowledges what all comers bring to the table and listens to the voices of Others as efforts are made to disconnect social violence from the practices of science and science education. Different lenses are needed to discern what can count as science and what roles science can play in shaping our future.

NOTES

1 An extensive bibliography concerning membership categorization analysis can be found at the URL www.paultenhave.nl/MCA-bib.htm. (Accessed March 02, 2007)

REFERENCES

Bhabha, H. K. (2006). *The location of culture*. New York: Routledge.
Derrida, J. (2006). *On cosmopolitanism and forgiveness*. New York: Routledge.
Foucault, M. (1979). *Discipline and punish: The birth of the prison*. New York: Vintage Books.
Hester, S., & Eglin, P. (Eds.). (1997). *Culture in action: Studies in membership categorization analysis*. Washington, DC: International Institute for Ethnomethodology and Conversation Analysis & University Press of America.
Roth, W.-M., & McGinn, M. K. (1998). >unDELETE science education: /lives/work/voices. *Journal of Research in Science Teaching, 35*, 399–421.
Smith, D. E. (1990). *Conceptual practices of power: A feminist sociology of knowledge*. Toronto: University of Toronto Press.
Smith, D. E. (2005). *Institutional ethnography: A sociology for people*. Lanham, MD: Altamira.
Waterman, J. (2001). *Artic crossing: A journey through the Northwest Passage and Inuit culture*. New York: A. A. Knopf.

INDEX

A

Activity: collective, 84, 163, 168, 248, 251, 254, 342; cultural-historical, vii, 9, 62, 101, 149, 150, 155, 163, 167, 168, 183, 184, 185, 188, 189, 190, 199, 242, 243, 244, 248, 250, 252, 253, 339, 342; cultural-historical activity theory, vii, 9, 62, 101, 149, 150, 155, 163, 167, 168, 183, 184, 185, 188, 189, 190, 199, 242, 243, 244, 248, 250, 252, 253, 339, 342; system, 13, 14, 149, 156, 164, 183, 191, 201, 207, 241, 244, 246, 249, 254, 342, 343; theory, 149, 150, 163, 183, 184, 188, 203, 204, 205, 206, 207, 243, 244, 245, 248, 252, 253
African American, 3, 13, 15, 16, 17, 18, 19, 20, 21, 27, 33, 34, 35, 38, 45, 60, 81, 100, 101, 121, 123, 125, 127, 128, 129, 131, 133, 143, 204, 253, 339
Agency, 2, 3, 4, 5, 14, 16, 18, 19, 24, 29, 37, 39, 40, 44, 46, 51, 52, 53, 62, 63, 71, 85, 89, 90, 91, 93, 95, 117, 118, 125, 126, 133, 135, 136, 140, 149, 155, 162, 163, 164, 169, 170, 173, 178, 179, 184, 192, 194, 248, 252, 253, 254, 255, 262, 285, 287, 314, 326, 327, 328, 329, 331, 332, 334, 335, 340, 341, 342
Alienating youth, 95
Anderson, E., 18, 20, 21, 25, 35, 40, 123, 133
Antaki, C., 263, 282
Anthropogenesis, 4, 5, 168
Anthropology: cultural, 103, 105
Anyon, J., 244
Aporia, 2, 172, 173
Appropriation, 14, 225, 246, 283, 294, 295, 296, 330
Artifact, 92, 138, 167, 203, 204, 206, 207, 211, 215, 216, 227, 228, 230, 231, 236, 239, 240, 241, 243, 247
Asian American, 45, 290
Authenticity, 46, 89
Authority, 22, 24, 30, 39, 55, 56, 59, 121, 214, 274, 306, 309, 310, 311, 314, 316, 317, 318, 320, 321, 322, 327, 329, 331, 332
Auto/biography, 1, 2, 4, 6, 9, 10, 175, 184, 202, 302

Auto/ethnography, 1, 10, 184, 189, 190, 191, 202

B

Bakhtin, M. M., 154, 163, 181, 183, 199, 202, 240, 242, 251, 255
Barton, A., 100, 102
Bhabha, H., 191, 202, 344, 345
Body, 2, 3, 4, 17, 28, 35, 36, 44, 126, 155, 163, 169, 172, 176, 177, 178, 186, 187, 188, 191, 192, 195, 196, 202, 207, 218, 219, 220, 245, 253, 254, 262, 280, 311, 320, 341
Boundary, 13, 21, 36, 40, 44, 51, 60, 82, 111, 125, 126, 131, 161, 162, 187, 188, 189, 190, 191, 192, 193, 199, 201, 207, 211, 222, 245, 246, 248, 253, 280, 333, 344; porous, 19, 26, 29, 122
Bourdieu, P., 3, 5, 9, 18, 52, 62, 133, 135, 149, 187, 247, 255
Brickhouse, N., 4, 63, 79, 259, 260, 261, 282, 301, 303, 325
Bricolage, 13, 14, 90, 102, 330, 337; bricoleur, 261, 277
Bucciarelli, L. L., 108, 118

C

Capital, 19, 22, 23, 26, 29, 33, 39, 62, 83, 85, 86, 87, 88, 101, 126, 127, 133, 262, 344; exchange, 23, 29, 33; production, 23; social, 23, 33, 54, 126, 133; social and symbolic capital, 23, 33, 52; symbolic, 19, 22, 23, 24, 39, 55, 86, 101, 123, 182, 328
Category, 5, 22, 25, 35, 44, 50, 64, 85, 88, 103, 105, 106, 111, 115, 118, 123, 232, 240, 266, 271, 273, 275, 282, 323, 344
Cixous, H., 101
Cogenerative dialogue, 121, 124, 127, 129, 130, 131, 132, 133, 134, 136, 137, 138, 141, 144, 146
Cognition: cognitive apprenticeship, 297, 298; cold, 188; hot, 187
Cole, M., 206, 242

INDEX

Collaboration, 39, 46, 101, 165, 207
Collins, P. H., 123, 133, 142
Collins, R., 40, 50, 51, 52, 55, 150
Communalism, 33
Communication, 4, 6, 32, 36, 89, 115, 172, 180, 181, 187, 188, 189, 191, 192, 196, 199, 205, 216, 243, 251, 263, 265, 283, 284, 285, 288, 289
Community (of practice), 26, 27, 32, 33, 36, 40, 43, 44, 51, 54, 61, 63, 64, 65, 66, 70, 72, 73, 78, 84, 85, 88, 90, 93, 95, 104, 105, 110, 114, 117, 125, 126, 127, 132, 138, 139, 150, 156, 158, 160, 161, 162, 165, 168, 170, 171, 172, 177, 189, 203, 204, 206, 207, 209, 216, 221, 226, 229, 242, 247, 253, 268, 282, 283, 284, 285, 286, 288, 289, 305, 327, 336, 344
Consciousness, 5, 15, 16, 17, 26, 34, 36, 37, 62, 86, 100, 114, 121, 123, 124, 133, 136, 140, 154, 164, 165, 166, 168, 169, 184, 186, 189, 192, 199, 200, 202, 205, 208, 245, 247, 250, 253, 254, 274, 325, 326, 328, 330, 334, 335, 340, 341, 343
Constructivism, 5, 149
Consumption, 254
Continuity, 2, 76, 187, 189, 285, 290, 295, 335
Contradiction, 13, 14, 25, 33, 34, 39, 88, 123, 135, 136, 137, 141, 172, 254, 340, 345
Coteaching, 17, 18, 25
Counter-hegemony, 113, 117, 143
Countertransference, 190, 254
Critical race theory, 144
Culture, 3, 4, 6, 9, 14, 15, 16, 17, 18, 20, 21, 24, 26, 28, 29, 30, 33, 34, 35, 36, 37, 38, 39, 81, 82, 85, 87, 88, 89, 93, 103, 104, 105, 106, 108, 112, 113, 114, 115, 116, 117, 118, 122, 123, 126, 127, 129, 131, 132, 135, 136, 137, 138, 139, 140, 141, 142, 143, 149, 160, 167, 168, 177, 185, 187, 189, 190, 191, 194, 201, 202, 246, 249, 250, 251, 259, 260, 261, 264, 265, 267, 269, 272, 273, 274, 276, 277, 281, 282, 286, 289, 297, 298, 325, 326, 330, 331, 332, 337, 340, 341, 342, 344, 345; Cultural studies, 247, 326; mainstream, 35, 36, 123; popular, 36, 37; reproduction, 6, 328; subculture, 26, 28, 36, 39, 201

D

Damasio, A., 169, 183, 188, 202
Deficit (lenses), 16, 17, 36, 78, 87, 94, 270
Deleuze, G., 100, 102
Demonstration, 41, 44, 46, 47, 62, 113, 114, 204

Derrida, J., 8, 9, 99, 251, 255, 344, 345
Descartes: Cartesian approaches, 188; Cartesian position, 99
Determinism, 20, 31, 34, 167, 185, 286, 341
Devereux, G., 190, 202
Dialectics, 1, 2, 3, 4, 14, 15, 37, 52, 53, 78, 115, 123, 125, 135, 141, 149, 150, 165, 166, 171, 172, 183, 187, 188, 189, 192, 206, 245, 249, 251, 252, 253, 254, 255, 326, 328, 329, 331, 332, 334, 340, 341
Diaspora, 14, 90, 330
Dis/continuity, 185, 187, 188, 189, 201
Discourse: analysis, 325; discursive achievement, 325; discursive device, 274, 275, 277, 280, 282; discursive moves, 42; discursive space, 223, 226; practices, 283, 285, 287, 288, 291
Disposition, 24, 62, 122, 124, 125, 133, 187, 195, 301, 315, 322, 333
Distinction: symbolic, 130
Division of labor, 149, 168, 172, 188, 203, 206, 212, 216, 243, 245, 247
Domination: symbolic, 131

E

Eckert, P., 116, 118, 284, 298
Economy: economic realities, 88, 92, 93
Education: engineering, 100, 103, 104, 106, 108, 115, 135, 139, 140, 142, 145; multicultural, 138
Egology, 99, 249
Eisenhart, M. A., 105, 117, 118, 142, 146
Emotion, 3, 15, 16, 17, 18, 19, 23, 33, 51, 94, 121, 150, 168, 169, 181, 185, 186, 187, 188, 189, 191, 192, 194, 195, 196, 198, 199, 200, 201, 246, 248, 251, 328, 329, 332, 333, 341, 344; affective stance, 192; bonds, 93; disconnectedness, 186; emotional-volitional, 3, 149, 153, 154, 155, 156, 162, 168, 170, 172, 178, 180, 181, 182, 183, 244, 245; energy (EE), 43, 44, 50, 51, 53, 54, 55, 57, 59, 60, 61, 85, 94; expressiveness, 130; negative emotional valence, 153, 155, 179, 181, 195; positive emotional valence, 153, 169, 189; valence, 121, 155, 168, 169, 179, 182, 192, 194, 195, 196, 199, 329
Empowerment, 91
Engeström, Y., 206, 207, 242, 252, 255
Enslavement, 35
Entity: collective, 189
Entrainment, 44, 50, 51, 52, 54, 60
Environment: environmentalism, 153, 155, 156, 157, 159, 160, 162, 163, 164, 167,

348

169, 170, 171, 172, 173, 175, 176, 177, 178, 179, 180, 181; safe, 83
Epistemology, 100, 101, 141, 329
Ethics: ethico-moral dimensions, 3, 149, 153, 154, 155, 162, 163, 172, 178, 179, 180, 181, 182, 183, 244, 245; ethico-moral sense, 154
Ethnicity, 36, 123, 138, 263, 341, 344
Ethnography, 17, 18, 30, 65, 190, 191; practices, 105; videoethnography, 24, 26, 27, 28, 29, 31, 36, 66
Ethnomethodology, 191, 281, 343
Ethos, 165, 178, 185, 187, 198, 201
Everyday life, 1, 7, 62, 69, 79, 84, 92, 103, 117, 139, 154, 182, 191, 204, 267, 325, 335
Expectation: cultural, 110
Experiential, 6

F

Face: face-to-face, 103, 104, 150, 192, 196, 199, 332; face-work, 43, 83
Facticity, 170, 172, 195, 199, 249
Favelas, 88
Field, vii, 2, 14, 18, 19, 20, 21, 23, 24, 26, 27, 29, 31, 33, 37, 39, 46, 54, 62, 78, 85, 91, 95, 99, 105, 110, 111, 112, 121, 122, 123, 124, 125, 131, 133, 136, 138, 139, 140, 141, 144, 146, 149, 150, 161, 165, 167, 191, 245, 246, 248, 249, 252, 261, 290, 304, 327, 328, 333, 341, 344; cultural, 17, 35, 121, 122, 127
Flesh, 2, 3, 4, 5, 6, 93, 178, 186, 187, 188, 190, 192, 194, 195, 200, 201, 202, 253, 264; fleshly manifestation, 192
Fluency, 13, 22, 24, 26, 29, 33, 34, 37, 41
Focus: mutual, 44, 50, 51, 54, 60
Foucault, M., 135, 326, 339, 345
Fox-Keller, E., 100

G

Garfinkel, H., 265, 282
Gaze, 139, 193, 195, 196, 207, 246
Gee, J., 206, 242, 286, 289, 298
Gender, 20, 43, 44, 52, 78, 85, 88, 94, 99, 100, 101, 102, 104, 105, 108, 109, 115, 116, 118, 121, 122, 123, 124, 125, 130, 131, 138, 142, 143, 144, 252, 262, 340, 341, 344; black feminist, 129, 144; femininity, 101, 108, 109, 113, 117, 122, 124, 125, 127, 131, 142, 322; feminism, 100, 102, 130, 133, 134, 136, 142, 143, 144, 145, 345; gender (and heterosexual) stereotypes, 130; inequality, 131
Ghettos, 81
Giddens, A., 1, 9, 149, 174, 183, 247, 255
Glaser, B., 298, 305, 323
Global, 19, 21, 36, 340
Goffman, E., 43, 56, 62, 63, 79, 83
Grounded theory, 305, 323
Guattari, F., 100, 102
Guba, E., 46, 105, 118

H

Habitus, 18, 62, 122, 123, 124, 125, 131, 133, 149, 187
Hall, S., 36, 37, 184, 202, 282
Halliday, M. K., 284
Hands-on, 41, 46, 105, 114, 115, 203, 204, 209, 227, 229, 267, 269, 270, 281, 318
Haraway, D., 100
Harding, S., 100
Harré, R., 206, 242, 260, 302, 323, 327, 337
Hegel, G.W.F, 167, 183, 201, 202, 249, 255
Hegemony, 121, 143, 246, 332, 339, 340
Heidegger, M., 191, 202, 250
Hermeneutics, 326
Heterogeneity, 13, 14, 40, 64, 248, 329
Heyerdhal, T., 261
History, vii, 10, 35, 37, 59, 64, 95, 117, 118, 139, 161, 164, 172, 175, 178, 226, 271, 282, 286, 318, 333, 345
Holland, D., 90, 95, 105, 115, 117, 118, 172, 184
Holzkamp, K., 168, 182, 184, 189, 202
Husserl, E., 99, 192, 249
Hybridity, 13, 14, 21, 30, 33, 34, 36, 39, 40, 92, 125, 129, 132, 133, 137, 331; hybrid cultures, 36; hybridization, 94, 138

I

Identity: affinity, 287, 295; categorical, 43, 52, 59, 60, 85; discourse-identity, 286; discursive, 205, 239, 240, 242, 289, 298, 325; idem, 2, 150; institutional, 286; ipse, 2; nature-identity, 286
Ideology, 32, 39, 89, 91, 104, 106, 109, 115, 116, 117, 121, 136, 140, 141, 246, 332, 340, 343
Immigrant, 38, 54; immigration, 14, 36, 185, 191
Immigration: migrant, 185
Institution: level, 244
Intention, 135, 143, 155, 168, 170, 231, 250, 252, 254, 329, 330, 341

INDEX

Interaction, vii, 10, 14, 16, 21, 23, 24, 25, 26, 27, 28, 29, 31, 33, 36, 37, 41, 42, 43, 44, 45, 47, 50, 51, 52, 53, 54, 55, 56, 57, 60, 61, 62, 82, 83, 103, 104, 114, 118, 121, 130, 132, 139, 160, 163, 173, 183, 188, 191, 192, 196, 198, 199, 204, 205, 206, 207, 208, 226, 233, 238, 244, 246, 248, 278, 281, 283, 285, 286, 287, 295, 302, 303, 307, 317, 320, 325, 333, 334, 335, 336; discursive, 205, 206; interactional, 43, 50, 53, 55, 61, 206, 209, 266, 268, 280, 307; ritual (IR), 14, 41, 44, 47, 50, 51, 52, 53, 54, 55, 57, 59, 60, 61, 83; successful, 30
Interrogation: critical, 101
Intersubjectivity, 8, 47, 189, 190, 192, 249, 251
IRE, 57, 83

K

Kant, I., 99
Knowledge: conceptual, 154, 297; cultural, 110, 113, 139; stocks of, 16, 18, 19, 26, 30, 37, 38, 39, 82, 123
Knowledgeability, 68, 70, 72, 73, 154, 160, 165, 166, 169, 181, 214, 220, 222, 235, 243, 262, 266, 270, 281, 310, 318, 321, 322, 342; knowledgeable participation, 154
Kuhn, T., 138, 140

L

Laboratory, 41, 44, 131, 149, 150, 154, 160, 171, 189, 191, 193, 194, 195, 196, 197, 198, 199, 200
Labov, W., 288, 289, 298
Languaculture, 325
Latino, 45, 204, 230
Lave, J., 6, 10, 71, 79, 109, 115, 118, 206, 242, 262, 282, 297, 298
Learning, vii, 1, 4, 8, 9, 10, 13, 14, 17, 19, 25, 29, 31, 38, 39, 41, 42, 46, 52, 60, 62, 63, 64, 65, 67, 70, 71, 72, 75, 77, 78, 79, 82, 84, 87, 93, 95, 100, 104, 109, 114, 115, 116, 118, 121, 122, 123, 124, 125, 126, 129, 130, 131, 132, 133, 138, 140, 144, 149, 150, 155, 156, 160, 169, 172, 174, 178, 181, 185, 187, 188, 190, 191, 200, 203, 204, 207, 209, 228, 231, 236, 237, 238, 240, 242, 243, 244, 245, 250, 251, 253, 255, 262, 263, 264, 265, 266, 267, 268, 269, 270, 271, 273, 274, 275, 276, 277, 278, 280, 281, 282, 283, 285, 286, 287, 288, 289, 290, 295, 296, 297, 298, 299, 301, 322, 325, 327, 330, 333, 334, 339, 342; classroom, 245, 285, 288; formal, 280
Lemke, J., 53, 62, 63, 77, 78, 79, 287, 298, 321, 322, 323, 336, 337
Leont'ev, A. A., 155, 189
Leont'ev, A. N., 149, 249
Lessing, D., 139
Level: collective, 1, 6, 81, 163, 168, 192, 204, 244, 245, 252
Levinas, E., 189, 200, 202, 255
Lifeworld, 19, 24, 25, 26, 33, 37, 38, 82, 83, 85, 88, 89, 94, 123, 141, 189, 191, 192, 246, 250, 260, 330, 333, 341, 343
Lincoln, Y., 46, 105, 118
Longino, H., 100
Luke, A., 302, 323, 327, 337

M

Macro, 38, 136, 246, 328, 332, 341; macroculture, 36; macrostructure, 37, 137
Margin, 254, 295, 296
Marx, K., 183, 201, 243, 245, 254, 255, 263
Masculinity, 100, 108, 124, 125, 130, 131, 143; masculine stereotype, 131
Materiality, 196, 201; material context, 189
Meaning, 95, 110, 117, 138, 143, 145, 167, 175, 203, 205, 207, 208, 209, 216, 218, 244, 249, 250, 262, 278, 279, 284, 286, 288, 299, 341
Member: membership categories, 266, 267, 269, 271, 273, 275, 276, 278, 280, 329
Meso, 38, 136, 137, 141, 328, 341, 343; mesostructure, 121, 122
Metalogue, 142, 247, 250, 327, 334
Micro, 38, 326, 328, 343
Middle class, 16, 25, 30, 34, 35, 36, 39, 82, 84, 85, 87, 92, 122, 163, 290
Mikhailov, F., 1
Minority students, 5, 283, 284, 285, 289, 295, 296, 297, 298, 299, 330
Model: cultural, 66, 69, 70, 78, 118
Mood, 44, 47, 192; collective, 192
Morality, 172
Motive, 155, 156, 162, 163, 164, 167, 170, 182, 189, 206, 237, 245, 249, 250, 252, 254, 340
MOVE, 13, 81, 90, 91, 221, 222

N

Nancy, J.-L., 4, 5, 10, 167, 184, 249, 250, 255, 301, 305, 325, 326, 327, 328, 331, 332, 333, 335, 336

Narrative, 1, 2, 6, 9, 37, 176, 177, 178, 187, 247, 248, 260, 341
National Research Council, 41, 62
National Science Foundation, 29, 40, 140, 241
National Standards in Science Education, 140
Nietzsche, F., 99
No Child Left Behind, 88, 140
Non-vernacular, 288, 289, 291, 293, 294

O

Objectivity, 284; objectivism, 190
Ontology, 15, 136, 141
Operation, 155, 164, 165, 166, 167, 170, 172, 178, 188, 189, 250
Oppression, 95
Other, 6, 9, 50, 61, 103, 123, 161, 162, 188, 190, 199, 200, 211, 214, 280, 287, 332, 341; othermother, 125, 126, 127, 131, 144; otherness, 4, 16, 99, 186, 193

P

Participation, 5, 10, 14, 21, 24, 25, 30, 31, 36, 41, 43, 46, 50, 60, 63, 64, 65, 67, 69, 72, 73, 75, 77, 78, 79, 83, 85, 86, 94, 95, 105, 118, 130, 132, 136, 138, 144, 149, 156, 159, 160, 162, 163, 167, 169, 172, 174, 176, 182, 205, 206, 218, 227, 235, 239, 242, 245, 247, 248, 250, 252, 262, 283, 289, 291, 295, 302, 303, 305, 314, 318, 322, 330, 332, 335, 341, 343, 344
Passivity, 2, 3, 4, 5, 124, 135, 140, 164, 170, 184, 192, 194, 243, 252, 253, 254, 255, 321, 329, 330, 331, 332, 335, 340, 341, 342
Pathos, 178, 201
Pedagogy, 88, 99, 130, 133, 134, 286
Persona, 39, 139, 336, 337
Phallogocentrism, 99, 100, 101, 102
Phenomenology, 2, 3, 202, 251
Plural, 6, 99, 138, 141, 143, 166, 177, 178, 341; plurality, 102, 141, 150, 162
Position, 4, 14, 37, 38, 43, 45, 51, 52, 56, 60, 63, 64, 67, 69, 70, 72, 73, 76, 78, 79, 82, 85, 88, 89, 90, 99, 100, 101, 115, 128, 133, 135, 142, 145, 149, 158, 174, 178, 179, 180, 182, 196, 203, 205, 206, 207, 215, 220, 221, 222, 229, 232, 235, 238, 252, 253, 254, 260, 262, 269, 277, 283, 284, 285, 288, 290, 291, 293, 296, 302, 309, 310, 313, 314, 315, 317, 318, 319, 320, 322, 326, 327, 331, 332, 333, 334; positioning, 5, 63, 64, 66, 93, 135, 206, 209, 216, 226, 227, 228, 248, 252, 253, 285, 302, 304, 305, 306, 307, 313, 314, 315, 318, 321, 325, 326, 327, 329, 331, 333, 334
Positivism, 140, 141
Possibility, vii, 3, 9, 13, 44, 47, 52, 59, 64, 72, 77, 78, 81, 94, 99, 109, 150, 167, 176, 177, 179, 192, 200, 246, 253, 266, 326, 335
Postmodernism, 1
Post-positivism, 141
Poverty, 15, 21, 22, 23, 33, 35, 36, 38, 65, 81, 85, 86, 87, 89, 90, 340; high-poverty, 45; low-income, 31, 45, 91; poor, 14, 20, 21, 22, 34, 45, 63, 64, 65, 78, 81, 85, 87, 89, 93, 142, 261, 318
Power, 1, 19, 24, 25, 27, 37, 42, 87, 94, 100, 102, 104, 105, 109, 110, 112, 114, 115, 118, 128, 135, 140, 142, 144, 145, 146, 163, 164, 170, 172, 179, 207, 279, 280, 281, 317, 318, 326, 329, 331, 339, 345; struggle, 317, 318
Practice, 5, 16, 17, 18, 64, 71, 83, 87, 100, 102, 106, 109, 112, 113, 116, 121, 122, 123, 124, 125, 126, 129, 130, 131, 132, 133, 136, 137, 138, 141, 144, 146, 154, 167, 171, 172, 207, 240, 244, 245, 246, 263, 265, 283, 284, 285, 286, 287, 288, 289, 290, 291, 296, 301, 326, 329, 330, 340, 342, 343, 345; cultural, 5, 114, 136, 187, 188, 191, 244, 245, 286, 287; mundane everyday, 262
Praxis, 14, 18, 19, 33, 35, 36, 62, 123, 149, 150, 155, 162, 163, 164, 165, 167, 168, 171, 173, 175, 180, 181, 182, 184, 192, 201, 202, 255, 339, 343; concrete, 165, 167, 171, 172, 182
Pregnancy, 65
Production, vii, 4, 6, 8, 19, 23, 38, 76, 99, 110, 114, 117, 118, 135, 139, 160, 162, 169, 173, 174, 189, 192, 193, 194, 201, 202, 238, 242, 243, 247, 248, 251, 252, 254, 259, 260, 264, 302, 323, 328, 337, 341, 342, 343, 344; cultural, 23, 36, 37, 71, 79, 118, 139, 142, 245
Prosody, 28, 196, 202

R

Race, 35, 36, 43, 44, 45, 52, 59, 83, 85, 88, 93, 94, 123, 145, 146, 185, 295; racialized, 252
Rap, 15, 27, 28, 33, 34, 36
Rawls, J., 143
Read-aloud, 218, 223, 226, 227, 228, 241, 244
Reference, 8, 28, 40, 44, 54, 59, 75, 104, 106, 117, 143, 166, 167, 236, 325

INDEX

Reflexivity, 190, 192, 255, 302, 307, 309, 310, 313, 314, 317, 321, 329; reflexive positioning, 307, 321
Repertoire, 260, 266, 267, 268, 269, 270, 271, 272, 273, 274, 275, 276, 277, 278, 280, 282, 283, 326, 328, 329, 330, 335, 336, 344
Reproduction: cultural, 137
Researcher: youth researcher, 29
Resistance: collective, 39
Resource, 1, 6, 9, 13, 14, 18, 19, 20, 21, 23, 24, 30, 33, 36, 37, 38, 39, 45, 50, 52, 59, 60, 64, 68, 81, 86, 87, 90, 91, 93, 101, 123, 138, 140, 141, 149, 156, 159, 163, 164, 167, 170, 174, 179, 187, 188, 189, 191, 192, 195, 196, 198, 201, 203, 204, 222, 238, 246, 253, 259, 260, 262, 268, 274, 276, 297, 317, 320, 321, 331, 340, 341, 342, 343, 344
Respect, 7, 20, 21, 22, 23, 24, 25, 26, 29, 30, 31, 33, 34, 37, 39, 64, 72, 78, 82, 86, 87, 88, 92, 108, 123, 127, 128, 135, 136, 138, 142, 160, 169, 170, 174, 181, 185, 187, 189, 192, 196, 254, 273, 277, 278, 296, 304, 320, 327, 328, 331, 335, 344
Responsibility, 65, 75, 129, 130, 132, 145, 154, 156, 160, 171, 172, 183, 194, 198, 200, 223, 244, 298; collective, 146
Ricœur, P., 2, 4, 8, 10, 144, 146, 171, 172, 182, 183, 184, 247, 332
Roth, W.-M., 1, 3, 6, 10, 14, 18, 19, 36, 38, 40, 41, 53, 62, 81, 100, 102, 135, 149, 153, 155, 156, 170, 175, 184, 185, 188, 189, 190, 202, 243, 253, 254, 255, 259, 262, 271, 282, 325, 344, 345
Rule, 24, 70, 126, 181, 188, 203, 206, 211, 213, 217, 221, 228, 229, 231, 240, 243, 246, 325, 326

S

Sacks, H., 265, 282
Saxe, G., 287, 299
Scaffolding, 297, 298
Schema, 3, 18, 39, 53, 125, 141, 246, 331, 332
School, 2, 3, 6, 13, 14, 15, 16, 18, 19, 20, 21, 22, 23, 24, 25, 26, 29, 30, 31, 32, 35, 36, 39, 44, 45, 46, 64, 65, 66, 67, 69, 70, 71, 72, 79, 81, 82, 83, 84, 86, 88, 89, 90, 92, 93, 94, 95, 100, 101, 108, 109, 113, 116, 118, 121, 122, 123, 124, 125, 126, 127, 130, 131, 133, 136, 137, 138, 139, 141, 142, 143, 154, 155, 156, 157, 158, 160, 162, 163, 174, 180, 181, 182, 203, 205, 222, 231, 241, 243, 244, 245, 249, 253, 259, 261, 262, 264, 267, 268, 269, 270, 271, 272, 273, 274, 275, 276, 277, 278, 280, 283, 286, 287, 290, 291, 295, 299, 301, 302, 303, 304, 305, 307, 308, 313, 314, 316, 318, 320, 321, 322, 328, 330, 334, 335, 336, 339, 344; school-based learning, 272, 276; schooling, 13, 29, 39, 62, 72, 84, 87, 89, 93, 101, 118, 121, 123, 129, 140, 143, 146, 163, 175, 185, 243, 244, 245, 263, 267, 269, 275, 276, 285, 339, 340, 342, 343
Schutz, A., 246, 247, 250, 255
Science: canonical, 94, 138, 276, 331, 344; community, 43, 56, 59, 127, 329; content, 42, 43, 44, 47, 55, 59, 61, 83, 84, 85, 94, 209, 294; creolized, 94, 138, 329, 330, 331, 344; curriculum, 142; discourse, 42, 46, 54, 83, 284, 285, 287, 293, 294, 295, 296, 297, 330; language, 42, 43, 44, 47, 49, 54, 55, 56, 57, 58, 59, 60, 61, 82, 83, 85; literacy, 101, 102, 154, 155, 162, 177, 182, 183, 184, 323, 337; protocol, 71; vocabulary, 42, 47, 49, 59, 62, 85
Self, 1, 2, 3, 4, 5, 6, 9, 13, 16, 26, 27, 37, 51, 62, 67, 69, 70, 72, 73, 76, 77, 79, 88, 90, 91, 99, 113, 115, 121, 135, 136, 154, 167, 172, 173, 174, 175, 176, 183, 187, 188, 200, 202, 229, 236, 247, 248, 249, 255, 302, 332, 333, 334, 339, 341, 342; selfhood, 240, 241
Sense, 3, 4, 5, 14, 16, 18, 26, 29, 33, 34, 36, 37, 38, 44, 47, 50, 51, 53, 54, 55, 56, 59, 60, 68, 72, 73, 74, 76, 77, 84, 88, 90, 91, 92, 103, 105, 109, 115, 121, 124, 127, 137, 138, 139, 140, 141, 142, 143, 150, 163, 165, 167, 171, 172, 174, 181, 183, 186, 187, 188, 189, 190, 191, 192, 194, 195, 196, 200, 201, 202, 203, 204, 207, 208, 209, 212, 213, 216, 218, 222, 223, 228, 229, 230, 232, 240, 245, 246, 247, 248, 249, 250, 251, 253, 254, 262, 275, 276, 301, 304, 312, 326, 327, 329, 334, 336, 337, 341; sense making, 121, 204, 216, 222, 228, 229, 230
Sewell, W., 149, 247
Sexuality, 122, 123; sexually active, 123, 133
Shantytowns, 81
Sign, 4, 16, 19, 22, 23, 33, 123, 162, 194, 202, 250, 288
Singular, 1, 3, 6, 9, 10, 38, 62, 89, 99, 138, 141, 162, 166, 172, 175, 176, 178, 182, 184, 200, 205, 206, 254, 284, 341; singular plural, 10, 62, 177, 184, 200
Slums, 81
Social: circumstances, 285; justice, 143, 145; life, 19, 29, 33, 34, 35, 36, 38, 86, 87, 94, 136, 138, 141, 200, 245, 246, 247, 248,

251, 253, 265, 326, 328, 332, 341; networks, 23; organization, 92; otherness, 17; position, 52, 90, 317, 321; practices, 64, 247, 263, 282, 283; situatedness, 42; socialization, 3, 283, 304

Society: societal motive, 164, 167, 170, 244; societally mediated, 163, 165, 188, 259; societally motivated, 185, 188

Sociology: sociology of emotion, 14, 18, 150

Solidarity, 14, 33, 34, 42, 43, 44, 47, 50, 51, 60, 61, 82, 83, 84, 85, 86, 87, 89, 133, 189, 249, 267, 344

Stereotype: cultural, 127

Strauss, A., 305, 323

Street: code, 21, 22, 27, 28, 39; culture, 18, 20, 24, 27, 29, 35

Structure, 18, 24, 29, 37, 40, 43, 45, 52, 53, 55, 61, 62, 72, 76, 85, 93, 108, 123, 125, 135, 149, 158, 163, 166, 216, 218, 240, 252, 254, 255, 260, 304, 314, 326, 328, 329, 331, 332, 334, 335, 340; structural factors, 44, 51, 83

Subjectivity, 7, 93, 189, 192, 203, 205, 206, 245, 247, 249, 251

Symbol, 22, 44, 47, 50, 51, 52, 53, 92, 204, 209, 231, 284, 285, 286, 288, 290, 291, 330; system, 285, 286, 288, 290, 291

T

Taylor, C., 143

Theory: sociocultural, 141, 142, 243, 341

Time: timescale, 50, 54, 63, 77, 78, 205, 286, 296, 321, 336

Tobin, K., 1, 14, 18, 19, 40, 46, 62, 81, 94, 95, 101, 133, 135, 150, 243, 325, 337

Tool, 5, 65, 78, 88, 94, 124, 132, 137, 165, 167, 168, 172, 188, 191, 204, 207, 231, 237, 240, 259, 260, 262, 271, 280, 328, 329, 342

Transaction, 1, 26, 27, 100, 149, 173, 174, 183, 195, 196, 198, 199, 200, 328, 332, 341, 342

Traweek, S., 116, 119

Turner, J. H., 150

Tutoring, 126, 131

U

Uexküll, J. von, 163, 184

Unconscious, 36, 37, 121, 122, 123, 124, 125, 131, 155, 165, 170, 189, 205, 250, 265, 276, 325, 326, 328, 330, 332, 341

Unemployment: frictional, 95

Urban, 11, 13, 15, 81, 121; education, 15, 16, 38; setting, 14, 46; youth, 13, 14, 15, 16, 17, 19, 30, 33, 34, 38, 39, 82, 83, 93, 94, 296

V

Varela, F., 188, 202

Vernacular, 288, 289, 291, 294, 295

Volition: volitional level, 170; volitional moments, 164

Vygotsky, L. S., 150, 172, 184, 206, 208, 242, 249, 250

W

Wenger, E., 6, 10, 42, 62, 71, 79, 109, 115, 118, 206, 242, 287, 299

Willis, P., 89, 116, 119, 243, 245, 255

Wisdom: practical, 171, 182, 183

Workplace, 84, 118, 192, 265, 266, 267, 268, 269, 270, 271, 275, 276, 278, 280, 330, 336; workplace-based learning, 272, 274

World: cultural, 95, 118, 135, 184, 287

Wortham, S., 54, 62, 64, 79

By the same authors:

Doing Educational Research: *A Handbook*
Kenneth Tobin, *The Graduate Center, CUNY, USA* and **Joe Kincheloe**, *McGill University, Montreal, Canada* (eds.)

Doing Educational Research explores a variety of important issues and methods in educational research. Contributors include some of the most important voices in educational research. In the handbook these scholars provide detailed insights into one dimension of the research process that engages both students as well as experienced researchers with key concepts and recent innovations in the domain. The editors and authors believe that there is a need for a handbook on educational research that is both practical as it introduces beginning scholars to the field and innovative as it pushes the boundaries of the conversation about educational research at this historical juncture.

In this collection the authors explore a variety of topics from methodologies such as ethnography, action research, hermeneutics, historiography, psychoanalysis, literary criticism to issues such as social theory, epistemology, and paradigms. The book addresses complex topics in an accessible and readable manner. The book will be very useful as a text in educational research at the graduate and the undergraduate level.

> September 2006, 480 pp
> paperback: ISBN:90-77874-48-8
> hardback: ISBN:90-77874-01-1
> SERIES: BOLD VISIONS IN EDUCATIONAL RESEARCH 1

Auto/Biography and Auto/Ethnography: *Praxis of Research Method*
W. -M. Roth, *University of Victoria, Canada* (ed.)

In a number of academic disciplines, auto/biography and auto/ethnography have become central means of critiquing of the ways in which research represents individuals and their cultures. The contributors to this volume explore, by means of examples, auto/biography and auto/ethnography as means for critical analysis and as tool kit for the different stakeholders in education.

The book was written to be used by upper undergraduate and graduate students taking courses in research design andd professors, who want to have a reference on design and methodology.

> July 2005, 448 pp
> paperback: ISBN:90-77874-04-6
> hardback: ISBN:90-77874-49-6
> SERIES: BOLD VISIONS IN EDUCATIONAL RESEARCH 2

Doing Qualitative Research: *Praxis of Method*
W. -M. Roth, *University of Victoria, Canada*

The author takes readers on a journey of a large number of issues in designing actual studies of knowing and learning in the classroom, exploring actual data, and putting readers face to face with problems that he actually or possibly encountered, and what he has done or possibly could have done. The reader subsequently sees the results of data collection in the different analyses provided. The book is organized around six major themes (sections), in the course of which it develops the practical problems an educational researcher might face in a large variety of settings.

The book was written to be used by upper undergraduate and graduate students taking courses in research design and professors who want to have a reference on design and methodology.

> August 2005, 508 pp
> **paperback:** ISBN:90-77874-05-4
> **hardback:** ISBN:90-77874-51-8
> SERIES: BOLD VISIONS IN EDUCATIONAL RESEARCH 3

Learning Science: *A Singular Plural Perspective*
W.-M. Roth, *University of Victoria, Canada*

How do you *intend* (to learn, know, see) something that you do not yet know? Given the theory-laden nature of perception, how do you *perceive* something in a science demonstration that requires knowing the very theory that you are to learn? In this book, the author provides answers to these and other (intractable) problems of learning in science. He uses both first-person, phenomenological methods, critically analyzing his own experiences of learning in unfamiliar situations *and* third-person, ethnographic methods, critically analyzing the learning of students involved in hands-on investigations concerning motion and static electricity.

This book, which employs the cognitive phenomenological method described in the recently published *Doing Qualitative Research: Praxis of Method* (See page 1 of this brochure), has been written for all those who are interested in learning science: undergraduate students preparing for a career in science teaching, graduate students interested in the problems of teaching and learning of science, and faculty members researching and teaching in science education.

> March 2006, 372 pp
> **paperback:** ISBN:90-77874-25-9
> **hardback:** ISBN:90-77874-26-7
> SERIES: NEW DIRECTIONS IN MATH AND SCIENCE EDUCATION 1

Teaching to Learn: *A View from the Field*
Kenneth Tobin, *The Graduate Center, CUNY, USA* and **W.-M. Roth,** *University of Victoria, Canada*

A recurrent trope in education is the gap that exists between theory, taught at the university, and praxis, what teachers do in classrooms. How might one bridge this inevitable gap if new teachers are asked to learn (to talk) about teaching rather than to teach? In response to this challenging question, the two authors of this book have developed coteaching and cogenerative dialoguing, two forms of praxis that allow very different stakeholders to teach and subsequently to reflect together about their teaching. The authors have developed these forms of praxis not by theorizing and then implementing them, but by working at the elbow of new and experienced teachers, students, supervisors, and department heads. Tobin and Roth describe the many ways coteaching and cogenerative dialogues are used to improve learning environments—dramatically improving teaching and learning across cultural borders defined by race, ethnicity, gender, and language. Teaching to Learn is written for science educators and teacher educators along the professional continuum: new and practicing teachers, graduate students, professors, researchers, curriculum developers, evaluation consultants, science supervisors, school administrators, and policy makers. Thick ethnographic descriptions and specific suggestions provide readers access to resources to get started and continue their journeys along a variety of professional trajectories.

> July 2006, 282 pp
> **paperback:** ISBN:90-77874-81-X
> **hardback:** ISBN:90-77874-91-7
> *SERIES: NEW DIRECTIONS IN MATH. AND SCIENCE EDUCATION 4*

For more information on these and our other titles go to
WWW.SENSEPUBLISHERS.COM

Printed in the United States
99052LV00001B/76-78/A